Alan Werner
139 Langdon
PH 255-5872

ELEMENTARY THEORY OF STRUCTURES

McGRAW-HILL CIVIL ENGINEERING SERIES

HARMER E. DAVIS, *Consulting Editor*

Elementary Theory of Structures

CHU-KIA WANG, Ph.D.

Professor of Civil Engineering
University of Wisconsin

CLARENCE LEWIS ECKEL, C.E.

Professor of Civil Engineering
Dean of the College of Engineering
University of Colorado

McGRAW-HILL BOOK COMPANY

New York   Toronto   London

1957

ELEMENTARY THEORY OF STRUCTURES

Library of Congress Catalog Card Number 56-11058

9 10 11 12 13 14 15 - MP - 1 0 9 8

68134

PREFACE

This text is designed to present the essential principles of structural analysis in a first course for architectural and civil engineering students. The analysis of statically determinate structures is based on the laws of statics, while that of statically indeterminate structures depends on both the principles of statics and the geometric conditions of the deformed structure. These principles are relatively simple; nevertheless experience shows that, in order to acquire proficiency and facility, a student must expect to work a considerable number of problems involving the appropriate conditions of statics and geometry.

Incidental to the presentation of basic principles in this text, special emphasis has been given to illustrative examples. It is hoped that this feature will relieve the teacher of undue blackboard routine and thereby permit time for lively and fruitful class discussion.

Statically determinate structures are discussed in Chaps. 1 through 10. Chapters 11 through 15 are devoted to an introduction to the analysis of statically indeterminate structures. With the exception of Chaps. 6, 7, and 10, which deal with applications to the analysis of structures such as roof trusses, building bents, and bridge trusses, the basic concern of this text is the use of general principles and methods of structural analysis.

In schools where "unified" courses in structural analysis and design are offered, this text may be used as a principal source book for the "analysis" portion of "design" assignments.

Teachers who prefer to give a "unified" treatment of statically determinate and statically indeterminate structures will find that Chaps. 3 and 11, Chaps. 4 and 12, and Chaps. 5 and 13 may be conveniently used in pairs.

Chapters 8 and 9, which deal with influence diagrams and criteria for moving loads, are general in nature. The topics discussed in these chapters are essential for an understanding of the structural analysis of bridge trusses or other structures carrying moving loads. Although students in architecture or architectural engineering are likely to be primarily interested in building structures, they will find these chapters, and perhaps Chap. 10, of value in adding to their over-all understanding of the procedures of structural analysis.

The methods of slope deflection and moment distribution are treated separately in Chaps. 14 and 15. Again a choice is permitted in that the slope-deflection method, and then the moment-distribution method, may be studied; or both methods may be discussed in relation to a given problem at the same time. If the latter choice is made, Chaps. 14 and 15 may be used together.

Although great care has been taken in checking calculations and manuscript, the authors will appreciate notices of errors and suggestions for improvement in future editions.

The authors wish to thank Mrs. C. K. Wang for her valuable assistance, especially in typing the final manuscript of this text.

C. K. WANG
C. L. ECKEL

CONTENTS

CHAPTER 1

GENERAL INTRODUCTION

1-1. Theory of Structures Defined. Engineers design structures such as bridges, buildings, ships, machine parts, as well as various kinds of equipment and other structural installations. Incident to design, the engineer must first determine the layout of the structure, its shape, and its constituent members. Then he must estimate or otherwise determine the loads which the structure is to carry. The theory of structures deals with the principles and methods by which the direct stress, the shear and bending moment, and the deflection at any section of each constituent member in the structure may be calculated. The next phase of the design is to proportion the members in accordance with the allowable working stresses of the material and other requirements for the proper

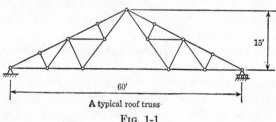

A typical roof truss

FIG. 1-1

functioning of the structure. This work is generally within the scope of texts on strength of materials or structural design and will not be discussed in this text. It may be well to point out that the process of design may have to be repeated a number of times before a satisfactory final design can be found.

Consider, for example, the design of a typical roof truss such as is shown in Fig. 1-1. The process of design involves four stages: (1) a layout of the truss is assumed; (2) the loading, which may consist of dead load (weight of the roofing material and the truss itself), snow load, wind load, or other loading, is estimated; (3) the direct stresses in the members of the truss are found; and (4) the sizes of the members are determined in accordance with the design specifications. This text will concern itself primarily with the third stage, but with occasional reference to the second

1

stage. The eventual reconciliation between the first and the fourth stages is largely a matter of experience.

1-2. Layout and Classification of Structures. The layout of any structure depends largely on the function of the structure, the loading conditions, and the properties of the material to be used. Except in routine situations, the determination of the layout of any structure requires knowledge, judgment, and experience. Usually after two or more layouts for the same structure are carried through the initial design stage, a comparison is made to determine the preferred design. Sometimes the preliminary layout has to be modified to meet unanticipated conditions encountered in the later stages of design.

Basically most structures may be classified as beams, rigid frames, or trusses or combinations of these elements. A beam is a structural member subjected to transverse loads only. It is completely analyzed when the shear and bending-moment values have been found. A rigid frame is a structure composed of members connected by rigid joints (welded joints, for instance). A rigid frame is completely analyzed when the variations in direct stress, shear, and bending moment along the lengths of all members have been found. A truss is a structure in which all members are usually assumed to be connected by frictionless hinges. A truss is completely analyzed when the direct stresses in all members have been determined. There are also structural members or machine parts which may be subjected to the action of direct stress, shear and bending moment, and twisting moment.

1-3. Loads on Structures. Generally, the loads on structures consist of dead load, live load, and the dynamic or impact effects of the live load. Dead load includes the weight of the structure itself; live load is the loading to be carried by the structure; and impact is the dynamic effect of the application of the live load. Thus, in building design, the weight of the flooring, beams, girders, and columns makes up the dead load; while the weight of movable partitions, furniture, etc., the snow load, and the wind load are considered as live load. Often the live load comes on a structure rather suddenly or as a moving or rolling load, as, for example, when a train passes over a bridge. In this case the live load is increased by an estimated percentage to include its dynamic effect. This increase is called the impact load.

It is obvious that most of the dead load, except such items as the roofing on roof trusses, ceiling plaster under floors, and handrails on bridges, cannot be determined until the members have actually been designed; therefore, dead load has to be first assumed and then checked after the sizes of the members have been determined. Except for unusual structures the dead-load stress normally constitutes only a relatively small percentage of the total stress in a member; so that in routine designs a modification of the first design is seldom necessary.

In its passage across the structure, the position which the live load assumes in order to cause a maximum direct stress, shear, or bending moment at a particular section in a member is of great importance and will receive comprehensive treatment in this text.

1-4. Methods of Analysis. In Art. 1-1 it was stated that the theory of structures deals with the principles and methods by which the direct stress, shear, and bending moment at any section of the member may be found under given conditions of loading. Because the forces acting on a structural member may usually be assumed to lie in the same plane and are in equilibrium, fundamental structural analysis involves the use of the three equations of equilibrium for a general coplanar-force system; viz., $\Sigma F_x = 0$, $\Sigma F_y = 0$, and $\Sigma M = 0$. These three equations, together with a good working knowledge of simple arithmetic, algebra, geometry, trigonometry, and some calculus, are the necessary prerequisites for studying the elementary theory of structures.

CHAPTER 2

EQUILIBRIUM OF COPLANAR-FORCE SYSTEMS

2-1. The Free Body. No matter how complicated a structure may be it may be assumed to be cut into various members, parts, or sections, each of which is under the action of a system of coplanar forces. Any one member, part, or section, thus set free from the whole structure, is called a free body. A free body, clearly drawn and complete with the magnitudes and directions (both known and unknown) of all the forces acting on it, is called a free-body diagram. The facility and ease with which the free-body diagrams are chosen and drawn are the key to the subject of structural analysis.

The free body, being at rest within the structure, must be in equilibrium under the action of all the coplanar forces acting on it. If the magnitude, or the direction, or both, of some of these forces are unknown, they can be found by the principles of statics, which are the three equations of equilibrium $\Sigma F_x = 0$, $\Sigma F_y = 0$, and $\Sigma M = 0$. In this chapter, the methods of solving for these unknown magnitudes or directions will be explained.

2-2. Equilibrium of Coplanar-concurrent-force Systems. If the free body happens to be a point (a pin, for instance), the forces acting on it are concurrent. The resultant of a coplanar-concurrent-force system must be a single force, the x component of which is ΣF_x of the component forces and the y component is ΣF_y. Thus the two equations $\Sigma F_x = 0$ and $\Sigma F_y = 0$ are necessary and sufficient to ensure that the resultant is zero or that the coplanar-concurrent-force system is in equilibrium. These two conditions for equilibrium permit the calculation of two unknowns, which may be the magnitudes of two forces with known directions, or the magnitude of one force with known direction and the direction of another force with known magnitude.

It should be noted that the x and y directions are purely arbitrary; thus, in applying the equation $\Sigma F_x = 0$, any direction may be considered as the x axis. Also, as long as the concurrent forces are in equilibrium and have no resultant, the sum of the moments of the component forces about any point in the plane must be zero. Thus in cases where they may be more conveniently applied, the moment equations $\Sigma M_A = 0$ and

$\Sigma M_B = 0$ may be substituted for either or both of the resolution equations $\Sigma F_x = 0$ and $\Sigma F_y = 0$.

In the graphic method of finding the resultant of a coplanar-concurrent-force system, a zigzag line is drawn connecting successively the component vectors taken in any convenient order; the resultant is then given by the vector extending from the starting point of the first component vector to the end point of the last component. Should the resultant be zero, the end point of the last component force must coincide with the starting point of the first component. Thus the graphic condition for the equilibrium of a coplanar-concurrent-force system is that the force polygon must close.

For example, if the four coplanar, concurrent forces ab, bc, cd, and de as shown in the space diagram of Fig. 2-1a are in equilibrium, the points A and E in the force polygon $ABCDE$ of Fig. 2-1b must coincide. Note that the forces in Fig. 2-1a may be designated in an irregular order, although they are normally named in alphabetical order around point O

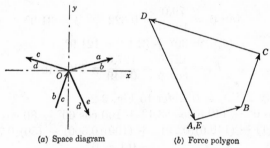

(a) Space diagram   (b) Force polygon

FIG. 2-1

in either the clockwise or counterclockwise direction. It should also be noted that the position and direction, but not the magnitude, of the forces must be plotted accurately in the space diagram; while both magnitude and direction of the forces, and not the position, are represented in the force polygon.

In the examples which follow, both the algebraic and graphic solutions are given.

Example 2-1. If the four coplanar, concurrent forces F_1, F_2, F_3, and F_4 shown in Fig. 2-2a are in equilibrium, find the magnitude and direction of F_4 which is arbitrarily assumed to act in the direction shown.

ALGEBRAIC SOLUTION. From $\Sigma F_x = 0$,

$$50 \cos 15° + 100 \cos 45° - 80 \sin 30° + (F_4)_x = 0$$
$$(F_4)_x = -(50)(0.966) - (100)(0.707) + (80)(0.500) = -79.0 \text{ lb}$$

Therefore $(F_4)_x$ acts to the left as shown in Fig. 2-2b and not to the right as assumed in Fig. 2-2a.

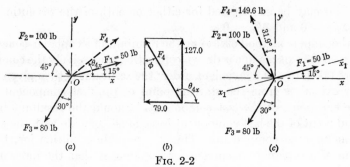

FIG. 2-2

From $\Sigma F_y = 0$,

$$50 \sin 15° - 100 \cos 45° - 80 \cos 30° + (F_4)_y = 0$$
$$(F_4)_y = -(50)(0.259) + (100)(0.707) + (80)(0.866) = +127.0 \text{ lb}$$

$(F_4)_y$ acts upward as assumed in Fig. 2-2a and as shown in Fig. 2-2b. Referring to Fig. 2-2b,

$$\tan \phi = \frac{79.0}{127.0} = 0.622 \qquad \phi = 31.9°$$
$$(\theta_4)_x = 90° + 31.9° = 121.9°$$
$$F_4 = \frac{127.0}{\cos \phi} = \frac{127.0}{0.849} = 149.6 \text{ lb}$$

CHECK. By $\Sigma(F_x)_1 = 0$ (refer to Fig. 2-2c),
$$+50 \cos 0° - 149.6 \cos 83.1° + 100 \cos 60° - 80 \cos 45° \approx 0$$
$$+(50)(1) - (149.6)(0.291) + (100)(0.500) - (80)(0.707) \approx 0$$
$$-0.1 \doteq 0$$

GRAPHIC SOLUTION. In Fig. 2-3, the forces AB, BC, and CD are drawn in succession to scale and in direction as given. The magnitude and direction of F_4 are then given by the vector DA.

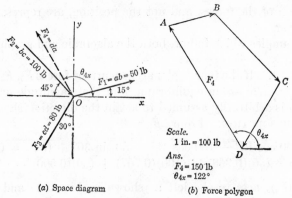

(a) Space diagram

(b) Force polygon

FIG. 2-3

Example 2-2.   If the four coplanar, concurrent forces F_1, F_2, F_3, and F_4 shown in Fig. 2-4a are in equilibrium, find the magnitudes of F_3 and F_4 acting in the directions shown.

ALGEBRAIC SOLUTION.   Since these forces are concurrent, this problem may be solved by resolving the forces into components parallel to the x and y axes and applying the conditions $\Sigma F_x = 0$ and $\Sigma F_y = 0$; however, this requires the use of simultaneous equations.   A simpler solution involving the solution of two separate equations is obtained by (1) finding F_3 from $\Sigma M_D = 0$, (2) finding F_4 from $\Sigma M_C = 0$, and (3) checking by $\Sigma F_x = 0$ and $\Sigma F_y = 0$.   It is usually easier to find the moment of

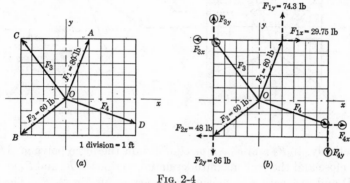

(a)   (b)

FIG. 2-4

forces such as these about a point D by use of the *principle of moments*, which states that the moment of a force about a point is equal to the sum of the moments of its two rectangular components obtained by resolution at any convenient point on the line of action of the force. Thus, from $\Sigma M_D = 0$ (Fig. 2-4b),

$$-(29.75)(7) - (74.3)(4) - (48)(1) + (36)(10) + 7(F_3)_x - 10(F_3)_y = 0$$
$$7(F_3)_x - 10(F_3)_y = +193.45$$

But
$$(F_3)_y = \tfrac{5}{4}(F_3)_x$$

so
$$(F_3)_x(7 - 10 \times \tfrac{5}{4}) = +193.45$$
$$(F_3)_x = -35.2 \text{ lb} \qquad (F_3)_y = -44.0 \text{ lb} \qquad F_3 = -56.3 \text{ lb}$$

Note that circles are drawn around the arrowheads of $(F_3)_x$, $(F_3)_y$, and F_3 in Fig. 2-4b to indicate that they are actually in directions opposite to those assumed in Fig. 2-4b.   From $\Sigma M_C = 0$,

$$(29.75)(0) + (74.3)(6) - (48)(8) + (36)(0) + 7(F_4)_x - 10(F_4)_y = 0$$
$$7(F_4)_x - 10(F_4)_y = -61.8$$

But
$$(F_4)_y = \tfrac{1}{3}(F_4)_x$$
$$(F_4)_x(7 - 10 \times \tfrac{1}{3}) = -61.8$$
$$(F_4)_x = -16.85 \text{ lb} \qquad (F_4)_y = -5.62 \text{ lb} \qquad F_4 = -17.76 \text{ lb}$$

CHECK
$$\Sigma F_x = 0 \quad +29.75 - 48 + 35.2 - 16.85 \approx 0$$
$$+0.10 \doteq 0$$
$$\Sigma F_y = 0 \quad +74.3 - 36 - 44.0 + 5.62 \approx 0$$
$$-0.08 \doteq 0$$

GRAPHIC SOLUTION. The space diagram and the force polygon are shown in Fig. 2-5. $AB = F_1$ and $BC = F_2$ are first plotted; D is the point of intersection of two lines drawn through C and A, parallel,

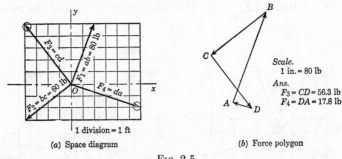

1 division = 1 ft

(a) Space diagram

(b) Force polygon

Scale.
1 in. = 80 lb
Ans.
$F_3 = CD = 56.3$ lb
$F_4 = DA = 17.8$ lb

FIG. 2-5

respectively, to F_3 and F_4. For equilibrium the force polygon $ABCDA$ must close and the true magnitudes and directions of F_3 and F_4 are equal to CD and DA acting as shown by the arrows in Fig. 2-5b. The circles drawn around the arrowheads of cd and da in the space diagram indicate that initially the directions of these forces were incorrectly assumed.

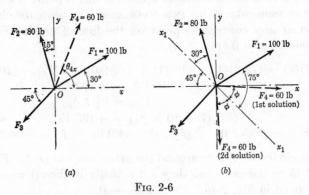

(a)

(b)

FIG. 2-6

Example 2-3. The four coplanar, concurrent forces F_1, F_2, F_3, and F_4 shown in Fig. 2-6a are in equilibrium. Find the magnitude of F_3 acting as shown, and the direction of F_4.

ALGEBRAIC SOLUTION. The x_1x_1 axis is drawn perpendicular to the line of action of F_3 in Fig. 2-6b. There are two positions of $F_4 = 60$ lb, each at an angle ϕ from the x_1x_1 axis, which will satisfy the condition $\Sigma(F_x)_1 = 0$. From $\Sigma(F_x)_1 = 0$,

$$-80 \cos 30° + 100 \cos 75° + 60 \cos \phi = 0$$
$$60 \cos \phi = (80)(0.866) - (100)(0.259) = 43.38$$
$$\phi = \pm 43.7°$$

FIRST SOLUTION $[(\theta_4)_x = 358.7°]$. From $\Sigma F_x = 0$,

$$+100 \cos 30° - 80 \sin 15° - F_3 \cos 45° + 60 \cos 1.3° = 0$$
$$F_3 \cos 45° = +(100)(0.866) - (80)(0.259) + (60)(1.000) = +125.9$$
$$F_3 = 178.1 \text{ lb}$$

CHECK. By $\Sigma F_y = 0$,

$$+100 \sin 30° + 80 \cos 15° - 178.1 \sin 45° - 60 \sin 1.3° \approx 0$$
$$+(100)(0.500) + (80)(0.966) - (178.1)(0.707) - (60)(0.023) \approx 0$$
$$0 = 0$$

SECOND SOLUTION $[(\theta_4)_x = 271.3°]$. From $\Sigma F_x = 0$,

$$+100 \cos 30° - 80 \sin 15° - F_3 \cos 45° + 60 \sin 1.3° = 0$$
$$F_3 \cos 45° = +(100)(0.866) - (80)(0.259) + (60)(0.023) = +67.3$$
$$F_3 = 95.2 \text{ lb}$$

CHECK. By $\Sigma F_y = 0$,

$$+100 \sin 30° + 80 \cos 15° - 95.2 \sin 45° - 60 \cos 1.3° \approx 0$$
$$+(100)(0.500) + (80)(0.966) - (95.2)(0.707) - (60)(1.000) \approx 0$$
$$0 = 0$$

GRAPHIC SOLUTION. In the force polygon of Fig. 2-7b, AB and BC are first plotted. With A as center and 60 lb as radius, an arc is drawn

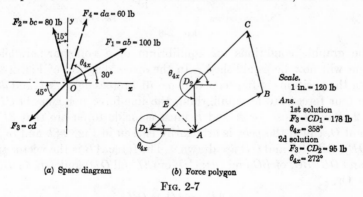

(a) Space diagram   (b) Force polygon

FIG. 2-7

to intersect a line through C and parallel to cd. The points of intersection are D_1 and D_2. As shown in Fig. 2-7b, the values of F_3 are CD_2 [$F_3 = 95$ lb and $(\theta_4)_x = 272°$] and CD_1 [$F_3 = 178$ lb and $(\theta_4)_x =$

358°]. Vectors D_2A and $D_1A = 60$ lb. A study of Fig. 2-7b shows that there may be no solution, one solution, or two solutions depending on whether AD is shorter than, equal to, or longer than the perpendicular AE.

2-3. Equilibrium of Coplanar-parallel-force Systems. The resultant of a coplanar-parallel-force system must be either a single force or a couple. The resultant cannot be a single force if the summation of the forces is zero and the resultant cannot be a couple if the summation of the moments of the forces about any point in the plane of the forces is zero. Thus the two equations $\Sigma F = 0$ and $\Sigma M = 0$ are necessary and sufficient to ensure that the resultant is zero or the coplanar-parallel-force system is in equilibrium. Two unknowns can be found from these two conditions of equilibrium. These unknowns may be the magnitude (including sense or direction) and position of the same force, the magnitudes of two forces (usually called the reactions), or the magnitude of one force with known position and the position of another force with known magnitude.

In the algebraic solution, one of the two equations used must be a moment equation, while the other may be either a resolution or a moment equation.

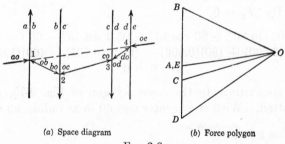

(a) Space diagram   (b) Force polygon

Fig. 2-8

The graphic conditions for equilibrium of a coplanar-parallel-force system will now be established. In the space diagram of Fig. 2-8a are shown the positions of four coplanar-parallel forces ab, bc, cd, and de. If these four forces are in equilibrium, the line-force polygon $ABCDE$ of Fig. 2-8b must close, or A and E must coincide to ensure that $\Sigma F = 0$. A point O, called the *pole*, is arbitrarily chosen in Fig. 2-8b and *rays OA* (or OE), OB, OC, and OD are drawn. The force AB is the vector sum of AO and OB; BC, of BO and OC; CD, of CO and OD; and DE, of DO and OE. Or,

$$AB = AO \nrightarrow OB$$
$$BC = BO \nrightarrow OC$$
$$CD = CO \nrightarrow OD$$
$$DE = DO \nrightarrow OE$$

Adding and noting that $OB \rightarrowtail BO = 0$, etc.,

$$AB \rightarrowtail BC \rightarrowtail CD \rightarrowtail DE = AO \rightarrowtail OE$$

Thus the four original forces are now replaced by two forces AO and OE, the resultant of which cannot be a single force because AO and OE are equal and opposite. To ensure that the resultant is not a couple the lines of action of AO and OE must be collinear. If in the space diagram of Fig. 2-8a, the *strings oa, ob, oc, od,* and *oe* are drawn in succession and respectively parallel to the rays in the force polygon, *oa* and *oe* must be collinear. The polygon 1-2-3-4-1 in Fig. 2-8a is usually called the *string polygon,* or the *equilibrium polygon.* Essentially the forces *ao* and *oe* replace the given force system and the fact that *ao* and *oe* are two collinear, equal, and opposite forces ensures that the given force system is in equilibrium. Thus the two graphic conditions for equilibrium of a coplanar-parallel-force system are: (1) the force polygon must close, and (2) the string polygon must close. Note that, in the actual solution of a problem, the force *de* would have been called *da* at the outset.

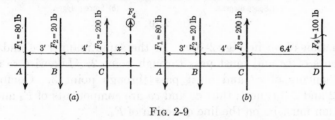

Fig. 2-9

Example 2-4. If the four coplanar, parallel forces F_1, F_2, F_3, and F_4 shown in Fig. 2-9a are in equilibrium, find the magnitude and position of F_4.

ALGEBRAIC SOLUTION. From $\Sigma F = 0$ (Fig. 2-9a),

$$80 + 20 = 200 + F_4$$
$$F_4 = -100 \text{ lb or } 100 \text{ lb downward}$$

Because a negative sign for F_4 is obtained, a circle is drawn around the arrowhead of F_4 in Fig. 2-9a to indicate that F_4 is actually 100 lb downward, as shown in Fig. 2-9b. From $\Sigma M_C = 0$ (Fig. 2-9a),

$$(80)(7) + (20)(4) = (100)(x)$$
$$x = 6.4 \text{ ft}$$

CHECK. By $\Sigma M_A = 0$,

$$(20)(3) + (100)(13.4) \approx (200)(7)$$
$$1,400 = 1,400$$

GRAPHIC SOLUTION. In the space diagram of Fig. 2-10a, the known forces F_1, F_2, and F_3 are called *ab, bc,* and *cd.* The unknown force F_4 is called *da.* The force polygon $ABCD$ of Fig. 2-10b is drawn; $F_4 = DA$ is scaled to be 100 lb downward. The pole O is arbitrarily chosen and rays OA, OB, etc., are drawn. The string oa is drawn parallel to the ray OA, intersecting ab at any point such as 1. From b to b draw the string ob or 1-2. From c to c draw the string oc or 2-3. Through point 3 draw the string od which intersects the string oa at point 4. Point 4 is on the line of action of da, which is scaled at $x = 6.4$ ft from cd.

It is to be noted that, in Fig. 2-10b, AO and OB are components of force AB and may be assumed to act through any point such as 1 on the

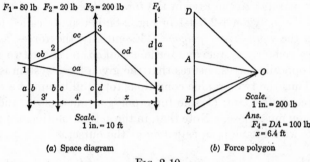

(a) Space diagram   (b) Force polygon

FIG. 2-10

line of action of ab in Fig. 2-10a. In the same manner bo and oc are components of bc and must pass through point 2. Likewise co and od are components of cd and must pass through point 3. Components along 1-2 and 2-3 cancel; thus do and oa are components of F_4 and their intersection must be on the line of action of F_4.

The three known forces have been named in the order ab, bc, and cd from the left toward the right. Actually any force can be called ab, and the other two bc and cd. For instance, F_3 may be called ab; F_1, bc; and F_2, cd. The same position for F_4 will be found by following the procedure described above.

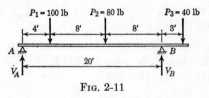

FIG. 2-11

Example 2-5. Three loads P_1, P_2, and P_3 act on the beam as shown in Fig. 2-11. Find the reactions V_A and V_B.

ALGEBRAIC SOLUTION. From $\Sigma M_A = 0$,

$$(100)(4) + (80)(12) + (40)(23) = 20V_B$$
$$2,280 = 20V_B$$
$$V_B = 114 \text{ lb}$$

From $\Sigma M_B = 0$,

$$20V_A + (40)(3) = (100)(16) + (80)(8)$$
$$20V_A = 2,120$$
$$V_A = 106 \text{ lb}$$

CHECK. By $\Sigma F = 0$,

$$100 + 80 + 40 \approx 106 + 114$$
$$220 = 220$$

GRAPHIC SOLUTION. The three known forces P_1, P_2, and P_3 are designated as ab, bc, and cd in the space diagram of Fig. 2-12a and plotted as AB, BC, and CD in the force polygon of Fig. 2-12b. The problem is to locate the point E which will divide the load line AD in segments proportional to the reactions V_A and V_B. The pole O is arbitrarily located and rays OA, OB, etc., are drawn. The unknown reactions in

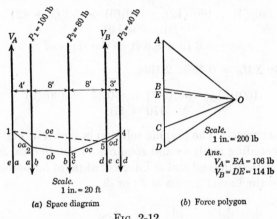

(a) Space diagram (b) Force polygon

Fig. 2-12

Fig. 2-12a are called de and ea. The strings oa, ob, etc., are drawn in succession, respectively parallel to the corresponding rays. It is important to note that the string oa, or 1-2, is drawn from a to a; ob, or 2-3, from b to b; oc, or 3-4, from c to c; and od, or 4-5, from d to d. By joining the points 1 and 5, the string oe from e to e is obtained. The ray OE is then drawn parallel to oe. By scaling, $V_A = EA = 106$ lb and $V_B = DE = 114$ lb.

It is to be noted that the designations ab, bc, and cd may be assigned to the three known forces in any order and similarly de and ea to the two unknown reactions V_A and V_B. The reader will get good practice in repeating this solution by designating P_2 as ab, P_1 as bc, P_3 as cd, V_A as de, and V_B as ea. In any event the unknowns must have a common letter, which is e in this case.

Example 2-6. The five coplanar, parallel forces shown in Fig. 2-13a are in equilibrium. Find the magnitude of F_4 and the position of F_5.

ALGEBRAIC SOLUTION. From $\Sigma F = 0$ (Fig. 2-13a),

$$50 + 80 + F_4 = 60 + 40$$
$$F_4 = -30 \text{ lb or } 30 \text{ lb downward}$$

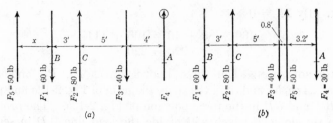

FIG. 2-13

From $\Sigma M_A = 0$ (Fig. 2-13a),

$$(40)(4) + (60)(12) = (80)(9) + (50)(x + 12)$$
$$50x = -440$$
$$x = -8.8 \text{ ft or 8.8 ft to the right of } B$$

CHECK. By $\Sigma M_C = 0$ (Fig. 2-13b),

$$(60)(3) + (50)(5.8) \approx (40)(5) + (30)(9)$$
$$470 = 470$$

GRAPHIC SOLUTION. The graphic solution is shown in Fig. 2-14. This follows the preceding analysis and requires no detailed explanation. It will be noted that strings od and oe intersect at point 5, thereby determining a point on the line of action of F_5 or de.

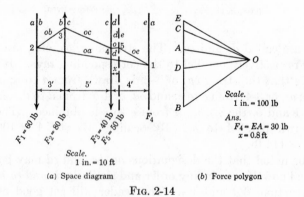

(a) Space diagram   (b) Force polygon

FIG. 2-14

2-4. Equilibrium of General Coplanar-force Systems. The resultant of a general coplanar (noncurrent and nonparallel) force system must be either a single force or a couple. The resultant cannot be a single force if ΣF_x and ΣF_y of the forces are both zero and the resultant cannot be a couple if ΣM of the forces about any point in the plane of the forces is zero. Thus the three equations $\Sigma F_x = 0$, $\Sigma F_y = 0$, and $\Sigma M = 0$ are necessary an dsufficient to ensure that the resultant is zero, or that the general coplanar-force system is in equilibrium. From these three conditions for equilibrium, three unknowns can be found. The more usual combinations of unknowns are: (1) the magnitude, direction, and position of the same force; (2) the magnitude of one force and both the magnitude and direction of another; and (3) three magnitudes.

Although it is obvious that not more than two of the equations used may be resolution equations, the resolution and moment equations may be used freely so that the unknowns can be solved as directly as possible without resorting to a system of simultaneous equations.

The two graphic conditions for equilibrium of a general coplanar-force system are: (1) the force polygon must close, and (2) the string polygon must close. Take, for instance, five coplanar forces F_1 through F_5, the positions of which are given by ab, bc, etc., in the space diagram of Fig. 2-15a and the magnitudes and directions are given by AB, BC, etc., in the force polygon of Fig. 2-15b. If the five forces are in equilibrium, the

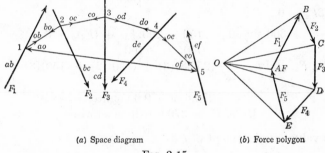

(a) Space diagram   (b) Force polygon

FIG. 2-15

first requirement is that points A and F must coincide in the force polygon. This is equivalent to the two algebraic conditions $\Sigma F_x = 0$ and $\Sigma F_y = 0$. In case the five forces are known to be in equilibrium initially, the force ef or EF would have been called ea or EA, thus eliminating the letters f and F. Next, any arbitrary point O, the pole, is chosen, and the rays OA, OB, etc., are drawn. By beginning at any point 1 on ab, the strings 1-2, 2-3, 3-4, and 4-5 are drawn in succession, respectively parallel to OB, OC, OD, and OE. (Note the rule, "from b to b draw ob parallel to OB, etc.") For equilibrium, the line joining points 1 and 5 must be parallel to OA or OF. The reason is that each of the five given forces is replaced by two forces (ab by ao and ob, for example), but the four intervening pairs (such as ob and bo) balance themselves. The two remaining forces ao and of, which are equivalent to the sum of F_1 to F_5, inclusive, must therefore be equal, opposite, and collinear (see string 1-5 in Fig. 2-15a).

Example 2-7. The three forces F_1, F_2, and F_3 shown in Fig. 2-16a and another force F_4 are in equilibrium. Find the magnitude, direction, and position of F_4.

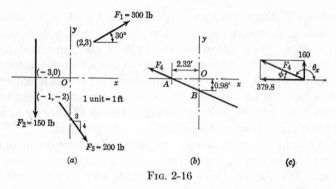

FIG. 2-16

ALGEBRAIC SOLUTION. From $\Sigma F_x = 0$ (Fig. 2-16a),

$$300 \cos 30° + 200(\tfrac{3}{5}) + (F_4)_x = 0$$
$$(F_4)_x = -379.8$$

From $\Sigma F_y = 0$,

$$300 \sin 30° - 150 - 200(\tfrac{4}{5}) + (F_4)_y = 0$$
$$(F_4)_y = +160$$

ΣM_O of F_1, F_2, and $F_3 = -(259.8)(3) + (150)(2) + (150)(3) + (120)(2)$
$$+ (160)(1)$$
$$= -370.6 \text{ or } 370.6 \text{ counterclockwise}$$

Thus M_O of $F_4 = 370.6$ ft-lb clockwise

Absolute value of x intercept of $F_4 = \dfrac{M_O \text{ of } F_4}{(F_4)_y}$

$$= \frac{370.6}{160} = 2.32 \text{ ft}$$

Absolute value of y intercept of $F_4 = \dfrac{M_O \text{ of } F_4}{(F_4)_x}$

$$= \frac{370.6}{379.8} = 0.98 \text{ ft}$$

Referring to Fig. 2-16b and c,

$$OA = 2.32 \text{ ft} \qquad OB = 0.98 \text{ ft}$$
$$\tan \phi = \frac{160}{379.8} = 0.421 \qquad \phi = 22.8° \qquad \theta_x = 157.2°$$
$$F_4 = \frac{379.8}{\cos 22.8°} = \frac{379.8}{0.922} = 412 \text{ lb}$$

Thus $F_4 = 412$ lb at $\theta_x = 157.2°$ passing through the point $(-2.32,0)$ or $(0,-0.98)$.

CHECK. A check can be made by verifying that ΣM of the four forces about each of the three points $(2,3)$, $(-3,0)$, or $(-1,-2)$ is zero.

GRAPHIC SOLUTION. In the force polygon of Fig. 2-17b, the forces are laid out and the closing force F_4 is determined. Next a pole such as

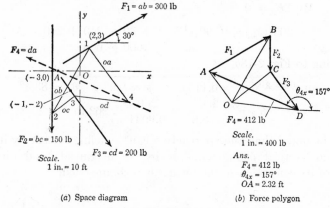

(a) Space diagram

(b) Force polygon

FIG. 2-17

O is selected and rays OA, OB, OC, and OD are drawn.   In the space diagram of Fig. 2-17a, the string oa is drawn parallel to the ray OA, intersecting ab at some convenient point 1.   Then strings ob (1-2), oc (2-3), and od are drawn in succession.   Strings oa and od intersect at point 4, which is on the line of action of da (F_4).

Example 2-8.   The beam shown in Fig. 2-18a has a roller support at A and a hinge support at B. Find the magnitude of R_A and the magnitude and direction of R_B due to the three loads shown.

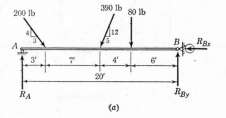

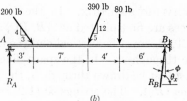

FIG. 2-18

ALGEBRAIC SOLUTION.   The horizontal and vertical components of the hinge reaction at B will be designated as $(R_B)_x$ and $(R_B)_y$.

From $\Sigma F_x = 0$ (Fig. 2-18a),

$$+(200)(\tfrac{3}{5}) - (390)(\tfrac{5}{13}) - (R_B)_x = 0$$
$$(R_B)_x = -30 \text{ lb}$$

From $\Sigma M_B = 0$,

$$20R_A = (200)(\tfrac{4}{5})(17) + (390)(\tfrac{12}{13})(10) + (80)(6)$$
$$R_A = 340 \text{ lb}$$

From $\Sigma M_A = 0$,

$$20(R_B)_y = (200)(\tfrac{4}{5})(3) + (390)(\tfrac{12}{13})(10) + (80)(14)$$
$$(R_B)_y = 260 \text{ lb}$$

CHECK.   By $\Sigma F_y = 0$,

$$(200)(\tfrac{4}{5}) + (390)(\tfrac{12}{13}) + 80 \approx 340 + 260$$
$$600 = 600$$

Referring to Fig. 2-18b,

$$\tan \phi = \tfrac{30}{260} = 0.1154 \qquad \phi = 6.6° \qquad \theta_x = 83.4°$$
$$R_B = \frac{260}{\cos 6.6°} = \frac{260}{0.9934} = 262 \text{ lb}$$

It will be noted that a hinge reaction can be defined either by its horizontal and vertical components or by its magnitude and direction. Because of its convenience, the former is recommended.

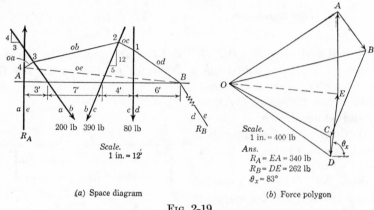

(a) Space diagram   (b) Force polygon

FIG. 2-19

GRAPHIC SOLUTION.   In the force polygon of Fig. 2-19b the known forces are first plotted as AB, BC, and CD.   Because a roller is provided at A, R_A must be vertical and the unknown point E must lie on a vertical line through A.   In the space diagram of Fig. 2-19a the string od (B-1) *must* first be drawn through B, the only known point on the line of action of de (R_B).   The strings oc (1-2), ob (2-3), and oa (3-4) are then drawn. The *closing line* B-4 must be the string oe.   From the pole a line is drawn parallel to oe, intersecting the vertical line through A at E.   Thus $R_B = DE$ and $R_A = EA$.

Example 2-9.   The five coplanar forces acting on the 4- by 8-ft board (Fig. 2-20a) are in equilibrium.   Find the magnitudes of F_3, F_4, and F_5.

ALGEBRAIC SOLUTION.   In order to avoid the solution of simultaneous equations, it will be desirable to find one of the unknowns by taking moments about the point of intersection of the lines of action of the other two unknown forces.   Thus from $\Sigma M_A = 0$ (Fig. 2-20a),

$$(200)(2) - (100)(2) - (\tfrac{4}{5}F_3)(5) + (\tfrac{3}{5}F_3)(1) = 0$$
$$F_3 = 58.8 \text{ lb} \qquad (F_3)_x = 35.3 \text{ lb} \qquad (F_3)_y = 47.0 \text{ lb}$$

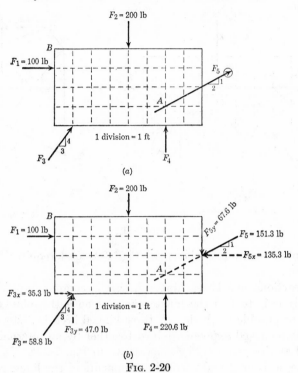

(a)

(b)

FIG. 2-20

From $\Sigma F_x = 0$,
$$+100 + 35.3 + (F_5)_x = 0$$
$$(F_5)_x = -135.3 \text{ lb} \qquad (F_5)_y = -67.6 \text{ lb} \qquad F_5 = -151.3 \text{ lb}$$

From $\Sigma F_y = 0$,
$$-200 + 47.0 + F_4 - 67.6 = 0$$
$$F_4 = 220.6 \text{ lb}$$

CHECK. The best check can be made by using as many of the previously unknown forces as possible in an independent equation. Thus from $\Sigma M_B = 0$ (see Fig. 2-20b),

$$+(100)(1) - (200)(4) + (35.3)(4) + (47.0)(1) + (220.6)(6) - (67.6)(8)$$
$$- (135.3)(2) \approx 0$$
$$+1,611.8 - 1,611.4 \doteq 0$$

GRAPHIC SOLUTION. The resultant of the forces F_4 and F_5 must pass through point A and will be designated R_{45}. The four forces F_1, F_2, F_3, and R_{45} are in equilibrium. The magnitude of F_3 and both the magnitude and direction of R_{45} can be found graphically as shown in Fig. 2-21. R_{45} may then be resolved into two components parallel to F_4 and F_5.

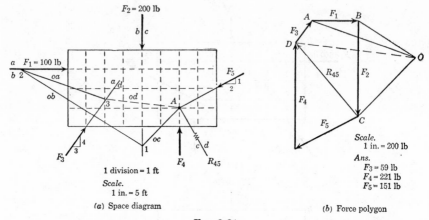

(a) Space diagram

(b) Force polygon

Fig. 2-21

The complete solution is shown in Fig. 2-21 and requires no detailed explanation.

2-5. Reactions on a Three-hinged Arch. A common problem in structural analysis is to find the hinge reactions on a three-hinged arch by algebraic or graphic methods. A three-hinged arch is a structure composed of two curved segments, joined together by an internal hinge and supported at two external hinges as shown in Fig. 2-22a. It is required to find the horizontal and vertical components of the hinge reactions at A, B, and C in this structure.

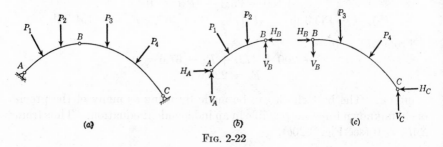

(a)   (b)   (c)

Fig. 2-22

ALGEBRAIC METHOD. When the supports A and C are at the same level, V_A and V_C can be solved from $\Sigma M_C = 0$ and $\Sigma M_A = 0$, respectively, by taking the whole arch as a free body. H_A or H_C may be found by taking moments about B, using either AB or BC as free body.

If the supports A and C are not on the same level, it is convenient to find H_B and V_B first. $\Sigma M_A = 0$ by taking the left segment AB as free body, and $\Sigma M_C = 0$ by taking the right segment BC as free body, will furnish the two simultaneous equations from which H_B and V_B can be solved.

GRAPHIC METHOD. The graphic method of finding the reactions on a three-hinged arch requires that a funicular (or string) polygon be passed through three given points. The general method will first be described and then proof will be given for the construction. The three-hinged arch in Fig. 2-23a is loaded as shown. The loads on the arch are plotted in the force polygon of Fig. 2-23b as $ABCDE$. It is required to find the location of the pole O such that the funicular (equilibrium) polygon X-1-2-3-4-Z in Fig. 2-23a (as will be found later) will pass through the three given points X, Y, and Z, which, in the present case, are the hinges of the arch. A random pole O' is arbitrarily chosen and the rays $O'A$, $O'B$, etc., are drawn. Lines mn and pq are drawn through X and Y and parallel to AC; lines st and uv are drawn through Y and Z and parallel to CE. Starting at any point X' on mn, a trial funicular polygon X'-$1'$-$2'$-$3'$-$4'$-Z' is drawn, with the strings respectively parallel to the

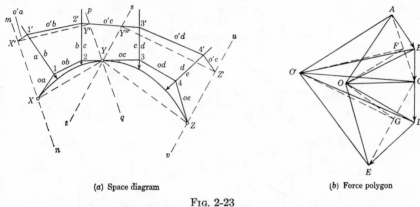

(a) Space diagram   (b) Force polygon

FIG. 2-23

rays radiating from the trial pole O'. The string $2'$-$3'$ ($o'c$) intersects pq and st at Y' and Y''. Join $X'Y'$ and $Y''Z'$. Draw $O'F$ parallel to $X'Y'$ and $O'G$ parallel to $Y''Z'$. Point O, the pole being sought, is the point of intersection of a line drawn parallel to XY at F and another line drawn parallel to YZ at G. The rays OA, OB, etc., are then drawn. The final funicular polygon X-1-2-3-4-Z will pass through the three given points X, Y, and Z.

It will be shown that EO and OA are the external hinge reactions at Z and X; and that CO and OC are the internal hinge reaction acting on the left and right segments, respectively.

PROOF. Suppose that the loads on the left segment are supported by two parallel reactions through the points X and Y. These reactions are FA and CF. Point F has been found by drawing $O'F$ parallel to the closing line $X'Y'$. Since point F must always take the same position regardless of the location of the pole, the required pole must lie on a line drawn through F and parallel to XY. Any funicular polygon, with a pole on FO, that passes through X, will also pass through Y. The same reasoning applies to the right segment, and any string polygon with a pole on GO will pass through Y and Z. Thus the required pole O is the point of intersection of FO and GO. The fact that OA and CO must be the hinge reactions on the left segment of the arch can be shown as follows:

$$AB = AO + OB$$
$$BC = BO + OC$$
$$\overline{AB + BC = AO + OC}$$

Since AO and OC are the two forces equivalent to all the loads acting on the left segment, OA and CO (the opposites of AO and OC) must be the supporting reactions. Similarly it can be shown that EO and OC are the hinge reactions on the right segment of the arch.

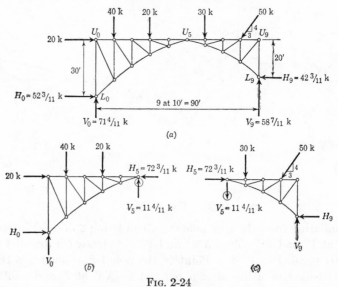

Fig. 2-24

The final funicular polygon is sometimes known as the pressure line in the arch because each string (such as ob in Fig. 2-23a) is the line of action of the resultant of all the forces (including the external hinge reaction) on either side of this string.

Example 2-10. Find the hinge reactions at L_0, U_5, and L_9 due to the loads acting on the three-hinged arch shown in Fig. 2-24a.

ALGEBRAIC SOLUTION. From Σ(moments about L_0) = 0 by taking the left segment as a free body (Fig. 2-24b),

$$30H_5 + 50V_5 = (20)(30) + (40)(10) + (20)(30)$$
$$3H_5 + 5V_5 = 160 \tag{a}$$

or, from Σ(moments about L_9) = 0 by taking the right segment as a free body (Fig. 2-24c),

$$20H_5 - 40V_5 = (30)(30) + (50)(\tfrac{4}{5})(10) + (50)(\tfrac{3}{5})(20)$$
$$2H_5 - 4V_5 = 190 \tag{b}$$

Solving (a) and (b),

$$H_5 = 72\tfrac{3}{11} \text{ kips}$$
$$V_5 = -114\tfrac{4}{11} \text{ kips}$$

From $\Sigma F_x = 0$ (Fig. 2-24b),

$$H_0 = 52\tfrac{3}{11} \text{ kips}$$

From $\Sigma F_y = 0$ (Fig. 2-24b),

$$V_0 = 71\tfrac{4}{11} \text{ kips}$$

From $\Sigma F_x = 0$ (Fig. 2-24c),

$$H_9 = 42\tfrac{3}{11} \text{ kips}$$

From $\Sigma F_y = 0$ (Fig. 2-24c)

$$V_9 = 58\tfrac{7}{11} \text{ kips}$$

CHECK.   A check can be made with $\Sigma F_x = 0$, $\Sigma F_y = 0$, and Σ(moments about L_0) = 0 by taking the whole arch (Fig. 2-24a) as the free body.

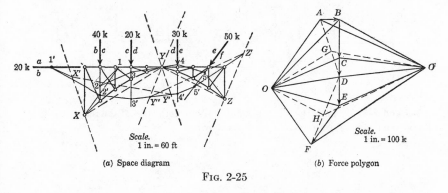

(a) Space diagram   (b) Force polygon

Fig. 2-25

GRAPHIC SOLUTION.   The graphic solution is shown in Fig. 2-25.   *DO* and *OA* are the hinge reactions acting on the left segment; *FO* and *OD* are the hinge reactions acting on the right segment.

PROBLEMS

2-1. By both the algebraic and graphic methods find the magnitude and direction of F_4 if the four coplanar, concurrent forces F_1, F_2, F_3, and F_4 are in equilibrium.

2-2. By both the algebraic and graphic methods find the magnitudes of F_3 and F_4

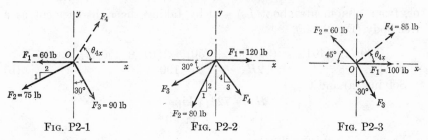

FIG. P2-1   FIG. P2-2   FIG. P2-3

acting in the directions shown if the four coplanar, concurrent forces F_1, F_2, F_3, and F_4 are in equilibrium.

2-3. By both the algebraic and graphic methods find the magnitude of F_3 and the direction of F_4 if the four coplanar, concurrent forces F_1, F_2, F_3, and F_4 are in equilibrium.

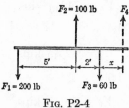

FIG. P2-4

2-4. By both the algebraic and graphic methods find the magnitude and position of F_4 if the four coplanar, parallel forces F_1, F_2, F_3, and F_4 are in equilibrium.

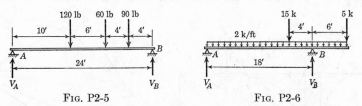

FIG. P2-5   FIG. P2-6

2-5 and 2-6. By both the algebraic and graphic methods find the reactions V_A and V_B on the beam subjected to the loads as shown.

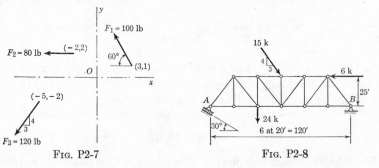

FIG. P2-7   FIG. P2-8

2-7. By both the algebraic and graphic methods find the magnitude, direction, and position of F_4 if the four coplanar forces F_1, F_2, F_3, and F_4 are in equilibrium.

2-8. By both the algebraic and graphic methods find the magnitude of R_A at the roller support and the magnitude and direction of R_B at the hinge support of the truss subjected to loads as shown.

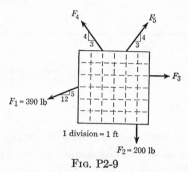

FIG. P2-9

2-9. By both the algebraic and graphic methods find the magnitudes of F_3, F_4, and F_5 if the five coplanar forces F_1 to F_5, inclusive, are in equilibrium.

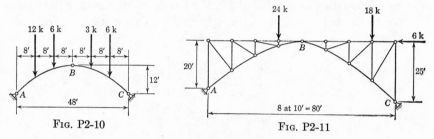

FIG. P2-10

FIG. P2-11

2-10 and 2-11. By both the algebraic and graphic methods find the horizontal and vertical components of each of the three hinge reactions.

CHAPTER 3

SHEARS AND BENDING MOMENTS IN BEAMS

3-1. Definition of Shears and Bending Moments. A beam has been defined as a structural member subjected to transverse loads only. Take, for example, the beam shown in Fig. 3-1a. The reactions R_A and R_B can be found from the principles of equilibrium of a coplanar-parallel-force system. The problem now is to investigate the *internal* forces (shears and bending moments) within the beam. Suppose that the beam is cut into the two segments by a right section at a distance x from A,

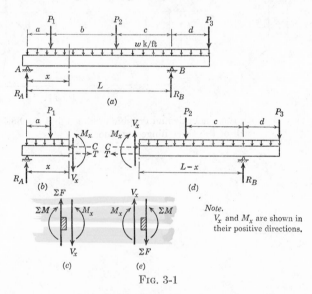

Fig. 3-1

and each segment is treated as a free body as shown in Fig. 3-1b and d. Assume the action of the left segment on the right segment to be a shearing force V_x upward and a bending moment M_x clockwise as shown in Fig. 3-1d. Consistent with the above assumption, the action of the right segment on the left segment must be a shearing force V_x downward and a bending moment M_x counterclockwise, as shown in Fig. 3-1b. In this text, these directions will hereafter be considered as those of positive

26

shear and positive bending moment. In other words, *positive shear* tends to rotate an element of the beam dx in length in the *clockwise* direction, and *positive bending moment* tends to *compress* the *upper* fibers of the element. (Check these definitions on both Fig. 3-1c and e.) V_x and M_x as above defined are known as the shear and bending moment on a section at a distance x from A.

By considering either the left or the right segment as the free body, V_x and M_x can be found from the two equations of equilibrium $\Sigma F = 0$ and $\Sigma M = 0$. The values as determined by either method should agree not only in magnitude but also in sign. Thus, from Fig. 3-1b,

$$\Sigma F = 0: \qquad V_x = R_A - P_1 - wx \qquad (3\text{-}1)$$

$$\Sigma M = 0: \qquad M_x = R_A x - P_1(x - a) - \frac{wx^2}{2} \qquad (3\text{-}2)$$

or, from Fig. 3-1d,

$$\Sigma F = 0: \quad V_x = P_2 + P_3 + w(L - x + d) - R_B \qquad (3\text{-}3)$$

$$\Sigma M = 0: \quad M_x = R_B(L - x) - P_2(L - x - c) - P_3(L - x + d)$$
$$- \frac{w(L - x + d)^2}{2} \qquad (3\text{-}4)$$

Thus, shear at a section is equal to the summation of all the upward forces minus the downward forces to the left of the section, or it is equal to the summation of all downward forces minus the upward forces on the right of the section. Bending moment at a section is equal to the summation of all clockwise moments minus the counterclockwise moments of all forces to the left of the section about the *section*, or it is equal to the summation of all counterclockwise moments minus the clockwise moments of all forces to the right of the section about the *section*. (The moments of *upward* forces, regardless of which side of the section the forces are on, are always positive according to the above definitions.)

If the definitions of shear and bending moment as described in the preceding paragraph are strictly followed, not only the numerical magnitude, but also the *sign*, of the shear or the bending moment at any section in a beam can be determined. Positive

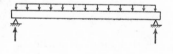

Positive shear

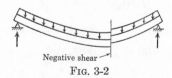

Negative shear

FIG. 3-2

shear and positive bending moment can be defined in another way on a more physical basis. Shear at a section is said to be positive if, by imagining the beam to be cut into two parts at the section, the left segment tends to slide upward relative to the right segment (see Fig. 3-2). Bending moment at a section is positive if the beam bends with the concave side on the top (see Fig. 3-3). The following examples will further emphasize the signs of shears and bending moments in beams.

Example 3-1.   Determine the magnitudes and signs of the shear and bending moment in the beam shown in Fig. 3-4 at a section whose distance from A is (a) 2 ft, (b) 13 ft, and (c) 22 ft.

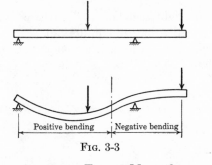

Positive bending   Negative bending

Fig. 3-3

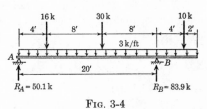

Fig. 3-4

SOLUTION.   From $\Sigma M_B = 0$,

$$20R_A + (10)(4) = (78)(7) + (30)(8) + (16)(16)$$
$$R_A = 50.1 \text{ kips}$$

From $\Sigma M_A = 0$,

$$20R_B = (78)(13) + (16)(4) + (30)(12) + (10)(24)$$
$$R_B = 83.9 \text{ kips}$$

CHECK.   By $\Sigma F = 0$,

$$50.1 + 83.9 \approx 78 + 16 + 30 + 10$$
$$134 = 134$$

(a) V_x and M_x, when $x = 2$ ft from A, are

$$V_x(\text{left}) = 50.1 - (3)(2) = +44.1 \text{ kips}$$
$$V_x(\text{right}) = 16 + 30 + 10 + (3)(24) - 83.9 = +44.1 \text{ kips}$$
$$M_x(\text{left}) = (50.1)(2) - \frac{(3)(2)^2}{2} = +94.2 \text{ kip-ft}$$
$$M_x(\text{right}) = (83.9)(18) - \frac{(3)(24)^2}{2} - (16)(2) - (30)(10) - (10)(22)$$
$$= +94.2 \text{ kip-ft}$$

(b) V_x and M_x, when $x = 13$ ft from A, are

$$V_x(\text{left}) = 50.1 - (3)(13) - 16 - 30 = -34.9 \text{ kips}$$
$$V_x(\text{right}) = 10 + (3)(13) - 83.9 = -34.9 \text{ kips}$$
$$M_x(\text{left}) = (50.1)(13) - (16)(9) - (30)(1) - \frac{(3)(13)^2}{2} = +223.8 \text{ kip-ft}$$
$$M_x(\text{right}) = (83.9)(7) - (10)(11) - \frac{(3)(13)^2}{2} = +223.8 \text{ kip-ft}$$

(c) V_x and M_x, when $x = 22$ ft from A, are

$V_x(\text{left}) = 50.1 + 83.9 - 16 - 30 - (3)(22) = +22.0$ kips
$V_x(\text{right}) = 10 + (3)(4) = +22.0$ kips

$M_x(\text{left}) = (50.1)(22) + (83.9)(2) - (16)(18) - (30)(10) - \dfrac{(3)(22)^2}{2}$

$\qquad = -44.0$ kip-ft

$M_x(\text{right}) = -(10)(2) - \dfrac{(3)(4)^2}{2} = -44.0$ kip-ft

Example 3-2. Determine the magnitudes and signs of the shear and bending moment in the beam shown in Fig. 3-5 at sections (a) 7 ft and (b) 17 ft from A.

SOLUTION. From $\Sigma F = 0$,

$$V_B = 6 + 9 + 24(0.8) = 34.2 \text{ kips}$$

From $\Sigma M_B = 0$,

$$M_B = (6)(20) + (9)(10) + \tfrac{1}{2}(0.8)(24)^2 = 440.4 \text{ kip-ft}$$

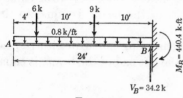

FIG. 3-5

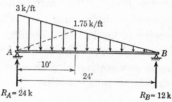

FIG. 3-6

CHECK. By $\Sigma M_A = 0$,

$$(6)(4) + (9)(14) + \tfrac{1}{2}(0.8)(24)^2 + 440.4 \approx (34.2)(24)$$
$$820.8 = 820.8$$

(a) V_x and M_x, when $x = 7$ ft from A, are

$V_x(\text{left}) = -6 - (0.8)(7) = -11.6$ kips
$V_x(\text{right}) = +9 + (0.8)(17) - 34.2 = -11.6$ kips
$M_x(\text{left}) = -(6)(3) - \tfrac{1}{2}(0.8)(7)^2 = -37.6$ kip-ft
$M_x(\text{right}) = +(34.2)(17) - 440.4 - (9)(7) - \tfrac{1}{2}(0.8)(17)^2$
$\qquad = -37.6$ kip-ft

(b) V_x and M_x, when $x = 17$ ft from A, are

$V_x(\text{left}) = -6 - 9 - (0.8)(17) = -28.6$ kips
$V_x(\text{right}) = -34.2 + (0.8)(7) = -28.6$ kips
$M_x(\text{left}) = -(6)(13) - (9)(3) - \tfrac{1}{2}(0.8)(17)^2 = -220.6$ kip-ft
$M_x(\text{right}) = -440.4 + (34.2)(7) - \tfrac{1}{2}(0.8)(7)^2 = -220.6$ kip-ft

Example 3-3. Determine the magnitude and sign of the shear and bending moment in the beam shown in Fig. 3-6 at a section 10 ft to the right of A.

SOLUTION

Total load on the beam $= \frac{1}{2}(3)(24) = 36$ kips

The center of gravity of the total load is at the third point of the span; thus,

$$R_A = 24 \text{ kips} \qquad R_B = 12 \text{ kips}$$

V_x and M_x, when $x = 10$ ft from A, are

$$V_x(\text{left}) = 24 - \frac{1}{2}(3 + 1.75)(10) = +0.25 \text{ kip}$$
$$V_x(\text{right}) = -12 + \frac{1}{2}(1.75)(14) = +0.25 \text{ kip}$$
$$M_x(\text{left}) = (24)(10) - \frac{1}{2}(3)(10)(2\frac{9}{3}) - \frac{1}{2}(1.75)(10)(1\frac{9}{3})$$
$$= +110.83 \text{ kip-ft}$$
$$M_x(\text{right}) = (12)(14) - \frac{1}{2}(1.75)(14)(1\frac{4}{3}) = +110.83 \text{ kip-ft}$$

3-2. Relationship between Load, Shear, and Bending Moment. The relationship between load, shear, and bending moment in a beam can be expressed in the following two propositions:

Proposition 1. The rate of decrease of shear with respect to x, on any section at a distance x from the left end of the beam, is equal to the intensity of load at the section.

Proposition 2. The rate of increase of bending moment with respect to x, on any section at a distance x from the left end of the beam, is equal to the shear at the section.

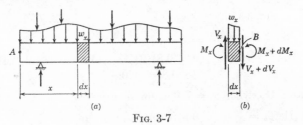

Fig. 3-7

PROOF. Let V_x and M_x be the shear and bending moment on a section at a distance x from A of the beam shown in Fig. 3-7a. Likewise $V_x + dV_x$ and $M_x + dM_x$ are the shear and bending moment on a section at a distance $x + dx$ from A. The free-body diagram of the segment of the beam which is dx in length is shown in Fig. 3-7b. Note that the load (w_x in intensity), shears, and bending moments are all shown in their positive directions.

From $\Sigma F_y = 0$ (Fig. 3-7b),

$$V_x - (V_x + dV_x) - w_x \, dx = 0$$
$$dV_x = -w_x \, dx$$
$$\frac{dV_x}{dx} = -w_x \qquad (3\text{-}5)$$

From $\Sigma M_B = 0$ (Fig. 3-7b),

$$V_x\, dx + M_x = M_x + dM_x + w_x\, dx \frac{dx}{2}$$

The term $\frac{1}{2}w_x(dx)^2$ approaches zero as a limit when compared with the other terms and can therefore be ignored. Simplifying,

$$dM_x = V_x\, dx$$
$$\frac{dM_x}{dx} = V_x \tag{3-6}$$

Equations (3-5) and (3-6) are symbolic expressions of Propositions 1 and 2 stated above. These relations will be further exemplified in the subsequent articles.

3-3. Shear and Bending-moment Equations. The shear and bending moment V_x and M_x on a section at a distance x from any arbitrary reference point on the beam can be expressed in terms of x. Such expressions are called shear and bending-moment equations. Any set of shear and bending-moment equations, however, is applicable only to a definite segment of the beam. This segment is usually between two adjacent concentrated loads or reactions. The reference point, from which x is measured, may be either inside or outside of the segment.

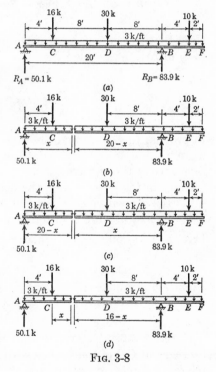

Example 3-4. Determine the shear and bending-moment equations for segment CD of the beam shown in Fig. 3-8a using A, B, and C as the reference points, respectively.

SOLUTION. From $\Sigma M_B = 0$ and $\Sigma M_A = 0$, R_A and R_B are found to be 50.1 kips and 83.9 kips, respectively (see also Example 3-1).

(a) V_x and M_x in segment CD in terms of x measured from A (Fig. 3-8b). Identical expressions for V_x and M_x should be obtained whether the left or the right side of the beam is considered. Although one seems to duplicate the other, both solutions will be shown for purpose of clarification.

Fig. 3-8

Left $\quad V_x = 50.1 - 16 - 3x = 34.1 - 3x$
$\quad\quad M_x = 50.1x - 16(x-4) - \frac{3}{2}x^2 = 64 + 34.1x - 1.5x^2$

Right $V_x = 30 + 10 + 3(26 - x) - 83.9 = 34.1 - 3x$
  $M_x = 83.9(20 - x) - 30(12 - x) - 10(24 - x) - \frac{3}{2}(26 - x)^2$
  $= 64 + 34.1x - 1.5x^2$

Propositions 1 and 2 in Art. 3-2 can be illustrated by noting that the derivative of the expression for $V_x = 34.1 - 3x$ with respect to x is -3 kips per ft, or $-w_x$, and the derivative of

$$M_x = 64 + 34.1x - 1.5x^2$$

is $dM_x/dx = V_x = 34.1 - 3x$.

  (b) V_x and M_x in segment CD in terms of x measured from B (Fig. 3-8c).

Left $V_x = 50.1 - 16 - 3(20 - x) = 3x - 25.9$
  $M_x = 50.1(20 - x) - 16(16 - x) - \frac{3}{2}(20 - x)^2$
  $= 146 + 25.9x - 1.5x^2$
Right $V_x = 30 + 10 + 3(x + 6) - 83.9 = 3x - 25.9$
  $M_x = 83.9x - 30(x - 8) - 10(x + 4) - \frac{3}{2}(x + 6)^2$
  $= 146 + 25.9x - 1.5x^2$

Note that, when x is increasing from right toward left, Eqs. (3-5) and (3-6) in Art. 3-2 become

$$\frac{dV_x}{dx} = +w_x \tag{3-7}$$

and

$$\frac{dM_x}{dx} = -V_x \tag{3-8}$$

In proving Eqs. (3-7) and (3-8), as shown in Fig. 3-7b, it is necessary to write $V_x + dV_x$ and $M_x + dM_x$ on the left face and V_x and M_x on the right face of the element.

  Thus, in the present case, the derivative of $V_x = 3x - 25.9$ is $+3$, or $+w_x$; the derivative of $M_x = 146 + 25.9x - 1.5x^2$ is $25.9 - 3x$, or $-V_x$.

  (c) V_x and M_x in CD in terms of x measured from C (Fig. 3-8d).

Left $V_x = 50.1 - 16 - 3(x + 4) = 22.1 - 3x$
  $M_x = 50.1(x + 4) - 16x - \frac{3}{2}(x + 4)^2$
  $= 176.4 + 22.1x - 1.5x^2$
Right $V_x = 30 + 10 + 3(22 - x) - 83.9 = 22.1 - 3x$
  $M_x = 83.9(16 - x) - 30(8 - x) - 10(20 - x) - \frac{3}{2}(22 - x)^2$
  $= 176.4 + 22.1x - 1.5x^2$

  Example 3-5.   Determine the shear and bending-moment equations for segment CD of the beam shown in Fig. 3-9a using A, B, and C as the reference points, respectively.

  SOLUTION.   From $\Sigma F = 0$ and $\Sigma M_B = 0$, V_B and M_B are found to be 34.2 kips and 440.4 kip-ft, respectively (see also Example 3-2).

  (a) V_x and M_x in segment CD in terms of x measured from A (Fig. 3-9b).
Left $V_x = -6 - 0.8x$
  $M_x = -6(x - 4) - \frac{1}{2}(0.8)x^2 \quad = 24 - 6x - 0.4x^2$

Right $\quad V_x = +9 + (0.8)(24 - x) - 34.2 = -6 - 0.8x$
$\quad\quad\ M_x = -440.4 + 34.2(24 - x) - 9(14 - x) - \frac{1}{2}(0.8)$
$\quad\quad\quad\quad\ (24 - x)^2 = 24 - 6x - 0.4x^2$

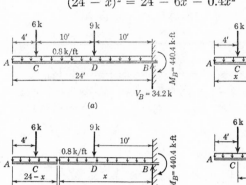

(a)

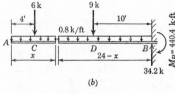

(b)

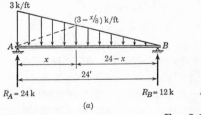

(c)

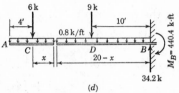

(d)

Fig. 3-9

(b) V_x and M_x in segment CD in terms of x measured from B (Fig. 3-9c).

Left $\quad\quad V_x = -6 - 0.8(24 - x) = -25.2 + 0.8x$
$\quad\quad\quad\ M_x = -6(20 - x) - \frac{1}{2}(0.8)(24 - x)^2$
$\quad\quad\quad\quad = -350.4 + 25.2x - 0.4x^2$
Right $\quad\quad V_x = +9 + 0.8x - 34.2 = -25.2 + 0.8x$
$\quad\quad\quad\ M_x = -440.4 + 34.2x - 9(x - 10) - \frac{1}{2}(0.8)x^2$
$\quad\quad\quad\quad = -350.4 + 25.2x - 0.4x^2$

(c) V_x and M_x in segment CD in terms of x measured from C (Fig. 3-9d).

Left $\quad\quad V_x = -6 - 0.8(x + 4) = -9.2 - 0.8x$
$\quad\quad\quad\ M_x = -6x - \frac{1}{2}(0.8)(x + 4)^2 = -6.4 - 9.2x - 0.4x^2$
Right $\quad V_x = +9 + 0.8(20 - x) - 34.2 = -9.2 - 0.8x$
$\quad\quad\quad\ M_x = -440.4 + 34.2(20 - x) - 9(10 - x) - \frac{1}{2}(0.8)(20 - x)^2$
$\quad\quad\quad\quad = -6.4 - 9.2x - 0.4x^2$

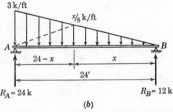

(a)   (b)

Fig. 3-10

Example 3-6.   Determine the shear and bending-moment equations for the beam shown in Fig. 3-10 using A and B as the reference points, respectively.

SOLUTION.   (a) V_x and M_x in the beam in terms of x measured from A (Fig. 3-10a).   The intensity of load w_x at a section distant x from A is

$$w_x = \frac{3(24 - x)}{24} = 3 - \tfrac{1}{8}x \text{ kips per ft}$$

Using the left free body,

$$V_x = 24 - \tfrac{1}{2}(3)x - \tfrac{1}{2}(3 - \tfrac{1}{8}x)x = 24 - 3x + \tfrac{1}{16}x^2$$
$$M_x = 24x - \tfrac{1}{2}(3x)(\tfrac{2}{3}x) - \tfrac{1}{2}x(3 - \tfrac{1}{8}x)(\tfrac{1}{3}x)$$
$$= 24x - \tfrac{3}{2}x^2 + \tfrac{1}{48}x^3$$

Using the right free body,

$$V_x = -12 + \tfrac{1}{2}(24 - x)(3 - \tfrac{1}{8}x) = 24 - 3x + \tfrac{1}{16}x^2$$
$$M_x = 12(24 - x) - \tfrac{1}{2}(24 - x)(3 - \tfrac{1}{8}x)\frac{(24 - x)}{3}$$
$$= 24x - \tfrac{3}{2}x^2 + \tfrac{1}{48}x^3$$

Note that the relations $dV_x/dx = -w_x$ and $dM_x/dx = V_x$ are satisfied by the above expressions for w_x, V_x, and M_x.

(b) V_x and M_x in the beam in terms of x measured from B (Fig. 3-10b). The intensity of load w_x at a section distant x from B is

$$w_x = \frac{3x}{24} = \frac{1}{8}x \text{ kips per ft}$$

Using the left free body,

$$V_x = 24 - \tfrac{1}{2}(3)(24 - x) - \tfrac{1}{2}(\tfrac{1}{8}x)(24 - x)$$
$$= -12 + \tfrac{1}{16}x^2$$
$$M_x = 24(24 - x) - \tfrac{1}{2}(3)(24 - x)[\tfrac{2}{3}(24 - x)]$$
$$\qquad\qquad\qquad - \tfrac{1}{2}(\tfrac{1}{8}x)(24 - x)[\tfrac{1}{3}(24 - x)]$$
$$= 12x - \tfrac{1}{48}x^3$$

Using the right free body,

$$V_x = -12 + \tfrac{1}{2}x(\tfrac{1}{8}x) = -12 + \tfrac{1}{16}x^2$$
$$M_x = 12x - \tfrac{1}{2}x(\tfrac{1}{8}x)(\tfrac{1}{3}x) = 12x - \tfrac{1}{48}x^3$$

In this case, x is increasing from right toward left; so the above expressions for w_x, V_x, and M_x should satisfy the relations $dV_x/dx = +w_x$ and $dM_x/dx = -V_x$.

3-4. Shear and Bending-moment Diagrams. Shear and bending-moment diagrams are the graphs showing the variations along the length of the beam in the values of the shear and bending moment due to a fixed loading condition.   The shear and bending-moment curves or diagrams can be plotted from the shear and bending-moment equations.   Since any set of shear and bending-moment equations is valid only within a definite range of values of x, they can be used only to plot the shear and bending-moment curves for the segment of the beam to which the equations apply. If the beam is subjected to concentrated loads only, the shear curves are horizontal and straight and the bending-moment curves are sloping and linear; but if the beam is subjected to both concentrated and uniformly distributed loads, the shear curves are linear and the bending-moment curves are parabolic.

Shear and bending-moment diagrams can usually be drawn without actually deriving the shear and bending-moment equations. Since the shear at any section is equal to the summation of all upward forces minus the downward forces to the left of the section, the shear ordinate can be obtained by summing all the forces to the *left* of the section. Thus the procedure is that of "stepping up and down with the forces from the left end to the right end of the beam." Inasmuch as the summation of all forces acting on the beam must equal zero, the shear diagram must *close*.

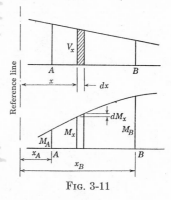

Fig. 3-11

The bending-moment ordinate at any section may be found by summing up the areas of the shear diagram to the left of the section. This is true because it can be shown that the change of moment between any two sections is equal to the area of the shear diagram between these two sections.

Thus, referring to Fig. 3-11 and Eq. (3-6),

$$\frac{dM_x}{dx} = V_x$$

from which $\qquad\qquad dM_x = V_x\,dx$

Integrating between sections A and B,

$$\int_{x_A}^{x} dM_x = \int_{x_A}^{x_B} V_x\,dx$$

or $\qquad\qquad M_B - M_A = \int_{x_A}^{x_B} V_x\,dx$ $\qquad\qquad$ (3-9)

With the known value of the bending moment at the left end of the beam, and by use of Eq. (3-9), the bending moment at all critical sections of the beam can be computed and plotted to obtain the bending-moment diagram. A check is always available because the bending moment at the left end plus the total area of the shear diagram is equal to the bending moment at the right end, or in other words, the bending-moment diagram must *close*. It will be noted that the bending moments and shear areas all carry signs, i.e., the bending moment at a section is positive if the upper part of the section is in compression, and the shear area is positive if it is above the horizontal reference line.

Since the derivative of the bending moment at a section with respect to x is equal to the shear at the same section (or the slope of the tangent to the bending-moment curve at any point is equal to the ordinate to the shear curve at the same point), it is obvious that the bending moment is maximum at a section where the shear equals zero. Such a section is sometimes called the dangerous section, but probably the term "section

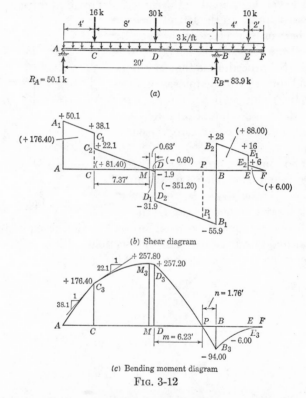

(a)

(b) Shear diagram

(c) Bending moment diagram

FIG. 3-12

of maximum bending moment" is more significant. The tangent to the bending-moment curve at the section of maximum bending moment is horizontal.

At a section where a concentrated load acts, there is an abrupt drop in shear and thus an abrupt change in the slope of the tangent to the bending-moment curve. The bending-moment curve is therefore *discontinuous* at such a section, or the tangent to the left branch of the bending-moment curve is not collinear with that to the right branch.

Example 3-7. Draw shear and bending-moment diagrams for the beam shown in Fig. 3-12a.

SOLUTION.   From $\Sigma M_B = 0$ and $\Sigma M_A = 0$, R_A and R_B are found to be 50.1 and 83.9 kips, respectively (see Example 3-1).

The shear diagram is plotted as $A A_1 C_1 C_2 D_1 D_2 B_1 B_2 E_1 E_2 F$, as shown in Fig. 3-12$b$. $A A_1 = 50.1$ kips upward. $A_1 C_1$ is a straight line which slopes down at the rate of 3 kips per ft.   Thus $CC_1 = 50.1 - 3(4) = 38.1$ kips. $C_1 C_2 =$ load at $C = 16$ kips. $CC_2 = 38.1 - 16 = 22.1$ kips. $DD_1 = 22.1 - 3(8) = -1.9$ kips. $DD_2 = -1.9 - 30 = -31.9$ kips. $BB_1 = -31.9 - 3(8) = -55.9$ kips. $BB_2 = -55.9 + 83.9 = +28$ kips. $EE_1 = +28 - 3(4) = 16$ kips. $EE_2 = 16 - 10 = 6$ kips.   The shear at F is found to be $6 - 2(3) = 0$.

The shear diagram shows that there are two values of shear at C; viz., $+38.1$ and $+22.1$ kips.   In fact, the $+38.1$ kips is the value of the shear at a section which is at an infinitesimal distance to the left of C where the load of 16 kips acts, and $+22.1$ kips is the value of shear at a section which is at an infinitesimal distance to the right of C.

The location of point M, the section of maximum bending moment, may be found as follows:

$$CM = \frac{CC_2}{w_x} = \frac{22.1}{3} = 7.37 \text{ ft}$$

or
$$MD = \frac{DD_1}{w_x} = \frac{1.9}{3} = 0.63 \text{ ft}$$

and
$$CM + MD = 7.37 + 0.63 = 8 \text{ ft} \qquad (check)$$

The shear areas are computed and entered in the shear diagram.   Thus

$$A A_1 C_1 C = +\tfrac{1}{2}(50.1 + 38.1)(4) = +176.40 \text{ kip-ft}$$
$$CC_2 M = +\tfrac{1}{2}(22.1)(7.37) = +81.40 \text{ kip-ft}$$
$$MDD_1 = -\tfrac{1}{2}(1.9)(0.63) = -0.60 \text{ kip-ft}$$
$$DD_2 B_1 B = -\tfrac{1}{2}(31.9 + 55.9)(8) = -351.20 \text{ kip-ft}$$
$$BB_2 E_1 E = +\tfrac{1}{2}(28 + 16)(4) = +88.00 \text{ kip-ft}$$
$$E_2 EF = +\tfrac{1}{2}(6)(2) = +6.00 \text{ kip-ft}$$

The total area of the shear diagram is found to be zero.   This verifies the statement that the "bending moment at the left end plus the total shear area is equal to the bending moment at the right end."   Note that in this problem the bending moment is zero at both ends.

The bending-moment diagram is plotted as $A C_3 M_3 D_3 B_3 E_3 F$ in Fig. 3-12c.

$$M_A = 0$$
$$M_C = M_A + 176.40 = 0 + 176.40 = +176.40$$
$$M_M = M_C + 81.40 = 176.40 + 81.40 = +257.80$$
$$M_D = 257.80 - 0.60 = +257.20$$
$$M_B = 257.20 - 351.20 = -94.00$$
$$M_E = -94.00 + 88.00 = -6.00$$
$$M_F = -6.00 + 6.00 = 0 \quad (check)$$

Upon determination of these controlling values a smooth curve may be drawn through the plotted points A, C_3, M_3, D_3, B_3, E_3, and F. The way in which the curvature occurs can best be visualized by inspecting the shear diagram. For instance, the shear curve A_1C_1 is linearly downward, thus indicating that the slope to the moment curve AC_3 is constantly decreasing from A to C_3. On the moment curve AC_3, the slope at A is 50.1 kips, the slope at C_3 is 38.1 kips, and the slope at other intermediate points is equal to the value of the shear ordinate at that point. From this consideration, it is apparent that the moment curve AC_3 is parabolic and it is convex at the top and concave at the bottom.

It has been stated that the slope of the tangent to the bending-moment curve at any point is equal to the ordinate to the shear curve at the same point. At point C_3 on the bending-moment curve of Fig. 3-12c, the slope of the tangent to curve C_3A is $+38.1$, and that to curve C_3M_3 is $+22.1$; $+38.1$ and $+22.1$ are, respectively, the shear at a section just to the left or right of C where a concentrated load of 16 kips acts.

The point at which bending moment is zero is called the point of inflection. The distances $DP = m$ or $BP = n$ in Fig. 3-12c locate the point of inflection P. The distances m and n, the sum of which must be 8 ft for a check, may be found as follows:

$$M_P - M_D = \text{shear area between } D \text{ and } P$$
$$0 - 257.20 = -\tfrac{1}{2}m(31.9 + 31.9 + 3m)$$
$$1.5m^2 + 31.9m = 257.20$$
$$m^2 + 21.27m + 10.64^2 = 171.47 + 10.64^2$$
$$m + 10.64 = \pm \sqrt{284.68} = \pm 16.87$$
$$m = 16.87 - 10.64 = 6.23 \text{ ft}$$
$$M_B - M_P = \text{shear area between } P \text{ and } B$$
$$-94.00 - 0 = -\tfrac{1}{2}n(55.9 - 3n + 55.9)$$
$$1.5n^2 - 55.9n = -94.00$$
$$n^2 - 37.27n + 18.64^2 = -62.67 + 18.64^2$$
$$n - 18.64 = \pm \sqrt{284.78} = \pm 16.88$$
$$n = -16.88 + 18.64 = 1.76 \text{ ft}$$
$$m + n = 6.23 + 1.76 = 7.99 \doteq 8 \text{ ft} \qquad (check)$$

Example 3-8. Draw shear and bending-moment diagrams for the beam shown in Fig. 3-13a.

SOLUTION. Considering the whole cantilever beam as a free body, from $\Sigma F_y = 0$ (Fig. 3-13a),

$$V_A = 8 + 15 + 4 + 6 = 33 \text{ kips upward}$$

From $\Sigma M_A = 0$,

$$M_A = (8)(9) + (15)(19) + (4)(24) + (6)(30)$$
$$= 633 \text{ kip-ft counterclockwise}$$

CHECK. By $\Sigma M_F = 0$,

$$(33)(32) = 633 + (6)(2) + (4)(8) + (15)(13) + (8)(23)$$
$$1{,}056 = 1{,}056$$

The shear diagram in Fig. 3-13b is plotted by summing the forces to the left of the section concerned. Note that the moment $M_A = 633$ kip-ft does not affect the shear diagram. The parts of the shear diagram are computed to be $+297$, $+250$, $+50$, and $+36$ kip-ft, respectively. The total shear area is $+297 + 250 + 50 + 36 = +633$ kip-ft, which checks the relation $M_E - M_A = 0 - (-633) = +633$ kip-ft. In other words, the increase in moment from A to E is equal to the shear area between A and E. Note that M_A is -633 kip-ft because at section A the upper fibers are in tension.

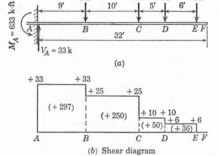

(a)

In plotting the moment diagram,

M_A is -633 kip-ft,
$M_B = -633 + 297 = -336$ kip-ft,
$M_C = -336 + 250 = -86$ kip-ft,
$M_D = -86 + 50 = -36$ kip-ft,
and $M_E = -36 + 36 = 0$ (check).

(b) Shear diagram

Note that the bending-moment diagram is bounded by straight lines because the shear, or the slope of the bending-moment curve, is constant between concentrated loads. The positive slope of the moment curve between A and B is constant at the

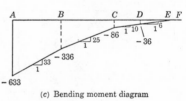

(c) Bending moment diagram
FIG. 3-13

rate of 33 kip-ft per ft, or 33 kips, which is the constant value of shear between A and B.

Example 3-9. Draw shear and bending-moment diagrams for the beam shown in Fig. 3-14a.

SOLUTION. From $\Sigma F = 0$ and $\Sigma M_B = 0$, V_B and M_B are found to be 34.2 kips and 440.4 kip-ft, respectively (see Example 3-2).

The shear diagram in Fig. 3-14b is plotted by summing the forces from the left toward the right. Since the summation of forces between A and B must be zero, the shear diagram must close. The sum of the shear areas is $-6.4 - 132 - 302 = -440.4$ kip-ft; thus the decrease in bending moment from A to B is 440.4 kip-ft. The bending-moment curve is plotted as shown in Fig. 3-14c. Note particularly the breaks in slope at sections C and D.

Example 3-10. Draw shear and bending-moment diagrams for the beam shown in Fig. 3-15a.

SOLUTION. From $\Sigma M_B = 0$ and $\Sigma M_A = 0$, R_A and R_B are found to be 24 and 12 kips, respectively (see Example 3-3).

A freehand sketch of the shear diagram can be drawn as in Fig. 3-15b. First, the shear ordinates at A and B must be $+24$ and -12. The slope of the second-degree curve at A is equal to the load intensity at A, or -3 kips per ft; while the slope at B is zero. The location of the section of zero shear may be determined by writing the expression for shear

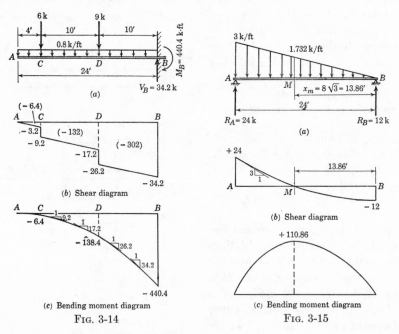

(a)

(b) Shear diagram

(c) Bending moment diagram

FIG. 3-14

FIG. 3-15

at point M in terms of its distance x_m from B, equating this shear to zero, and then solving for x_m. Thus,

$$V_m = 12 - \frac{1}{2}x_m \frac{3x_m}{24} = 0$$

$$x_m = 8\sqrt{3} = 13.86 \text{ ft}$$

In this particular case, the bending moment at section M can be more conveniently found by using the left or right free bodies rather than the shear areas. Considering AM as a free body,

Bending moment at $M = (24)(10.14) - \frac{1}{2}(3)(10.14)(\frac{2}{3})(10.14)$
$$- \frac{1}{2}(1.732)(10.14)(\frac{1}{3})(10.14)$$
$$= 110.86 \text{ kip-ft}$$

Considering BM as a free body,

Bending moment at $M = (12)(13.86) - \frac{1}{2}(1.732)(13.86)(\frac{1}{3})(13.86)$
$$= 110.87 \text{ kip-ft} \quad (check)$$

A sketch of the bending-moment diagram is shown in Fig. 3-15c.

It is interesting to note that, if the total load of 36 kips is uniformly distributed on a 24-ft span, the maximum bending moment at the mid-span is $\frac{1}{8}(36)(24) = 108$ kip-ft.

3-5. Bending-moment Diagram by the Graphic Method. At times when the span is long and the loading is complicated, it may be more convenient to draw the bending-moment diagram by the graphic method, which will now be explained.

PROOF. Let it be required to find the moment of the coplanar, parallel forces $ab(AB)$, $bc(BC)$, $cd(CD)$, and $de(DE)$ shown in Fig. 3-16, about

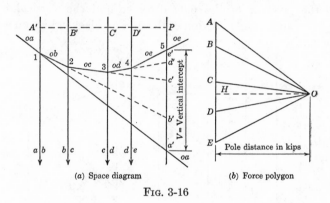

(a) Space diagram   (b) Force polygon

FIG. 3-16

point P. A pole O is so chosen that its distance from the line $ABCDE$ is a convenient number, such as 10 kips, 50 kips, etc. Note that the *pole distance H*, although called a "distance," is expressed in terms of units of force, such as pounds or kips. The strings oa, ob, oc, od, and oe are then drawn in succession. The prolongations of strings oa and oe intersect a vertical line through P at points a' and e'. The distance $a'e'$ is known as the *vertical intercept V*. It will be shown that the moment of forces AB, BC, CD, and DE, about P is equal to the product of H and V, or

$$M_P = HV \tag{3-10}$$

Note that H is a force usually expressed in kips, and V is a distance commonly measured in feet.

By definition (Fig. 3-16),

$$M_P = (AB)(A'P) + (BC)(B'P) + (CD)(C'P) + (DE)(D'P) \tag{3-11}$$

From similar triangles 1-a'-b' and OAB,

$$\frac{a'b'}{A'P} = \frac{AB}{H} \quad \text{(ratio of base to altitude)}$$

or
$$(AB)(A'P) = (a'b')H \qquad (3\text{-}12a)$$

Likewise, from similar triangles 2-b'-c' and OBC, 3-c'-d' and OCD, and 4-d'-e' and ODE,

$$(BC)(B'P) = (b'c')H \qquad (3\text{-}12b)$$
$$(CD)(C'P) = (c'd')H \qquad (3\text{-}12c)$$
$$(DE)(D'P) = (d'e')H \qquad (3\text{-}12d)$$

Substituting (3-12) in (3-11),

$$M_P = (a'b' + b'c' + c'd' + d'e')H = (a'e')H = HV$$

which is Eq. (3-10).

Thus the moment of a group of vertical forces about any point is equal to the product of the pole distance and the vertical intercept between the points of intersection of the first and last strings on a vertical line through the moment center.

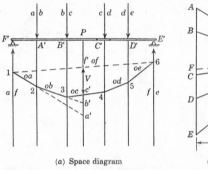 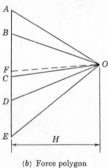

(a) Space diagram   (b) Force polygon

FIG. 3-17

Now consider the simple beam $F'E'$ (Fig. 3-17) carrying loads AB, BC, CD, and DE. The force polygon and the string polygon are drawn as shown. It will be shown that to some scale the string polygon 1-2-3-4-5-6-1 is the bending-moment diagram for the given loads. The bending moment at P is equal to HV, wherein H is the conveniently chosen pole distance and V is the vertical intercept within the string polygon under the point P. By definition of bending moment,

$$M_P = (FA)(F'P) - (AB)(A'P) - (BC)(B'P) \qquad (3\text{-}13)$$

From similar triangles 1-f'-a' and OFA,

$$\frac{f'a'}{F'P} = \frac{FA}{H}$$

or
$$(FA)(F'P) = (f'a')H \qquad (3\text{-}14a)$$

Likewise, from similar triangles 2-$a'b'$ and OAB, and 3-b'-c' and OBC,

$$\frac{a'b'}{A'P} = \frac{AB}{H} \quad \text{or} \quad (AB)(A'P) = (a'b')H \qquad (3\text{-}14b)$$

$$\frac{b'c'}{B'P} = \frac{BC}{H} \quad \text{or} \quad (BC)(B'P) = (b'c')H \qquad (3\text{-}14c)$$

Substituting (3-14) in (3-13),

$$M_P = (f'a' - a'b' - b'c')H = (f'c')H = HV$$

Thus it is seen that the funicular, equilibrium, or string polygon represents to some scale the bending-moment diagram.

It is to be noted that, if a horizontal base line for the bending-moment diagram is desired, it will be necessary to redraw the string polygon by relocating the pole on the horizontal line through point F in Fig. 3-17b. Also note that the bending-moment diagram shown in Fig. 3-17a will be inverted if the pole O is chosen to the left of the load line AE.

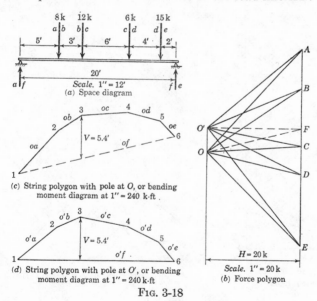

FIG. 3-18

Example 3-11.   Construct the bending-moment diagram for the beam shown in Fig. 3-18a by the graphic method.

SOLUTION.   As shown in the space diagram of Fig. 3-18a, the loads are called ab, bc, cd, and de.   Likewise the reactions are designated ef and fa.   The force polygon $ABCDE$ (Fig. 3-18b) is drawn with a convenient pole distance of 20 kips.   The string polygon 1-2-3-4-5-6-1 is constructed as shown in Fig. 3-18c and then the closing line 1-6 or of is drawn.   The point F is located on the force polygon by drawing OF parallel to of.

In fact, EF and FA are, respectively, the reactions at the right and left supports of the beam. Figure 3-18c is the required bending-moment diagram. Bending-moment ordinates may be measured using a scale of 1 in. = 240 kip-ft, in which 240 kip-ft is the product of the pole distance (20 kips) and the scale of the space diagram (1 in. = 12 ft). For instance

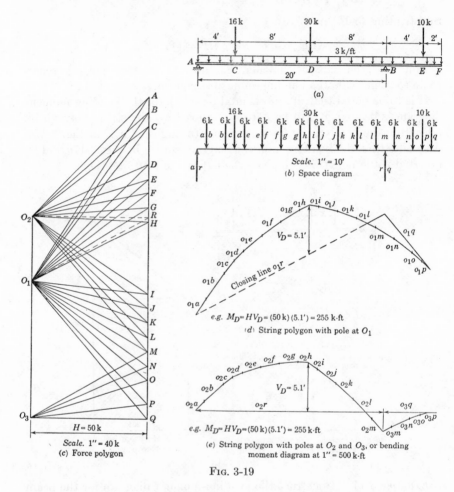

Fig. 3-19

the bending moment at the 12-kip load is measured to be 108 kip-ft, or the product of $V = 5.4$ ft and $H = 20$ kips. If the bending-moment diagram with a horizontal base line is required, it is necessary to redraw the string polygon as shown in Fig. 3-18d, for which the pole O' ($O'F$ is horizontal) has been used.

Example 3-12. Construct the bending-moment diagram for the beam shown in Fig. 3-19a by the graphic method.

SOLUTION. The uniformly distributed load is replaced by a series of evenly spaced concentrated loads. In this case, 13 loads of 6 kips each will be used as shown in Fig. 3-19b. (The larger the number of equivalent concentrated loads, the more accurate will be the results.) The loads are designated from a to q in the space diagram and the reactions are called qr and ra. The force polygon AQ (Fig. 3-19c) is drawn and a pole O is chosen such that the pole distance H is a convenient number, which is 50 kips in this case. The string polygon of Fig. 3-19d is then drawn and the closing line o_1r is found. The point R is located on the force polygon by drawing O_1R parallel to o_1r. In fact, QR and RA are, respectively, the reactions at the right and left supports of the beam. Figure 3-19d is the required bending-moment diagram, and the bending moment

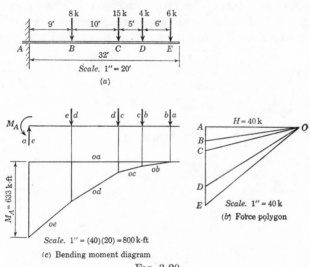

Scale. $1'' = 20'$

(a)

$M_A = 633$ k-ft

$H = 40$ k

Scale. $1'' = 40$ k

(b) Force polygon

Scale. $1'' = (40)(20) = 800$ k-ft

(c) Bending moment diagram

FIG. 3-20

at any point is the product of the *vertical* intercept V in feet and the pole distance H in kips. For instance, the bending moment at D is equal to $HV_D = (50 \text{ kips})(5.1 \text{ ft}) = 255$ kip-ft. Note that in this diagram o_1r and o_1q are the base lines from which the vertical intercepts are measured either upward or downward, a fact which determines whether the bending moment is positive or negative (this is always true when the pole is on the left side of the force polygon). If a bending-moment diagram with *horizontal* base lines (Fig. 3-19e) is desired, it is necessary to use the poles O_2 and O_3 (O_2R and O_3Q are horizontal) for the strings o_2a to o_2m and o_3p to o_3m, respectively (note that o_2m and o_3m meet at the right support in Fig. 3-19e). Figure 3-19d or e is the bending-moment diagram with a scale of 1 in. = 500 kip-ft, wherein 500 is the product of the pole distance (50 kips) and the scale of the space diagram (1 in. = 10 ft).

It is to be noted that the bending-moment diagram thus plotted is theoretically exact for the loading of Fig. 3-19*b*, but only approximate for the loading of Fig. 3-19*a*.

Example 3-13.   Construct the bending-moment diagram for the beam shown in Fig. 3-20*a* by the graphic method.

SOLUTION.   Starting at the free end (Fig. 3-20*c*), the loads on the cantilever beam are designated as *ab*, *bc*, *cd*, and *de*.   The force polygon *ABCDE* in Fig. 3-20*b* is drawn with pole *O* at *H* = 40 kips horizontally to the right of *AE*.   This is done for convenience so that the string polygon will look like the familiar bending-moment diagram.   Had the loads been designated *ab*, *bc*, *cd*, and *de* starting from the left, the pole would be located horizontally to the left of the lower end of the force polygon *ABCDE*.   The string polygon is drawn as shown in Fig. 3-20*c*.   This is the bending-moment diagram required with a scale of 1 in. equal to *H* times the scale of the space diagram, or

$$1 \text{ in.} = (40 \text{ kips})(20 \text{ ft}) = 800 \text{ kip-ft}$$

With this scale, the moment at the fixed end is measured to be 635 kip-ft, which checks with the result of Example 3-8.

PROBLEMS

3-1. Determine the magnitudes and signs of the shear and bending moment in the beam as shown at a section whose distance from *A* is (*a*) 2 ft, (*b*) 10 ft, and (*c*) 15 ft. Check each result by using both the left and right sides of the section as free bodies.

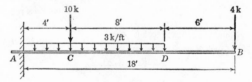

FIG. P3-1

3-2. Determine the magnitudes and signs of the shear and bending moment in the beam as shown at a section whose distance from *A* is (*a*) 5 ft, (*b*) 10 ft, and (*c*) 20 ft. Check each result by using both the left and right sides of the section as free bodies.

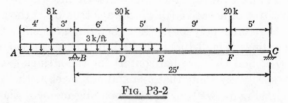

FIG. P3-2

3-3. Determine the magnitude and sign of the shear and bending moment in the beam as shown at a section 8 ft to the right of *A*.

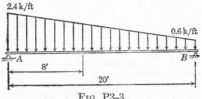

Fig. P3-3

3-4. Determine the shear and bending-moment equations for segment CD of the beam of Prob. 3-1 using A and B as the reference points, respectively. Check each equation by using both the left and right sides of the section as free bodies.

3-5. Determine the shear and bending-moment equations for segment DE of the beam of Prob. 3-2 using A, B, and C as the reference points, respectively. Check each equation by using both the left and right sides of the section as free bodies.

3-6. Determine the shear and bending-moment equations for the beam of Prob. 3-3 using A and B as the reference points, respectively.

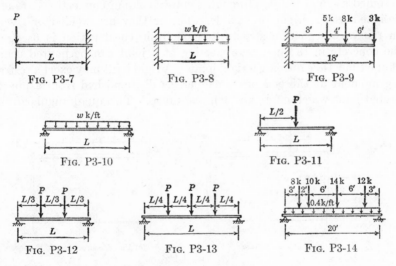

Fig. P3-7   Fig. P3-8   Fig. P3-9

Fig. P3-10   Fig. P3-11

Fig. P3-12   Fig. P3-13   Fig. P3-14

3-7 to 3-14. Draw shear and bending-moment diagrams for the beams shown.

3-15. Draw shear and bending-moment diagrams for the beam of Prob. 3-1.

3-16. Draw shear and bending-moment diagrams for the beam of Prob. 3-2.

3-17. Draw shear and bending-moment diagrams for the beam of Prob. 3-3.

3-18. Construct the bending-moment diagram for the beam of Prob. 3-9 by the graphic method.

3-19. Construct the bending-moment diagram for the beam of Prob. 3-1 by the graphic method.

3-20. Construct the bending-moment diagram for the beam of Prob. 3-2 by the graphic method.

CHAPTER 4

ANALYSIS OF STATICALLY DETERMINATE
RIGID FRAMES AND COMPOSITE STRUCTURES

4-1. Analysis of Statically Determinate Rigid Frames.[1]   A rigid frame
is a structure composed of members which are connected by relatively
stiff or rigid joints.   In steel structures welded or riveted joints may be
designed as rigid joints, although beam-to-column connections are often
designed as semirigid or flexible, whether they are welded or riveted.
In reinforced-concrete structures monolithic construction is used, and
the concrete in the members meeting at a joint can be poured usually
at one time, thus forming a rigid joint.   A rigid joint may exert a restraining moment at the end of a member, as distinguished from a pin-connected joint which offers no such resistance.   Two quadrangular frames

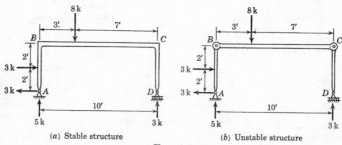

(a) Stable structure   (b) Unstable structure

Fig. 4-1

are shown in Fig. 4-1.   One of these has rigid connections (Fig. 4-1a)
at joints B and C, and the other has pinned connections (Fig. 4-1b) at
these two joints.   It is obvious that the structure shown in Fig. 4-1a is
stable, while that in Fig. 4-1b is unstable.   If the principles of static
equilibrium are applied to the whole frame ABCD, the external reaction
components at A and D for the two frames are identical; however, for
equilibrium of member AB, the joint B must exert a thrust of 5 kips
downward and a restraining moment of 6 kip-ft counterclockwise on
member BA (see Fig. 4-2a).   The restraining moment of 6 kip-ft counterclockwise on member BA and of 6 kip-ft clockwise on member BC can

[1] It is more common for rigid frames to be statically indeterminate.   For the difference between statically determinate and indeterminate rigid frames, see Chap. 13.

be developed by the rigid joint at B in Fig. 4-1a, but not by the pin-connected joint in Fig. 4-1b. The free-body diagrams of joint B and member BC are shown in Fig. 4-2b and c.   Joint C must be regarded as a rigid joint, although it happens in this case that there is no transverse load acting on member CD and consequently no moment is required of joint C.   The free-body diagrams of joint C and of member CD are shown in Fig. 4-2d and e.   Thus one way of finding out whether a structure is stable or not is to draw the free-body diagrams of each member or joint and see that the moment-resisting joints are rigid.

In the analysis of rigid frames, it is first necessary to find the external reaction components.   Free-body diagrams of all members are then drawn, and from these the variation in direct stress, shear, and bending moment in each individual member may be readily computed.   For the beginner it is advisable to sketch separate free-body diagrams of the rigid joints themselves.   It should be noted that the forces and moments (or couples) acting on each free-body diagram, whether it is that of a joint, a

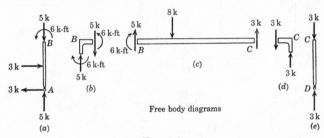

Free body diagrams

FIG. 4-2

member, or the whole frame, must satisfy the three fundamental equations of equilibrium of a general coplanar-force system; i.e., $\Sigma F_x = 0$, $\Sigma F_y = 0$, and $\Sigma M = 0$.   It should also be noted that the analysis is self-checking, because by progressive use of the law that actions and reactions (forces and moments) are equal and opposite, one completely defined free body can be obtained in the end, and the equilibrium of this last free body ensures the correctness of the solution.

It must be emphasized that only statically *determinate* structures with rigid joints are discussed in this chapter.   The definition of statically indeterminate rigid frames and their methods of analysis will be considered in the latter part of this text.

Example 4-1.   Draw free-body, shear, and bending-moment diagrams for all members and joints of the rigid frame shown in Fig. 4-3a.

SOLUTION.   The three unknown reaction components at the fixed support are M_A, H_A, and V_A assumed to act as shown.   By applying the equations $\Sigma F_x = 0$, $\Sigma F_y = 0$, and $\Sigma M_A = 0$ to the whole frame as a free body, M_A, H_A, and V_A are found to be 66 kip-ft counterclockwise, 5 kips to the right, and 24 kips upward, respectively.   The free-body, shear, and moment diagrams of members AB, BC, and CD are drawn in succession as shown in Fig. 4-3b to d.   The reader is advised to sketch these diagrams independently and check the results shown in Fig. 4-3.   Note that plus or minus signs as used in connection with the shear diagrams follow the convention of Art. 3-1; vertical members are treated as beams by looking from the right side of the page.   It is not necessary to indicate moment signs because all bending-moment diagrams are plotted on the *compression* side of the member, which means that the bending moment at any one section is of such a direction that it causes compression on the side of the member where it is plotted.

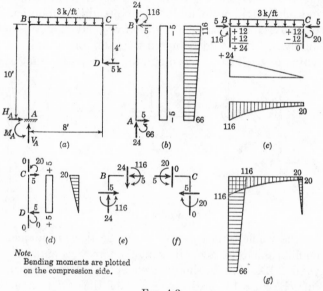

Fig. 4-3

It should be remembered that the change in the bending moment between any two sections is equal to the area of the shear diagram between those two sections.   For example, consider member BC.   The bending moment of 116 kip-ft is first plotted on the lower side of the bending-moment diagram at end B (Fig. 4-3c).   By adding the shear area of $+\frac{1}{2}(24)(8) = +96$ kip-ft between B and C to the moment at B, the moment curve *rises* from 116 kip-ft on the lower side at B to 20 kip-ft on the lower side at C, which checks with the known value of the end moment at C on the free-body diagram of member BC.   The free-body diagrams of joints B and C are shown in Fig. 4-3e and f.   As shown in Fig. 4-3g the bending-moment diagram is plotted on the compression side throughout the whole frame.

Example 4-2.   Draw free-body, shear, and bending-moment diagrams for all members and joints of the rigid frame shown in Fig. 4-4*a*.

SOLUTION.   By applying the equations $\Sigma F_x = 0$, $\Sigma F_y = 0$, and $\Sigma M_D = 0$ to the whole frame as a free body, M_D, H_D, and V_D are found to be 51 kip-ft clockwise, 5 kips to the left, and 24 kips upward, respectively. The free-body, shear, and bending-moment diagrams of members AB, BC, and CD are drawn in succession as shown in Fig. 4-4*b* to *d*.   Particular attention should be called to the manipulation of the forces and end moments acting on the member BC.   It is convenient to first find

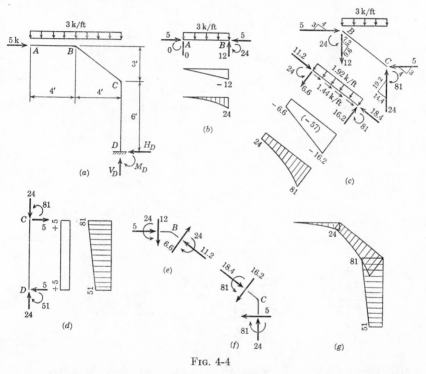

FIG. 4-4

the horizontal and vertical forces acting at B and C by the equations of statics, and then replace these forces by two forces parallel and perpendicular to the axis of the member as shown in Fig. 4-4*c*.   The total vertical uniform load on the member is $(3)(4) = 12$ kips, of which the component in direction of BC is $(12)(\frac{3}{5}) = 7.2$ kips and the component perpendicular to BC is $(12)(\frac{4}{5}) = 9.6$ kips.   Thus the uniform load perpendicular to BC is $9.6/5 = 1.92$ kips per ft and the axial load along BC is $7.2/5 = 1.44$ kips per ft.   The total axial stress in the member thus increases from a compression of 11.2 kips at B to $11.2 + 7.2 = 18.4$ kips at C.   The shear and bending-moment diagrams for member BC are then constructed in the usual manner. The free-body diagrams of joints B and C are shown separately in Fig. 4-4e and f. The complete bending-moment diagram, plotted on the compression side of the frame, is shown in Fig. 4-4g.

Example 4-3. Draw free-body, shear, and bending-moment diagrams for all members and joints of the rigid frame shown in Fig. 4-5a.

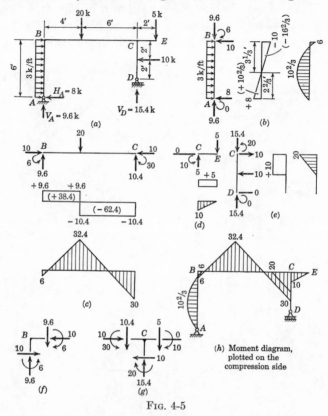

FIG. 4-5

SOLUTION. Taking the whole frame as a free body (Fig. 4-5a),

$$\Sigma M_A = 0: \quad (18)(3) + (20)(4) + (5)(12) = (10)(4) + 10V_D$$
$$V_D = 15.4 \text{ kips upward}$$
$$\Sigma F_x = 0: \quad 18 = 10 + H_A$$
$$H_A = 8 \text{ kips to the left}$$
$$\Sigma F_y = 0: \quad V_A + 15.4 = 20 + 5$$
$$V_A = 9.6 \text{ kips upward}$$

CHECK. By $\Sigma M_D = 0$,

$$(9.6)(10) + (8)(2) + (18)(1) + (5)(2) \approx (20)(6) + (10)(2)$$
$$140 = 140$$

The free-body, shear, and bending-moment diagrams for members AB, BC, CE, and CD are shown in Fig. 4-5b to e.   The free-body diagrams of joints B and C are shown in Fig. 4-5f and g.   The free-body diagram of joint C at which three members are joined should be particularly noted.

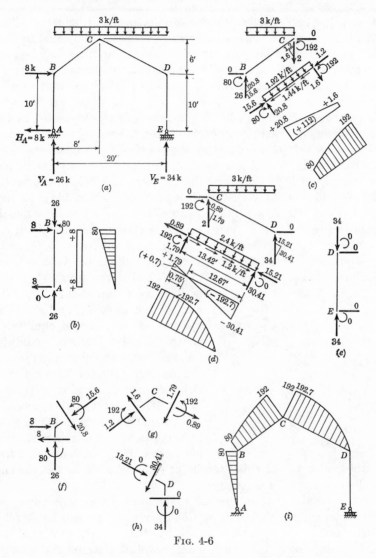

FIG. 4-6

The bending-moment diagram for the whole frame is plotted on the compression side as shown in Fig. 4-5h.

Example 4-4.   Draw free-body, shear, and bending-moment diagrams for all members and joints of the rigid frame shown in Fig. 4-6a.

SOLUTION. The solution of this problem is fully shown in Fig. 4-6. The reader is advised to sketch these diagrams independently on a separate piece of paper and check the results shown in Fig. 4-6. The free-body diagram of joint B, on which an *external* horizontal load of 8 kips acts, should be particularly noted.

4-2. Analysis of Statically Determinate Composite Structures.[1] There are structures in which some members are primarily subjected to direct stresses and others to bending stresses. Such structures may be called "beam trusses" or "truss beams" but are commonly known as composite structures. In analyzing composite structures, it is important to recognize which members are *two-force* members and which are *three-force* members. A two-force member is one which is pin-connected at both ends and is not subjected to any load between the end joints. A member that does not satisfy these requirements is a three-force member. Thus, when a member is treated as a two-force member, its own weight must not be considered in the analysis (unless the member is in the vertical position, in which case the member is subjected to a variable axial stress only). For example, a typical free-body diagram of a two-force member is shown in Fig. 4-7. Here member AB is pin-connected at both ends and is subjected only to forces F_1, F_2, and F_3 at A and forces F_4, F_5, and F_6 at B. Naturally the six forces F_1 through F_6 acting on AB must be in equilibrium. Let R_1 replace F_1, F_2, and F_3 and R_2 replace F_4, F_5, and F_6; or R_1 is the resultant of F_1, F_2, and F_3 and R_2 that of F_4, F_5, and F_6. Then R_1 and R_2 must be in equilibrium. Inasmuch as two forces in equilibrium must be collinear, equal, and opposite, the member AB must be subjected to either direct axial tension, as is shown in Fig. 4-7, or direct axial compression. Thus members which satisfy the definition of two-force members as stated above can always be considered to be under the action of *two* equal tensile or compressive forces at the ends, hence the name two-force members.

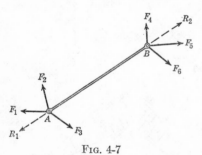

FIG. 4-7

Once the two-force members are differentiated from the three-force members in a composite structure, free-body diagrams can be sketched. The forces and moments acting on each free-body diagram may be determined by using freely the three equations of statics and the fact that actions and reactions are always equal and opposite.

[1] For treatment of statically indeterminate composite structures, see C. K. Wang, "Statically Indeterminate Structures," chap. 12, McGraw-Hill Book Company Inc., New York, 1953.

Example 4-5.   Analyze completely the composite structure shown in Fig. 4-8a.

SOLUTION.   The external reaction components at A and C are shown in Fig. 4-8a as H_A, V_A, H_C, and V_C.   The free-body diagrams of members BC and AB are shown in Fig. 4-8b and c.   Member BC is recognized as a two-force member.   Obviously the resultant of H_C and V_C must be T

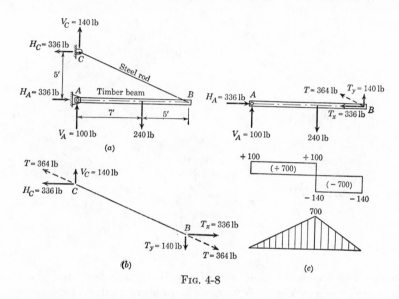

FIG. 4-8

in the direction of BC.   Thus the ratio of V_C to H_C must be $\frac{5}{12}$.   The computations are as follows:

Fig. 4-8a: $\Sigma M_A = 0$ $(240)(7) = 5H_C$ $H_C = 336$ lb
$V_C = \frac{5}{12}H_C = \frac{5}{12}(336) = 140$ lb
$\Sigma F_x = 0$ $H_A = H_C = 336$ lb
$\Sigma F_y = 0$ $V_A = 240 - V_C = 240 - 140 = 100$ lb
Fig. 4-8b: $\Sigma F_x = 0$ $T_x = H_C = 336$ lb
$\Sigma F_y = 0$ $T_y = V_C = 140$ lb

Fig. 4-8c.   The forces acting on this free-body diagram have been determined above.   As a check, these forces should be tested for equilibrium before the shear and bending-moment diagrams are drawn.

Example 4-6.   Analyze completely the composite structure shown in Fig. 4-9a.

SOLUTION.   In this structure AC is a two-force member and AB and BC are three-force members.   By treating the whole structure as a free body (Fig. 4-9a), H_A, V_A, and V_B are found to be 0, 10, and 6 kips, respec-

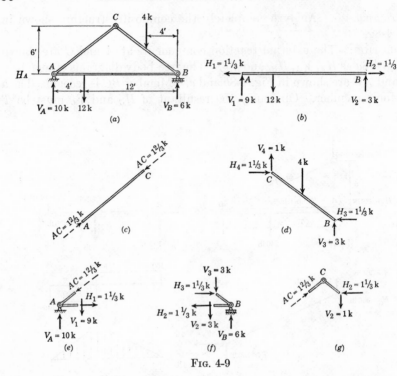

FIG. 4-9

tively. The free-body diagrams of the three members and the three joints are shown in Fig. 4-9b to g. The order of analysis follows:

1. Fig. 4-9b: $V_1 = 9$ kips $V_2 = 3$ kips
2. Fig. 4-9e: $AC = 1\frac{2}{3}$ kips $H_1 = 1\frac{1}{3}$ kips
3. Fig. 4-9b: $H_2 = 1\frac{1}{3}$ kips
4. Fig. 4-9f: $H_3 = 1\frac{1}{3}$ kips $V_3 = 3$ kips
5. Fig. 4-9d: $H_4 = 1\frac{1}{3}$ kips $V_4 = 1$ kip
6. Fig. 4-9d: Check by $\Sigma M = 0$
7. Fig. 4-9g: Check by $\Sigma F_x = 0$ and $\Sigma F_y = 0$

It must be noted that the reactions V_A and V_B are *external* forces acting on the joints A and B. The shear and bending-moment diagrams for members AB and BC are not shown but may be supplied by the reader. Note that it is necessary to replace the horizontal and vertical forces on member BC by forces parallel and perpendicular to the axis of the member before shear and bending-moment diagrams can be drawn. The reader is again advised to repeat the analysis and to note that the solution is entirely self-checking.

Example 4-7. Analyze completely the composite structure shown in Fig. 4-10a.

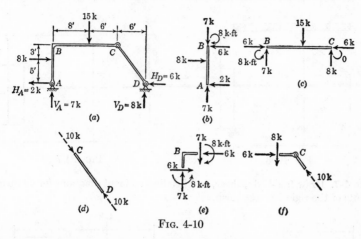

Fig. 4-10

SOLUTION. CD is recognized to be the only two-force member. (Thus the ratio of V_D to H_D must be $\frac{8}{6}$ or $\frac{4}{3}$ in order that their resultant may be in the direction of CD.) The remainder of the solution is shown in Fig. 4-10 and will not be explained in detail. It will be interesting to solve a problem similar to this one, but with the addition of a load in some direction on member CD (see Prob. 4-11).

PROBLEMS

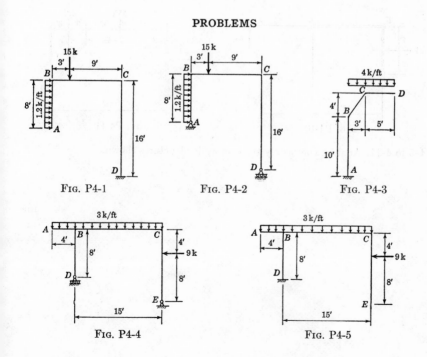

FIG. P4-1   FIG. P4-2   FIG. P4-3

FIG. P4-4   FIG. P4-5

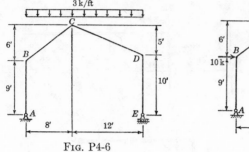

Fig. P4-6

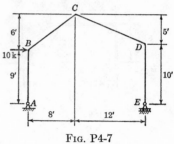

Fig. P4-7

4-1 to 4-7. Draw free-body, shear, and bending-moment diagrams for all members and joints of the rigid frames shown.

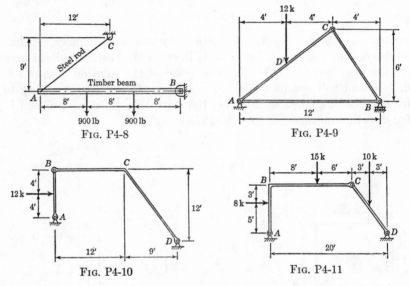

Fig. P4-8

Fig. P4-9

Fig. P4-10

Fig. P4-11

4-8 to 4-11. Analyze completely the composite structures shown.

CHAPTER 5

STRESSES IN TRUSSES

5-1. Stress Analysis of Trusses. A truss is a structure composed of individual members joined together so as to form a series of triangles. The joints are usually assumed to act as frictionless hinges or pins, but they are often riveted or otherwise connected so that some end restraint is developed. Generally all the loads on a truss act or are assumed to act only at the joints. Thus it follows that all members of a truss are assumed to be *two-force* members which are subjected only to direct axial stress (tension or compression). A truss is completely analyzed when the direct axial stresses (*kind* and *amount*) in all members have been determined.

Either algebraic or graphic methods may be used in the stress analysis of trusses. The algebraic method consists essentially of isolating certain parts of the truss by cutting through some members and treating the stresses in these members as external forces acting on the free body. If the force system acting on the free body is concurrent, as in a joint, *two* unknown stresses can be found. If the force system acting on the free body is nonconcurrent, *three* unknown stresses can be found. Such a free body is usually obtained by cutting a section through *three* members and treating either side of the truss as a free body. The former procedure is commonly known as the method of joints and the latter as the method of sections. One variation of the method of sections is called the method of moments and shears. This procedure will be illustrated in Art. 5-4. The graphic method of stress analysis is equivalent to the algebraic method of joints. As a matter of convenience, the force polygons for the concurrent-force system acting on each joint are compounded to form a combined stress diagram, and from this, all unknown stresses may be measured or scaled. As will be explained later, the "closing" of the combined stress diagram provides a valuable check in the graphic analysis.

The three methods of stress analysis of trusses, viz., the algebraic methods of joints and of sections (or of moments and shears) and the graphic method, will be treated in the following articles.

59

5-2. Method of Joints. The external reaction components are first determined by taking the whole truss as a free body. The two equations of equilibrium are then applied to the free-body diagrams of all joints in succession, so that not more than two unknown stresses are involved at each joint. It is advisable to start first with the joint at the left and proceed joint by joint to the middle of the structure and then start with the joint at the right and work back to the middle, thereby obtaining *three* checks at the junction. These three checks ensure the correctness of the solution. It is also advisable to indicate on the truss diagram not only the total stress in each member, but also the horizontal and vertical components of this stress, so that the equilibrium of any one joint can be checked at a glance.

To explain why there are always three independent checks in the method of joints, the relation between the number of joints and number of members in a truss will be developed. Since a truss is an assemblage of triangles, it takes three members and three joints to form the first triangle, and each additional triangle requires two additional members but only one additional joint. Let m and j, respectively, equal the total number of members and joints in a truss; then $(m - 3)$ and $(j - 3)$ are the number of additional members and joints beyond the first triangle. From the previous discussion, the number of additional members is always twice the number of additional joints, or

$$(m - 3) = 2(j - 3)$$

Simplifying,
$$m = 2j - 3 \tag{5-1}$$

The free-body diagram of each joint furnishes two conditions of statics, so there are altogether $2j$ conditions, but only $m = 2j - 3$ unknown stresses; the three extra conditions will therefore be available for checking the solution.

Example 5-1. Using the method of joints, determine the kind and amount of stress in each member of the truss shown in Fig. 5-1a.

SOLUTION. By applying the equations of statics to Fig. 5-1a, V_0 and V_3 are found to be 19.6 and 22.4 kips, respectively. The free-body diagrams of joints L_0, U_1, L_1, L_3, L_2, and U_2 are drawn in succession in Fig. 5-1b to g. Two unknown bar stresses can be solved at each joint; thus,

1. Joint L_0 (Fig. 5-1b): $L_0U_1 = 24.5$ kips compression
                               $L_0L_1 = 14.7$ kips tension
2. Joint U_1 (Fig. 5-1c): $U_1U_2 = 14.7$ kips compression
                               $U_1L_1 = 1.6$ kips tension
3. Joint L_1 (Fig. 5-1d): $L_1U_2 = 2.7$ kips compression
                               $L_1L_2 = 16.8$ kips tension

When the middle of the truss has been reached, it is usually preferable to start with the right end and work back to the middle where three checks may be obtained.

4. Joint L_3 (Fig. 5-1e): $L_2L_3 = 16.8$ kips tension

$U_2L_3 = 28.0$ kips compression

5. Joint L_2 (Fig. 5-1f): $U_2L_2 = 24.0$ kips tension

$\Sigma F_x = 0$ (check)

6. Joint U_2 (Fig. 5-1g): $\Sigma F_x = 0$ (check)

$\Sigma F_y = 0$ (check)

In applying the resolution equations, it is helpful to show on all free-body diagrams the horizontal and vertical projections of the resultant

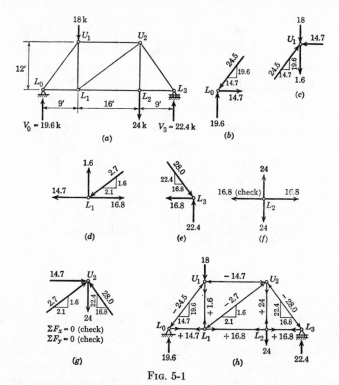

Fig. 5-1

stress in each member. In Fig. 5-1h is shown the *answer diagram* on which the stresses with their horizontal and vertical components in all members are indicated. Note that a minus sign indicates a compressive stress and a positive sign indicates tension. The arrows in Fig. 5-1h represent the direction of the action of the member *on the joint*, not the action of the joint on the member. Take, for example, member L_0U_1.

The member is pushing on the joints L_0 and U_1; thus the joints are pushing on the ends of the member. This indicates a compressive stress. Usually the experienced computer can omit diagrams such as Fig. 5-1b to g and work only with a diagram like Fig. 5-1h.

Example 5-2. Using the method of joints, determine the kind and amount of stress in each member of the truss shown in Fig. 5-2a.

SOLUTION. The solution by the method of joints is shown in Fig. 5-2b. The reactions at L_0 and L_1 are 2,000 lb each. First take joint L_0 as a free body. To satisfy $\Sigma F_y = 0$ at this joint, L_0U_1 must push on the joint with an effective vertical component of 1,500 lb. By proportion, the corresponding horizontal component is 3,000 lb and the total stress

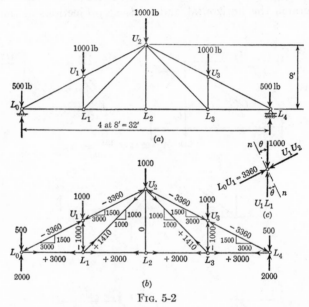

FIG. 5-2

in member L_0U_1 is 3,360 lb (compression). To satisfy the condition that $\Sigma F_x = 0$ at L_0, member L_0L_1 must pull on the joint with 3,000 lb (tension). Because only two unknowns can be involved at each joint, consideration is next given to joint U_1 (there are three unknowns at L_1), which is shown as a free body in Fig. 5-2c. To satisfy $\Sigma F_n = 0$, $1,000 \cos \theta = U_1L_1 \cos \theta$, and $U_1L_1 = 1,000$ lb compression. By resolving forces along the member it is seen that stress in U_1U_2 is equal to that in L_0U_1 (compression). Joints L_1, L_2, and U_2 are taken as free bodies in order and the condition $\Sigma F_y = 0$ for joint U_2 is used as a check. Because of symmetry, this is the only independent check available.

5-3. Method of Sections. The method of sections involves the passing of a section through three members and applying the three equations of statics to the free body on either side of the section. Usually the force system acting on such a free body is nonconcurrent and nonparallel; thus the unknown stresses in the three cut members can be found. Take, for example, the truss shown in Fig. 5-3a. The stresses in members U_2U_3, U_2L_3, and L_2L_3 can be found by passing a section through these three members and treating either side (Fig. 5-3b or c) as a free body. In Fig. 5-3b and c, the stresses in members U_2L_3 and L_2L_3 are assumed to be tensile and that in U_2U_3 compressive. Positive results from subsequent computations will confirm the assumption. In case of a negative result, a circle should be drawn around the original arrowhead. This will call

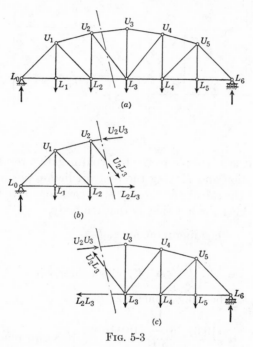

Fig. 5-3

attention to the fact that the assumed direction of the stress in this member was incorrect. Note that, in Fig. 5-3b or c, any one unknown stress may be found independently from a moment equation about the point of intersection of the two other unknown stresses.

The method of sections is particularly useful when only the stresses in some members are desired, or it can be used for solving stresses in all members. In this event, the equilibrium at each joint can be used to check the correctness of the solution. The method of joints is convenient when at least one of the two unknowns at each joint is either horizontal or vertical in direction. When both unknowns are inclined in direction, it may be more convenient to apply the moment, instead of resolution equations to the free body of the joint. This procedure will be illustrated in Example 5-5.

Example 5-3.   Using the method of sections, determine the kind and amount of stress in members U_1U_2, L_1U_2, and L_1L_2 of the truss shown in Fig. 5-4a.

SOLUTION.   By taking the whole truss as a free body, V_0 and V_3 are found to be 19.6 and 22.4 kips, respectively.   Next, a section is passed through members U_1U_2, L_1U_2, and L_1L_2 as shown in Fig. 5-4b.   Directions of the unknown forces or stresses are assumed as indicated on the cut members.   Although it is not necessary to anticipate correctly the

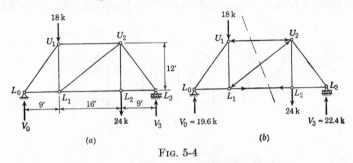

FIG. 5-4

kind of stress, it is advisable to try to avoid negative results by visualizing the truss action and making the best prediction possible (a negative sign means that the calculated force or stress acts opposite to the assumed direction).   Taking the left side as the free body,

$$U_1U_2 = \frac{\text{bending moment at } L_1}{12}$$

$$= \frac{(19.6)(9)}{12} = 14.7 \text{ kips compression}$$

$$L_1L_2 = \frac{\text{bending moment at } U_2}{12}$$

$$= \frac{(19.6)(25) - (18)(16)}{12} = 16.8 \text{ kips tension}$$

$$\Sigma F_y = 0: \qquad +19.6 - 18 - (L_1U_2)(\tfrac{3}{5}) = 0$$
$$L_1U_2 = 2.7 \text{ kips compression}$$

Taking the right side as the free body,

$$U_1U_2 = \frac{\text{bending moment at } L_1}{12}$$

$$= \frac{(22.4)(25) - (24)(16)}{12}$$

$$= 14.7 \text{ kips compression} \qquad (check)$$

$$L_1L_2 = \frac{\text{bending moment at } U_2}{12}$$

$$= \frac{(22.4)(9)}{12} = 16.8 \text{ kips tension} \qquad (check)$$

$\Sigma F_y = 0:$ $\qquad +22.4 - 24 + (L_1U_2)(\tfrac{3}{5}) = 0$

$\qquad\qquad L_1U_2 = 2.7$ kips compression $\qquad$ (*check*)

Example 5-4. Using the method of sections, determine the kind and amount of stress in members U_1U_2, L_1U_2, and L_1L_2 of the truss shown in Fig. 5-5a.

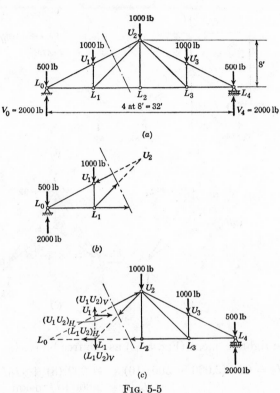

(a)

(b)

(c)

FIG. 5-5

SOLUTION. By passing a plane through members U_1U_2, L_1U_2, and L_1L_2 of the truss in Fig. 5-5a, the free-body diagrams of the left and right sides are obtained as shown in Fig. 5-5b and c. Taking the left segment (Fig. 5-5b) as the free body,

ΣM about $U_2 = 0:$ $\quad (2,000 - 500)(16) = (1,000)(8) + (L_1L_2)(8)$

$\qquad\qquad\qquad\qquad\qquad L_1L_2 = 2,000$ lb tension

ΣM about $L_1 = 0:$ $\quad (2,000 - 500)(8) = (U_1U_2)_H(4)$

$\qquad\qquad\qquad\qquad\quad (U_1U_2)_H = 3,000$ lb

$\qquad\qquad\qquad\qquad\quad (U_1U_2)_V = 1,500$ lb

$\qquad\qquad\qquad\qquad\quad U_1U_2 = 3,360$ lb compression

Note that the stress in U_1U_2 has been resolved at U_1 into its horizontal and vertical components and that the moment of these two components about L_1 is equal to the moment of their resultant about L_1.

ΣM about $L_0 = 0$: $(1,000)(8) = (L_1U_2)_V(8)$

$$(L_1U_2)_V = 1,000 \text{ lb}$$
$$(L_1U_2)_H = 1,000 \text{ lb}$$
$$L_1U_2 = 1,410 \text{ lb tension}$$

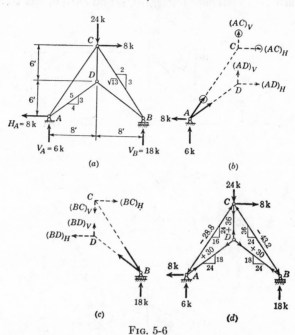

FIG. 5-6

Taking the right segment (Fig. 5-5c) as the free body,

ΣM about $U_2 = 0$: $(2,000 - 500)(16) = (1,000)(8) + (L_1L_2)(8)$
$$L_1L_2 = 2,000 \text{ lb tension} \qquad (check)$$

ΣM about $L_1 = 0$:
$$(2,000 - 500)(24) = (1,000)(16) + (1,000)(8) + (U_1U_2)_H(4)$$
$$(U_1U_2)_H = 3,000 \text{ lb}$$
$$(U_1U_2)_V = 1,500 \text{ lb}$$
$$U_1U_2 = 3,360 \text{ lb compression} \qquad (check)$$

ΣM about $L_0 = 0$:
$$(2,000 - 500)(32) = (1,000)(24) + (1,000)(16) + (L_1U_2)_V(8)$$
$$(L_1U_2)_V = 1,000 \text{ lb}$$
$$(L_1U_2)_H = 1,000 \text{ lb}$$
$$L_1U_2 = 1,410 \text{ lb tension} \qquad (check)$$

Example 5-5. Determine the kind and amount of stress in each member of the truss shown in Fig. 5-6a. Use either the method of joints or the method of sections.

SOLUTION. The external reaction components at A and B must first be determined. From $\Sigma F_x = 0$ on the whole free body, H_A must be equal to 8 kips to the left. The 24-kip vertical load will be reacted by 12 kips each at A and B, and the 8-kip horizontal load will produce a clockwise moment of 96 kip-ft about A or B, which can be balanced by a counterclockwise reaction couple of $96/16 = 6$ kips upward at B and 6 kips downward at A. Thus,

$$V_A = 12 - \frac{(8)(12)}{16} = 12 - 6 = 6 \text{ kips upward}$$

$$V_B = 12 + \frac{(8)(12)}{16} = 12 + 6 = 18 \text{ kips upward}$$

The free-body diagrams of joints A and B are shown in Fig. 5-6b and c. Here the resolution equations of equilibrium are not readily applicable, but the moment equations can be used advantageously. Taking joint A (Fig. 5-6b) as the free body,

$\Sigma M_C = 0$: $\qquad (6)(8) + (8)(12) = (AD)_H(6)$
$\qquad\qquad (AD)_H = 24 \text{ kips}$
$\qquad\qquad (AD)_V = (AD)_H(\tfrac{3}{4}) = 18 \text{ kips}$
$\qquad\qquad AD = (AD)_H(\tfrac{5}{4}) = 30 \text{ kips tension}$
$\Sigma M_D = 0$: $\qquad (6)(8) + (8)(6) + (AC)_H(6) = 0$
$(AC)_H = -16$ kips [draw circles around arrowheads of $(AC)_H$, $(AC)_V$, and AC]
$(AC)_V = (AC)_H(\tfrac{3}{2}) = 24$ kips (numerically only)

$$AC = (AC)_H\left(\frac{\sqrt{13}}{2}\right) = 28.8 \text{ kips compression}$$

Taking joint B (Fig. 5-6c) as the free body,

$\Sigma M_C = 0$: $\qquad (18)(8) = (BD)_H(6)$
$\qquad\qquad (BD)_H = 24 \text{ kips}$
$\qquad\qquad (BD)_V = (24)(\tfrac{3}{4}) = 18 \text{ kips}$
$\qquad\qquad BD = (24)(\tfrac{5}{4}) = 30 \text{ kips tension}$
$\Sigma M_D = 0$: $\qquad (18)(8) = (BC)_H(6)$
$\qquad\qquad (BC)_H = 24 \text{ kips}$
$\qquad\qquad (BC)_V = (24)(\tfrac{3}{2}) = 36 \text{ kips}$

$$BC = (36)\left(\frac{\sqrt{13}}{3}\right) = 43.2 \text{ kips compression}$$

As soon as they are calculated, the values of the horizontal and vertical components as well as the resultant stress in each member are entered in the answer diagram shown in Fig. 5-6d. By inspection of the equilibrium of joint C, the stress in CD is found to be 36 kips tension. The equilibrium of all joints can be readily checked on the answer diagram.

5-4. Method of Moments and Shears. The method of moments and shears may be considered as merely a variation of the method of sections. A horizontal truss subjected to vertical loads resembles a beam in its structural action. The top chord of the truss takes compression and the bottom chord takes tension. In a parallel chord truss the web member (vertical or diagonal) resists the shearing force at the section. Consider, for example, the truss shown in Fig. 5-7. The stress in member U_4U_5, by reason of the method of sections, is equal to the bending moment at L_4 divided by the height of the truss. Similarly, the stress in member L_4L_5 is equal to the bending moment at U_5 divided by the height of the truss. Note that the bending moment at any joint of a truss may be computed by using either the left or the right free body. It is also obvious that, in this simply supported truss, the entire top chord is in compression and the entire bottom chord in tension. The stress in the diagonal L_4U_5, by observing $\Sigma F_y = 0$ in the method of sections, is equal

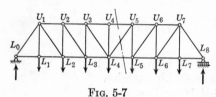

FIG. 5-7

to the shear at the section multiplied by the secant of the angle which the member makes with the vertical. If the shear at the section is positive, the stress in L_4U_5 is compressive, and if the shear is negative, the stress is tensile. The same procedure applies to vertical members, except the stress in a vertical member of a horizontal-chord truss is numerically equal to the shear at the section, and the kind of stress may be determined by inspection. If the upper chord or the lower chord (or sometimes both) cut by a section is inclined in direction, the vertical component of the stress in the web member is equal to the shear at the section modified by the vertical component of the stress in the inclined chord member.

The conception of having the chord members resist the bending moment and the web members take the shearing force at a section may be conveniently applied to finding the stresses in all members of a truss. After the stresses have thus been found, each joint should be checked for equilibrium; i.e., the forces acting on the joint must satisfy the two conditions $\Sigma F_x = 0$ and $\Sigma F_y = 0$. If this check is applied at every joint, the correctness of the solution is assured.

Example 5-6. Determine the kind and amount of stress in each member of the truss shown in Fig. 5-8 by the method of moments and shears.

SOLUTION. Let

d = panel length
h = height of truss
s = length of diagonal
θ = angle between diagonal and vertical

Then $\qquad \tan \theta = \dfrac{d}{h} \qquad \sec \theta = \dfrac{s}{h}$

It is noted that the top chord is in compression and the bottom chord is in tension; so the kind of stress in chord members is no longer in question. The amount of the chord stresses may be found as shown below.

$$L_0L_1 = L_1L_2 = \frac{\text{bending moment at } U_1}{h} = \frac{3\tfrac{1}{2}Pd}{h} = 3\tfrac{1}{2}P \tan \theta$$

$$U_1U_2 = L_2L_3 = \frac{\text{bending moment at } L_2 \text{ or } U_2}{h}$$

$$= \frac{(3\tfrac{1}{2}P)(2d) - P(d)}{h} = 6P \tan \theta$$

$$U_2U_3 = L_3L_4 = \frac{\text{bending moment at } L_3 \text{ or } U_3}{h}$$

$$= \frac{(3\tfrac{1}{2}P)(3d) - P(d + 2d)}{h} = 7\tfrac{1}{2}P \tan \theta$$

$$U_3U_4 = \frac{\text{bending moment at } L_4}{h}$$

$$= \frac{(3\tfrac{1}{2}P)(4d) - P(d + 2d + 3d)}{h} = 8P \tan \theta$$

By the method of shears, the stresses in the diagonals L_0U_1, U_1L_2, U_2L_3, and U_3L_4 are found to be $3\tfrac{1}{2}P \sec \theta$, $2\tfrac{1}{2}P \sec \theta$, $1\tfrac{1}{2}P \sec \theta$, and

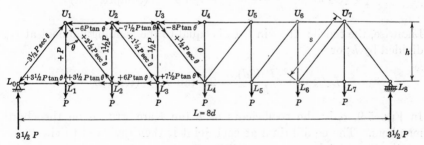

FIG. 5-8

$\tfrac{1}{2}P \sec \theta$; those in the verticals U_2L_2 and U_3L_3 are $1\tfrac{1}{2}P$ and $\tfrac{1}{2}P$, respectively. The kind of stress in these web members is determined by inspection. The stress in the vertical hanger U_1L_1 is $+P$, and that in the center vertical U_4L_4 is zero. Since the truss is symmetrically loaded, the stresses in the members on the right side of the center line are the same as those in the corresponding members on the left segment of the truss.

In this solution it will be noted that the stresses in the chord members are all expressed in terms of $P \tan \theta$; in the diagonals, $P \sec \theta$; and in the verticals, P. The coefficients themselves are sometimes known as index stresses, and the method here treated may be called the method of index stresses.

Again, the correctness of the solution may be checked by observing that the two resolution equations are satisfied at each joint.

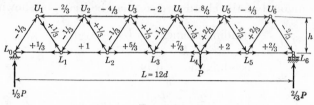

Fig. 5-9

Example 5-7.   Determine the kind and amount of stress in each member of the truss shown in Fig. 5-9 by the method of moments and shears.

SOLUTION.   The reactions are noted to be $\frac{1}{3}P$ and $\frac{2}{3}P$. The stresses in the diagonals are first found by the method of shears. Note that only the coefficients of $P \sec \theta$ are entered on the diagonals of Fig. 5-9. The stresses in the chord members may be found by taking moments. For instance, the compressive stress in U_2U_3 is equal to the bending moment at L_2 divided by h; or

$$U_2U_3 = \frac{(\frac{1}{3}P)(4d)}{h} = \frac{4}{3}P \tan \theta \qquad \text{(compression)}$$

Likewise, the tensile stress in L_4L_5 is equal to the bending moment at U_5 divided by h; or

$$L_4L_5 = \frac{(\frac{2}{3}P)(3d)}{h} = 2P \tan \theta \qquad \text{(tension)}$$

In Fig. 5-9, only the coefficients of $P \tan \theta$ are written on the chord members. The equilibrium at each joint is then reviewed to check the solution completely.

Example 5-8.   Determine the kind and amount of stress in each member of the truss shown in Fig. 5-10 by the method of moments and shears.

SOLUTION.   Finding the reactions,

$$\Sigma M_7 = 0: \qquad V_0 = \frac{9(1) + 12(2) + 10(3)}{7} = 9 \text{ kips}$$

$$\Sigma M_0 = 0: \qquad V_7 = \frac{10(4) + 12(5) + 9(6)}{7} = 22 \text{ kips}$$

CHECK.   By $\Sigma F_y = 0$, $9 + 22 = 10 + 12 + 9$

$$31 = 31$$

The stresses in the lower chord members and the horizontal components of the stresses in the upper chord members are determined by taking moments. Thus,

$$L_0L_1 = L_1L_2 = \frac{\text{bending moment at 1}}{24} = \frac{(9)(24)}{24} = 9 \text{ kips}$$

$$L_2L_3 = (U_1U_2)_H = \frac{\text{bending moment at 2}}{32} = \frac{(9)(2)(24)}{32} = 13.50 \text{ kips}$$

$$L_3L_4 = (U_2U_3)_H = \frac{\text{bending moment at 3}}{36} = \frac{(9)(3)(24)}{36} = 18 \text{ kips}$$

$$U_3U_4 = (U_4U_5)_H = \frac{\text{bending moment at 4}}{36} = \frac{(9)(4)(24)}{36} = 24 \text{ kips}$$

$$L_4L_5 = (U_5U_6)_H = \frac{\text{bending moment at 5}}{32} = \frac{[(22)(2) - (9)(1)]24}{32}$$
$$= 26.25 \text{ kips}$$

$$L_5L_6 = L_6L_7 = \frac{\text{bending moment at 6}}{24} = \frac{(22)(24)}{24} = 22 \text{ kips}$$

The vertical components of the stresses in the upper chord members are found from the horizontal components computed above and the slopes of the upper chords.

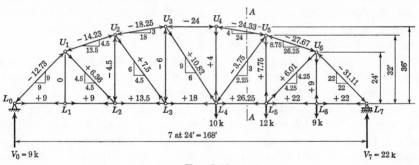

Fig. 5-10

The stresses in the vertical members exclusive of U_1L_1, U_4L_4, and U_6L_6 and the vertical components of the stresses in the diagonals are determined by the method of shears. When a section is cut through an inclined upper chord, however, the vertical component of the stress in this upper chord must be considered in applying the equation $\Sigma F_y = 0$ to the free body on either side of the section. For instance, considering the free body to the right of section AA (Fig. 5-10) and remembering that the vertical component of the stress in U_4U_5 is 4 kips,

$$(L_4U_5)_V + 22 - 9 - 12 - 4 = 0$$
$$(L_4U_5)_V = 3 \text{ kips}$$

After the vertical components of the stresses in the diagonals are computed, the horizontal components can easily be found by simple proportion using the slopes of the diagonals. The stresses in U_1L_1, U_4L_4, and U_6L_6 are found by considering $\Sigma F_y = 0$ at joints L_1, U_4, and L_6.

Equilibrium at each joint can now be reviewed and the total stresses in the inclined members can be found from their horizontal and vertical components. The complete answer diagram is shown in Fig. 5-10.

5-5. The Graphic Method. Two steps are involved in the graphic method of stress analysis of trusses: (1) the external reaction components must be determined by either the algebraic or graphic methods previously discussed in Chap. 2, and (2) the internal stresses in all members can be obtained from the *stress diagram*, which is the superposition of all the individual force polygons for the concurrent-force systems acting on each joint.

In order that the second phase of this work can be systematically performed, a definite scheme of notation and procedure has been suggested, as shown in Fig. 5-11a. The external reaction components may be determined by one of the methods discussed in Chap. 2. Thus the balanced external-force system acting on the truss includes P_4, P_2, V_4, P_3, P_1, V_0, and H_0, named in the *clockwise* order around the truss. Each space between two adjacent external forces is labeled with a capital letter in consecutive order, by starting with the letter A and then proceeding in the *clockwise* direction. Thus the force P_4 can be called AB, the force V_4 called CD; etc. (see Fig. 5-11a). Each triangle is labeled with a numeral by starting with the number 1 and proceeding from left to right. The two numbers, or one letter and one number, on opposite sides of each member are used to represent the *magnitude* and *direction* of the stress in the member. By referring to joint L_0, and reading around the joint in the *clockwise* direction, the stress in member L_0U_1 will be called A-1, not 1-A; but when referring to joint U_1, the stress in the same member L_0U_1 should be called 1-A, not A-1. It will be found that this distinction is of great importance in determining the kind of stress in a member.

The procedure by which the stress diagram in Fig. 5-11b is drawn will now be given. The force polygon of the external-force system is first plotted as $ABCDEFGA$. The points 1, 2, 3, 4, 5, and 6 are then to be located. Point 1 is at the intersection of a line through A parallel to member A-1 (or L_0U_1) and a line through F parallel to member 1-F (or L_0L_1). With the locations of points 1 and E known in Fig. 5-11b, point 2 is the intersection of two lines through 1 and E, respectively, parallel to members 1-2 (or L_1U_1) and 2-E (or L_1L_2). In this case it may be advisable to locate the points 1, 2, and 3 first and then points 6, 5, and 4. The fact that line 3-4 should be vertical (or parallel to member L_2U_2)

serves as a check on the accuracy of the graphic solution. By proceeding from each end and *closing up* near the middle, the cumulative effect of graphic errors can be largely eliminated.

The amount of the stress in member 2-3 (or U_1L_2) is given by the magnitude of the vector 2-3 in the stress diagram. The *kind* of stress

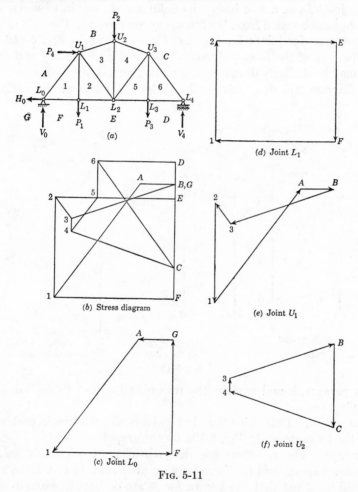

(a)

(d) Joint L_1

(b) Stress diagram

(e) Joint U_1

(c) Joint L_0

(f) Joint U_2

FIG. 5-11

can be determined from the following consideration. The stress in member U_1L_2 is called 2-3 when read clockwise around joint L_2. The direction of 2-3 in Fig. 5-11b is downward to the right, which, when referred to joint L_2, means that the member exerts compression on the joint. The stress in member U_1L_2 is therefore compressive. The same conclusion can be reached if the stress in member U_1L_2 is called 3-2 when referred to joint U_1. The direction of 3-2 in Fig. 5-11b is upward to the left, which, with reference to joint U_1, means compression on the joint. Thus the amount and kind of the stress in each member can be readily found from the stress diagram.

In fact, the stress diagram is merely the superposition of the individual force polygons for the concurrent-force systems at each joint. By considering joint L_0 as a free body, the unknown stresses in members L_0U_1 and L_0L_1 can be scaled from the force polygon shown in Fig. 5-11c. The force polygons for joints L_1, U_1, and U_2 are shown in Fig. 5-11d to f. A comparison of the force polygons shown in Fig. 5-11c to f and others which may be similarly drawn for the remaining joints with the combined stress diagram will demonstrate the statement made at the beginning

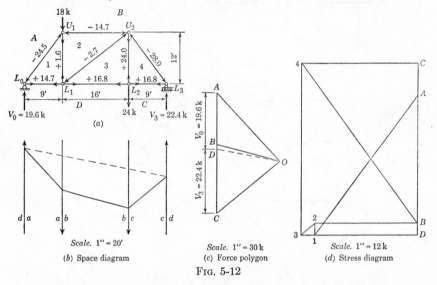

Scale. $1'' = 20'$

(b) Space diagram

Scale. $1'' = 30k$

(c) Force polygon

Scale. $1'' = 12k$

(d) Stress diagram

Fig. 5-12

of this paragraph and explain the reasoning incident to the combined stress diagram.

Example 5-9. Determine the kind and amount of stress in each member of the truss shown in Fig. 5-12a by the graphic method.

SOLUTION. The reactions are determined by the graphic method. The space diagram and the force polygon are shown in Fig. 5-12b and c. It should be noted that the known forces are designated ab and bc and the unknown reactions are indicated by cd and da. By scaling the force polygon in Fig. 5-12c, it is found that $V_3 = CD = 22.4$ kips and $V_0 = BA = 19.6$ kips.

The previously mentioned notation for the graphic solution of the internal stresses is now inserted in Fig. 5-12a; A, B, C, and D around the truss and 1, 2, 3, and 4 in the internal triangles. The stress diagram shown in Fig. 5-12d is drawn in accordance with this notation. The stress in each member is scaled in Fig. 5-12d and then written on the proper member in Fig. 5-12a.

Example 5-10. Determine the kind and amount of stress in each member of the truss shown in Fig. 5-13a by the graphic method.

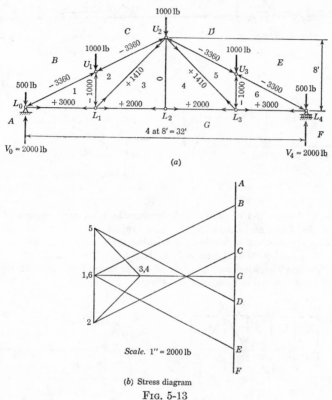

(a)

(b) Stress diagram

FIG. 5-13

SOLUTION. Because of symmetry, the reactions are **2,000** lb each. The stress diagram is shown in Fig. 5-13b, and the scaled stresses are written on the truss members in Fig. 5-13a. It may be interesting to note that, in the case of this simple and symmetrical stress diagram, it is relatively simple to compute the stresses from the geometry of the stress diagram.

PROBLEMS

5-1 to 5-4. Determine the kind and amount of stress in each member of the truss shown by the method of joints.

5-5. Using the method of sections, determine the kind and amount of stress in members U_2U_3, L_2U_3, and L_2L_3 of the truss in Prob. 5-1.

5-6. Using the method of sections, determine the kind and amount of stress in members U_2U_3, U_2L_3, and L_2L_3 of the truss in Prob. 5-2.

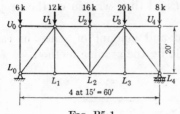

Fig. P5-1

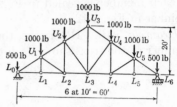

Fig. P5-2

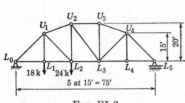

Fig. P5-3

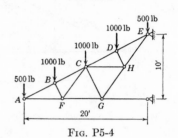

Fig. P5-4

5-7. Using the method of sections, determine the kind and amount of stress in members U_1U_2, U_2L_2, and L_2L_3 of the truss in Prob. 5-3.

5-8. Using the method of sections, determine the kind and amount of stress in members BC, CF, and FG of the truss in Prob. 5-4.

5-9. Determine the kind and amount of stress in each member of the truss in Prob. 5-3 by the method of moments and shears.

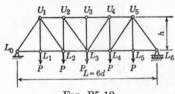

Fig. P5-10

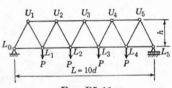

Fig. P5-11

5-10 and 5-11. Determine the kind and amount of stress in each member of the truss shown by the method of moments and shears.

5-12 to 5-15. Determine the kind and amount of stress in each member of the trusses shown in Probs. 5-1 to 5-4, inclusive, by the graphic method.

CHAPTER 6

ANALYSIS OF ROOF TRUSSES

6-1. General Description. In building design wherein an open, unobstructed space with a width of more than 40 or 50 ft is to be provided, the roof is commonly supported by roof trusses spaced from about 15 to 25 ft apart. These trusses may rest on columns or on masonry walls along the sides of the building. A roof truss attached to its supporting columns is commonly called a *bent*, the analysis of which will be treated in Chap. 7. The discussion in the present chapter will be limited to roof trusses supported on masonry walls. If the span of the roof truss is

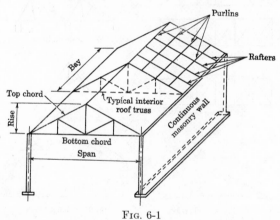

Fig. 6-1

small, say less than 40 or 50 ft, the truss may usually be anchored to the wall at both ends, or the anchor bolts at one end may pass through slotted holes in the bearing plate to make some provision for expansion or contraction due to temperature changes. For longer spans, trusses should be hinged at one end and supported on rollers or rockers at the other end.

A sketch showing the typical roof construction is shown in Fig. 6-1. Here the roof trusses are supported on continuous masonry walls. The distance between adjacent roof trusses is called the *bay*. The *purlins* are longitudinal beams which rest on the top chord, and preferably at the joints of the truss. Unless the purlins are placed at the joints, the top chord will be subjected to combined bending and direct stresses. The roof covering (with or without *sheathing*) may rest directly on the purlins, or on *rafters* which are in turn supported by the purlins. A typical interior roof truss receives purlin loads from both sides; so it supports roof loads on the equivalent of one whole bay length. The *span* of the roof truss is the horizontal distance between the supports; the *rise* is the vertical distance from the *ridge* to the *eaves;* the *pitch* is the ratio of the rise to the span. Although the walls and the purlins assist in maintaining longitudinal stability, additional bracing is usually necessary. The

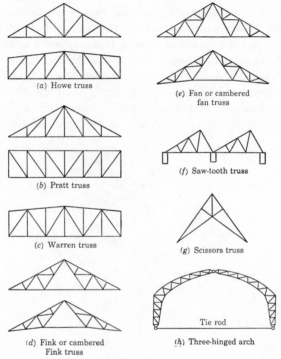

(a) Howe truss

(e) Fan or cambered fan truss

(b) Pratt truss

(f) Saw-tooth truss

(c) Warren truss

(g) Scissors truss

Tie rod

(d) Fink or cambered Fink truss

(h) Three-hinged arch

Fig. 6-2

longitudinal or diagonal bracing members which run from truss to truss may be in the plane of the top chord, the bottom chord, or both; thus the names *top-chord bracing* or *bottom-chord bracing*. Although approximate stress calculations may be made, actual design of the bracing system is largely a matter of experience and judgment.

The selection of the type of the roof truss to be used generally depends on the length of span, the amount of loading, and the kind of materials to be used. Eight common types of roof trusses are shown in Fig. 6-2a to h. The vertical members of the Howe truss or the diagonal members of the Pratt truss are normally in tension and may economically be steel rods if the rest of the truss is built of wood. The top chords of Howe, Pratt, and Warren trusses may be horizontal (or with sufficient slope to provide drainage) or with pronounced slopes as indicated. The Fink and Fan trusses, with or without camber in the lower chord, are usually built of steel. The sawtooth truss is suitable for mill buildings where light from windows facing to the north is desired. The scissors truss is often used in church structures. Three-hinged arches, with or without tie rods, are often used in buildings with long spans, such as armories or gymnasiums.

6-2. Dead, Snow, and Wind Loads. A truss carries its own weight, the weight of the bracing and ceiling and other suspended loads, and the loads from the purlins. One of the requirements for true truss action is that the loads be applied at the joints only. Sometimes purlins are placed on the top chord between panel points. In this case the loads are distributed to the adjacent joints during the truss analysis, but both the direct stresses as determined from the truss analysis and the bending stresses due to the intermediate purlin loads between panel points must be considered in the design of the top chords. The weight of the truss and the bracing system may be assumed to be divided among the joints on the top chord, while ceiling and suspended loads are assumed to be carried by the appropriate lower chord joints.

The loads on roof trusses generally consist of dead, snow, and wind loads. The dead load includes (1) the weight of roof covering (with or without sheathing), rafters if any, and purlins; (2) the weight of the bracing system; (3) the weight of the truss itself; and (4) ceiling and other suspended loads. If desired, (1) and (4) may be ascertained before the beginning of the truss analysis; items (2) and (3) must be first assumed and then reviewed after design calculations have been made. Fortunately items (2) and (3) are usually a small part of the total load; so even a rather large error in their assumed values may have a relatively insignificant effect on the resultant maximum stresses. Roof coverings are commonly corrugated steel, asphalt or asbestos, various types of shingles, tiles, slates, or thin concrete slabs, and tar and gravel. Sheathing or rafters may or may not be used, depending on whether the roofing material is self-supporting. The weight of item (1) above can easily vary from 5 to 25 lb per square foot of roof surface. The weight of a plastered ceiling may be 8 to 10 lb per square foot of horizontal surface. The weight of the bracing system may vary from $\frac{1}{2}$ to $1\frac{1}{2}$ lb per square foot of roof surface.

The weight of the roof truss is usually estimated by use of an appropriate empirical formula. Two of these are given here; others are available in handbooks.

For wooden roof trusses,

$$w = 0.5 + 0.075L \qquad \text{(H. S. Jacoby)} \qquad (6\text{-}1)$$

For steel roof trusses,

$$w = 0.4 + 0.04L \qquad \text{(C. E. Fowler)} \qquad (6\text{-}2)$$

In the above formulas, w is the weight of truss in pounds per square foot of horizontal surface, and L is the span in feet. It should be noted that any empirical formula should be used with discretion, or with adapting modifications. Some preliminary estimates are often advisable when empirical formulas are used.

The snow load which may come to the roof depends on the climate of the locality and on the pitch of the roof truss. The density of dry snow may be taken at 8 lb per cu ft, and that of wet snow at 12 lb per cu ft. Snow tends to pile up on flat roof surface. The snow load per square foot of inclined roof surface may be assumed as $(1 - \theta/60)$ times the estimated load per square foot of flat surface, wherein θ is the angle in degrees between the inclined roof surface and the horizontal. If appropriate for the climate, the snow load may be assumed to be 15, 20, or 25 lb per square foot of roof surface.

The wind load on a roof surface depends on the pitch of the roof truss and on the velocity of wind, which in turn is a function of the height of the building. The wind pressure p in pounds per square foot of vertical surface due to a wind velocity V in miles per hour is usually assumed to vary from $p = 0.003V^2$ to $p = 0.004V^2$. Thus a provision for wind pressures of 20, 25, or 30 lb per square foot of vertical surface may provide for wind velocities of 75, 85, or 95 mph. The wind pressure normal to an inclined roof surface is usually found by use of the Duchemin empirical formula (1829),

$$p_n = p \, \frac{2 \sin \theta}{1 + \sin^2 \theta} \qquad (6\text{-}3)$$

in which p_n is the normal pressure on an inclined roof surface at an angle θ with the horizontal, and p is the assumed pressure on the vertical surface.

Recent investigations have shown that wind may not only exert pressure on the windward side; it may actually exert suction on the leeward side. Suction is not likely to affect the design of the members of the roof truss supported on masonry walls, but its effect is of enough importance to point up the necessity for sufficient anchorage to prevent lifting of the truss. Also, if the windows on the windward or leeward side are open or broken, pressure or suction may come to the inside of the roof. Although the use of the Duchemin formula as discussed above is generally considered to be conservative, these newer conceptions are noteworthy.

6-3. Wind Loads as Recommended in the 1940 ASCE Final Report.
Sub-committee 31 of the American Society of Civil Engineers made some definite recommendations in regard to wind forces in its 1940 final report.[1] Although the report prescribes wind forces for both plane and round roof surfaces, only those for plane surfaces have been abstracted from the above-mentioned source as follows:

1. A uniformly distributed force of 20 psf for the first 300 ft above ground level, increased above this level by 2.5 psf for each additional

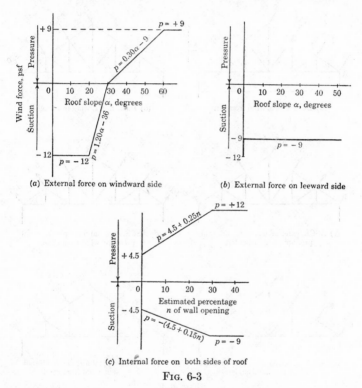

(a) External force on windward side

(b) External force on leeward side

(c) Internal force on both sides of roof

FIG. 6-3

100 ft of height, is recommended as a standard wind load for the United States and Canada.

2. For plane surfaces inclined to the wind and not more than 300 ft above the ground, the external wind force may be pressure or suction, depending on the exposure and the slope. For a windward slope inclined at not more than 20° to the horizontal, a suction of 12 psf is recommended; for slopes between 20 and 30°, a suction uniformly diminishing from 12 psf to 0 ($p = 1.20\alpha - 36$); and for slopes between 30 and 60°, a pressure increasing uniformly from 0 to 9 psf ($p = 0.30\alpha - 9$). On the lee-

[1] Final Report of Sub-committee 31, *Trans. ASCE*, vol. 105, pp. 1713–1720, 1940.

ward slope, for all inclinations in excess of zero, a suction of 9 psf is recommended.

3. For a flat roof a normal external suction of not less than 12 psf should be considered as applied to the entire roof surface.

4. For buildings that are normally airtight an internal pressure or suction of 4.5 psf should be considered as acting normal to the walls and the roof. For buildings with 30 per cent or more of the wall surfaces open, or subject to being open, an internal pressure of 12 psf, or an

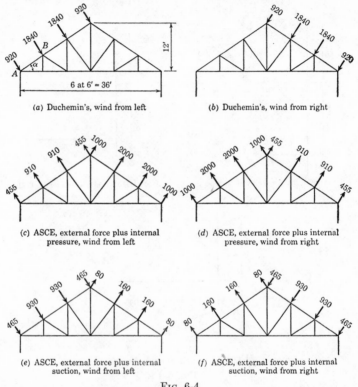

(a) Duchemin's, wind from left   (b) Duchemin's, wind from right

(c) ASCE, external force plus internal
pressure, wind from left

(d) ASCE, external force plus internal
pressure, wind from right

(e) ASCE, external force plus internal
suction, wind from left

(f) ASCE, external force plus internal
suction, wind from right

Fig. 6-4

internal suction of 9 psf, is recommended; for buildings with wall openings varying from 0 to 30 per cent of the wall space, an internal pressure varying uniformly from 4.5 to 12 psf ($p = 4.5 + 0.25n$, n = percentage of opening), or an internal suction varying uniformly from 4.5 to 9 psf ($p = 4.5 + 0.15n$) is recommended.

5. The design wind force applied to any surface of a building is to be a combination of the afore-mentioned appropriate external and internal wind forces.

6. When wind surfaces are more than 300 ft above the ground, the external and internal wind forces should be scaled up in the proportion

that the prescribed wind force on plane surfaces normal to the wind at the level under consideration bears to 20 psf.

The external and internal wind forces on inclined-plane surfaces at not more than 300 ft above the ground, as described in items 2, 3, and 4 above, may be indicated graphically as in Fig. 6-3. The determination of wind loads on roof trusses on the basis of the 1940 ASCE final report is illustrated in the following examples.

Example 6-1. Compare the wind loads on the roof truss shown in Fig. 6-4a (a) by the use of the Duchemin formula and (b) in accordance with the 1940 ASCE recommendation. The bay distance between trusses is 15 ft and the normal pressure in the direction of the wind is 20 psf. Assume 20 per cent wall opening.

SOLUTION

$$\text{Length of } AB \text{ (Fig. 6-4a)} = \sqrt{4^2 + 6^2} = 7.21 \text{ ft}$$
$$\text{Tributary area of roof surface per panel} = (15)(7.21)$$
$$= 108.2 \text{ sq ft}$$

(a) *Duchemin formula*

$$p_n = p\,\frac{2\sin\theta}{1 + \sin^2\theta} = 20\,\frac{2(2/\sqrt{13})}{1 + (2/\sqrt{13})^2} = 16.97 \text{ psf}$$

Wind panel load = $(16.97)(108.2) = 1,836$ or $1,840$ lb

(b) *ASCE recommendation*

$$\alpha = \arctan \tfrac{2}{3} = 33.7°$$
$$\text{External force, windward side} = 0.30(33.7) - 9$$
$$= 1.11 \text{ psf (pressure)}$$
$$\text{External force, leeward side} = 9 \text{ psf (suction)}$$
$$\text{Internal force, both sides} = 4.5 + 0.25(20)$$
$$= 9.5 \text{ psf (pressure)}$$
$$\text{Internal force, both sides} = 4.5 + 0.15(20)$$
$$= 7.5 \text{ psf (suction)}$$

Thus, taking the inward panel load as positive and the outward as negative, for the case of external force plus internal pressure,

$$\text{Wind panel load, windward side} = (+1.11 - 9.5)(108.2)$$
$$= -908 \text{ or } -910 \text{ lb}$$
$$\text{Wind panel load, leeward side} = (-9 - 9.5)(108.2)$$
$$= -2,002 \text{ or } -2,000 \text{ lb}$$

Considering external force plus internal suction,

$$\text{Wind panel load, windward side} = (+1.11 + 7.5)(108.2)$$
$$= +932 \text{ or } +930 \text{ lb}$$
$$\text{Wind panel load, leeward side} = (-9 + 7.5)(108.2)$$
$$= -162 \text{ or } -160 \text{ lb}$$

The wind panel loads as computed above are summarized in Fig. 6-4. It is apparent that the use of the Duchemin formula is conservative as far as maximum combined stresses (with the same sign as the dead-load stress in the member) are concerned. But the ASCE loading of Fig. 6-4c or d should be carefully considered because it may cause a stress of the opposite sign numerically larger than the dead-load stress so that the member must be designed to take both tension and compression. Also the required anchorage of the truss must be investigated. The loading shown in Fig. 6-4e and f, being less than that shown in Fig. 6-4a and b, may not have any significance. The reader is advised to familiarize himself with the ASCE recommendation, especially for use in unusual situations.

Example 6-2.   Compare the wind loads on the roof truss shown in Fig. 6-5a (a) by the use of the Duchemin formula and (b) in accordance with the 1940 ASCE recommendation. The bay distance between trusses is 18 ft and the normal pressure in the direction of the wind is 20 psf. Assume more than 30 per cent wall opening.

SOLUTION

$$\text{Length of } AB \text{ (Fig. 6-5}a) = \sqrt{7.5^2 + 3.75^2} = 8.385 \text{ ft}$$
$$\text{Tributary area of roof surface per panel} = (18)(8.385)$$
$$= 150.93 \text{ sq ft}$$

(a) *Duchemin formula*

$$p_n = p \frac{2 \sin \theta}{1 + \sin^2 \theta} = 20 \frac{2(1/\sqrt{5})}{1 + (1/\sqrt{5})^2} = 14.91 \text{ psf}$$
$$\text{Wind panel load} = (14.91)(150.93) = 2,250 \text{ lb}$$

(b) *ASCE recommendation*

$$\alpha = \arctan \tfrac{1}{2} = 26.57°$$
$$\text{External force, windward side} = 36 - 1.20(26.57)$$
$$= 4.12 \text{ psf (suction)}$$
$$\text{External force, leeward side} = 9 \text{ psf (suction)}$$
$$\text{Internal force, both sides} = 12 \text{ psf (pressure)}$$
$$\text{Internal force, both sides} = 9 \text{ psf (suction)}$$

Taking the inward panel load as positive and the outward as negative, for the case of external force plus internal pressure,

$$\text{Wind panel load, windward side} = (-4.12 - 12)(150.93)$$
$$= -2,433 \text{ or } -2,430 \text{ lb}$$
$$\text{Wind panel load, leeward side} = (-9 - 12)(150.93) = -3,170 \text{ lb}$$

Considering external force plus internal suction,

$$\text{Wind panel load, windward side} = (-4.12 + 9)(150.93)$$
$$= -773 \text{ or } +770 \text{ lb}$$
$$\text{Wind panel load, leeward side} = (-9 + 9)(150.93) = 0$$

The wind panel loads as computed above are summarized in Fig. 6-5. Again it is apparent that the use of the Duchemin formula is conservative with regard to the maximum combined stresses due to dead, snow, and wind loads. But, because of the possibility of more than 30 per cent wall opening, the internal pressure, together with external suction, yields large outward panel loads as shown in Fig. 6-5c and d. In some members these loads may produce stresses numerically larger than the

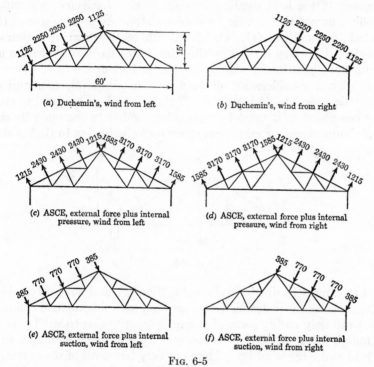

(a) Duchemin's, wind from left

(b) Duchemin's, wind from right

(c) ASCE, external force plus internal pressure, wind from left

(d) ASCE, external force plus internal pressure, wind from right

(e) ASCE, external force plus internal suction, wind from left

(f) ASCE, external force plus internal suction, wind from right

Fig. 6-5

dead-load stresses. Inasmuch as these dead and wind stresses are of the opposite sign, there will be stress reversals for which provision must be made.

6-4. Combinations of Loads. The function of stress analysis is to provide the designer with the most probable maximum or minimum (reversal) stresses to which any truss member may be subjected; consequently consideration must be given to the combinations of dead, snow, and wind loads which are to be accommodated in the design. The usual combinations are: (1) dead plus full snow on both sides, (2) dead plus wind on either side, (3) dead plus half snow on both sides plus wind from either side, (4) dead plus full snow on the leeward side and wind on the windward side, (5) dead plus ice (which may be 5 to 10 lb per square foot of roof surface) on both sides plus wind on either side, and (6) dead plus ice on both sides plus full snow on the leeward side plus wind on the windward side. Much depends on the judgment of the designer as to which load combinations should be used.

Because maximum winds may come only occasionally and are usually of short duration, most specifications allow a 33⅓ per cent increase in the unit working stress in cases where wind effect is included. In such a case, any of the load combinations (2) to (6), if selected, will not be controlling unless the resulting combined stress exceeds ⁴⁄₃ times that indicated in condition (1). Or stated differently, load condition (1) will control the design if the resulting stress exceeds ¾ of that from any other load combination in which wind load is included.

Ordinarily a consideration of load combinations (1), (2), and (3) will provide adequate design data. If there may be a reversal of stress in any one member, it must be caused by wind from the opposite side. The minimum stress, or maximum stress opposite in sign to that of dead

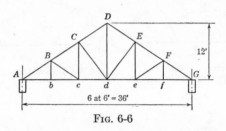

FIG. 6-6

load, if any, must be due to load combination (2), and not load combination (3), because the inclusion of half snow on both sides in load combination (3) will only nullify some of the reverse stress due to wind. It has been found, however, that with the exception of cases involving small dead load and large wind load, there are very few cases of stress reversal in roof trusses supported on masonry walls. In the usual cases then, it seems desirable to devise some sort of "equivalent" vertical loading to simulate the combined effect of snow and wind on the basis of normal working stress. It is logical to conclude that this "equivalent" vertical loading should be either that of full snow only or half snow plus a certain fraction of the wind pressure *normal* to the roof surface. Here again, only experience and judgment can help to decide what to use as "equivalent" loading for snow and wind.

The following two examples will serve to illustrate the general procedure of stress analysis of roof trusses. The reader will note that the assumed data are purely arbitrary.

Example 6-3. On the basis of normal working stress for vertical loading determine the maximum and minimum (if opposite in sign) stresses in all members of the roof truss shown in Fig. 6-6. The data for the analysis are:

Bay length = 15 ft   Span = 36 ft   Rise = 12 ft

The dead load consists of

Weight of roofing, rafters, and purlins
$$= 16 \text{ lb per square foot of roof surface}$$
Weight of bracing system = 1 lb per square foot of roof surface
Weight of truss = 0.5 + 0.075L
$$= 3.2 \text{ lb per square foot of horizontal surface}$$
Weight of ceiling = 10 lb per square foot of horizontal surface

The snow load is 10 lb per square foot of roof surface.
The wind load is 30 lb per square foot of vertical surface.

$$\text{Duchemin formula, } p_n = p \frac{2 \sin \theta}{1 + \sin^2 \theta}$$

The truss is anchored at both ends; wind reactions to be assumed parallel; $33\frac{1}{3}$ per cent increase in working stress allowed when wind effect is included.

Load combinations are

(1) Dead + full snow on both sides
(2) Dead + wind from either side
(3) Dead + half snow on both sides + wind from either side

SOLUTION.   (a) *Dead-load stresses.*

Length of AB (Fig. 6-6) = $\sqrt{4^2 + 6^2}$ = 7.21 ft

Tributary area of roof surface per panel = (15)(7.21) = 108.2 sq ft
Tributary area of horizontal surface per panel = (15)(6) = 90 sq ft
Panel load on top chord = (16 + 1)(108.2) + (3.2)(90) = 2,130 lb
Panel load on bottom chord = (10)(90) = 900 lb

The dead panel loads as computed above are shown in Fig. 6-7a. The weight of truss and bracing has been assumed to act at the panel points on the top chord. Some designers assume this load to be equally divided between the top and bottom panel points, but this is generally an unnecessary refinement. The stresses in all members of the truss may be determined by either the algebraic or the graphic method. The graphic solution has always been considered as standard procedure, but it is often found that the algebraic method actually takes much less time, especially when a computing machine is available. It should again be noted that each method is self-checking. Because both the algebraic

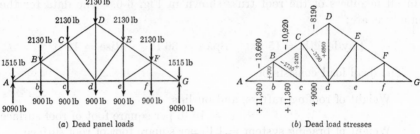

(a) Dead panel loads   (b) Dead load stresses

FIG. 6-7

and the graphic methods have been treated at length in Chap. 5, they will not be shown here. The dead-load stresses are shown in Fig. 6-7b.

(b) *Snow-load stresses.*

$$\text{Panel load on top chord} = (10)(108.2) = 1,080 \text{ lb}$$

The snow panel loads are shown in Fig. 6-8a and the snow-load stresses in all members of the truss are shown in Fig. 6-8b.

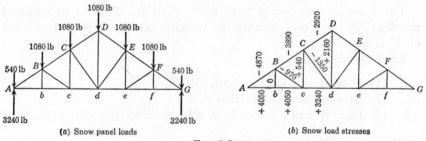

(a) Snow panel loads   (b) Snow load stresses

FIG. 6-8

(c) *Wind-load stresses.* From the Duchemin formula,

$$p_n = p \frac{2 \sin \theta}{1 + \sin^2 \theta} = 30 \frac{2(2/\sqrt{13})}{1 + (2/\sqrt{13})^2}$$
$$= 25.45 \text{ lb per square foot of roof surface}$$
$$\text{Panel load} = (25.45)(108.2) = 2,750 \text{ lb}$$

Since the truss is anchored at both ends and wind reactions are assumed to be parallel to the loads, only the stresses due to wind from the left need to be determined because the stresses due to wind from the right will be opposite-handed to those due to wind from left. The stresses due to wind from left are shown in Fig. 6-9b.

(d) *Combination of stresses.* In comparing Figs. 6-7b, 6-8b, and 6-9b, it is seen that wind from either side will not cause stress in any member opposite in sign to that of dead or snow loads. Inasmuch as there can be no reversal of stress in any member, load condition (2) (dead plus wind from either side) need not be considered at all. The maximum stress in a member on the basis of normal working stress for vertical loading is, then, the larger of the two values: (1) dead plus full snow on both sides, or (2) three-fourths of dead, plus half snow on both sides,

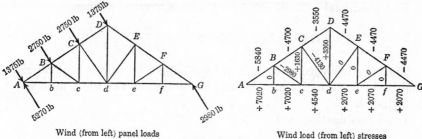

Wind (from left) panel loads   Wind load (from left) stresses

Fig. 6-9

plus wind from either side. Table 6-1 shows the combination of stresses. In examining this table, it is found that the $\frac{3}{4}(D + S/2 + W_L \text{ or } W_R)$ condition controls the maximum stress in members Bc, Cc, and Cd only.

TABLE 6-1. COMBINATION OF STRESSES

Member	D	S	W_L	W_R	$D + S$	$D + \frac{S}{2} + W_L \text{ or } W_R$	$\frac{3}{4}\left(D + \frac{S}{2} + W_L \text{ or } W_R\right)$	Max stress
AB	$-13,660$	$-4,870$	$-5,840$	$-4,470$	$-18,530$	$-21,935$	$-16,450$	$-18,530$
BC	$-10,920$	$-3,890$	$-4,700$	$-4,470$	$-14,810$	$-17,565$	$-13,170$	$-14,810$
CD	$-8,190$	$-2,920$	$-3,550$	$-4,470$	$-11,110$	$-14,120$	$-10,590$	$-11,110$
Ab	$+11,360$	$+4,050$	$+7,020$	$+2,070$	$+15,410$	$+20,405$	$+15,300$	$+15,410$
bc	$+11,360$	$+4,050$	$+7,020$	$+2,070$	$+15,410$	$+20,405$	$+15,300$	$+15,410$
cd	$+9,090$	$+3,240$	$+4,540$	$+2,070$	$+12,330$	$+15,250$	$+11,440$	$+12,330$
Bb	$+900$	0	0	0	$+900$	....	....	$+900$
Bc	$-2,730$	-970	$-2,980$	0	$-3,700$	$-6,195$	$-4,650$	$-4,650$
Cc	$+2,420$	$+540$	$+1,650$	0	$+2,960$	$+4,340$	$+3,260$	$+3,260$
Cd	$-3,790$	$-1,350$	$-4,130$	0	$-5,140$	$-8,595$	$-6,450$	$-6,450$
Dd	$+6,960$	$+2,160$	$+3,300$	$+3,300$	$+9,120$	$+11,340$	$+8,510$	$+9,120$

Example 6-4. On the basis of normal working stress for vertical loading determine the maximum and minimum (if opposite in sign) stresses in all members of the Fink roof truss shown in Fig. 6-10. The data for the analysis are:

Span = 60 ft   Rise = 15 ft   Bay length = 18 ft

The dead load consists of

Weight of roofing, rafters, and purlins
= 14 lb per square foot of roof surface
Weight of bracing system = 1 lb per square foot of roof surface

Weight of truss $= 0.4 + 0.04L$
$\qquad\qquad\quad = 2.8$ lb per square foot of horizontal surface
No ceiling or other suspended loads

The snow load is 15 lb per square foot of roof surface.
The wind load is 25 lb per square foot of vertical surface.

$$\text{Duchemin formula, } p_n = p\,\frac{2\sin\theta}{1+\sin^2\theta}$$

The truss is hinged at the left end and supported on rollers at the right end; $33\frac{1}{3}$ per cent increase in working stress allowed when wind effect is included.

Load combinations are

    (1) dead + full snow on both sides
    (2) dead + wind from either side
    (3) dead + half snow on both sides + wind from either side

Equivalent loading for snow and wind: 20 lb per square foot of roof surface on the basis of normal working stress for vertical loading. (Compare maximum stresses due to load combinations stated above with

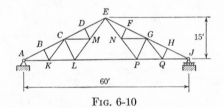

Fig. 6-10

stresses due to dead load plus equivalent loading for snow and wind condition.)

SOLUTION.   (a) *Dead-load stresses*

$$\text{Length of } AB \text{ (Fig. 6-10)} = \sqrt{(7.5)^2 + (3.75)^2} = 8.385 \text{ ft}$$

Tributary area of roof surface per panel $= (18)(8.385) = 150.93$ sq ft
Tributary area of horizontal surface per panel $= (18)(7.5) = 135$ sq ft
Panel load on top chord $= (14 + 1)(150.93) + (2.8)(135) = 2{,}640$ lb

The dead panel loads and stresses are shown in Fig. 6-11. The stresses are obtained by multiplying those in Fig. 6-12 by 2.64, because the dead-load panel loads are 2.64 times the 1,000-lb panel loads indicated in Fig. 6-12. Although a graphic solution might have been used, the stresses shown in Fig. 6-12 are the results of an algebraic solution.

In the algebraic solution of the stresses shown in Fig. 6-12, it should be noted that the slope of the top chord is 1 on 2, the slope of members BK, CL, and DM is 2 on 1, the slope of members CK, EM, and ML is 4 on 3,

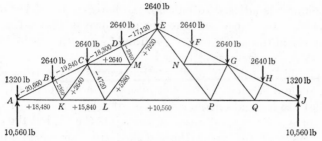

Dead panel loads and stresses

FIG. 6-11

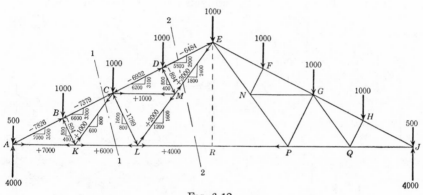

FIG. 6-12

and member CM is horizontal. The slope of EML is 4 on 3 because

$$\text{Length of } AC = 15\left(\frac{\sqrt{5}}{2}\right) = \frac{15}{2}\sqrt{5} \text{ ft}$$

$$\text{Length of } AL = AC\left(\frac{\sqrt{5}}{2}\right) = 18.75 \text{ ft}$$

$$\text{tan angle } ELR = \frac{ER}{LR} = \frac{15}{30 - 18.75} = \frac{4}{3}$$

The methods of joints or sections may be used interchangeably. The stresses may be solved in the following order:

1. Joint A: members AB and AK
2. Section 1-1: member CK
3. Joint K: members BK and KL
4. Joint B: member BC (one check available)
5. Section 2-2: member DE, ME, and LP
6. Joint D: members CD and DM (first use $\Sigma M_C = 0$ to find the vertical component of DM)
7. Joint M: members CM and LM
8. Joint C: member CL (one check available)
9. Joint E: check $\Sigma F_y = 0$

The computer is advised to indicate the horizontal and vertical components of the stress in every member as shown in Fig. 6-12. A final review to see that the conditions $\Sigma F_x = 0$ and $\Sigma F_y = 0$ are satisfied at *every* joint checks the correctness of the solution.

If the graphic solution is used for the stresses in the truss shown in Fig. 6-12, some difficulty will be encountered after the points 1, 2, and 3 (Fig. 6-13a) are located on the *stress diagram* (Fig. 6-13c). Apparently there are three unknown stresses at either joints C or L, but the difficulty can be surmounted by use of the modified truss shown in Fig. 6-13b. From a comparison of the original Fink truss in Fig. 6-13a with the modified Fink truss of Fig. 6-13b, it is obvious that the stress in member E'-6

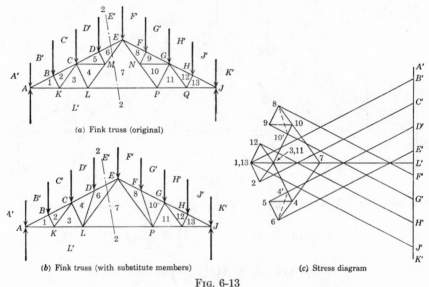

(a) Fink truss (original)

(b) Fink truss (with substitute members)   (c) Stress diagram

Fig. 6-13

has been unchanged by the alteration (in the algebraic method, stress in member E'-6 can be found by cutting section 2-2 and taking moments about L). Thus points 1, 2, 3, and 6 (or 13, 12, 11, and 8) must assume the same positions in the stress diagrams for the trusses shown in either Fig. 6-13a or b. The graphic solution, then, consists of (1) determining points 1, 2, and 3 with reference to Fig. 6-13a or b; (2) determining points 4' and 6 with reference to Fig. 6-13b only; and (3) with point 6 known, determining 5, 4, and then 7 with reference to Fig. 6-13a. This procedure applies equally well to the right half of the truss.

(b) *Snow-load stresses.*

Panel load on top chord = (15)(150.93) = 2,260 lb

The snow panel loads and stresses due to these loads are shown in

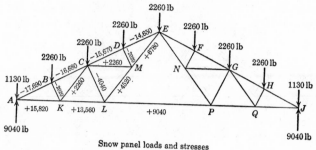

Snow panel loads and stresses

Fig. 6-14

Fig. 6-14. The stresses are obtained by multiplying those in Fig. 6-12 by 2.26.

(c) *Wind-load stresses.* From the Duchemin formula,

$$p_n = p\,\frac{2\sin\theta}{1+\sin^2\theta} = 25\,\frac{2(1/\sqrt{5})}{1+(1/\sqrt{5})^2}$$

$$= 18.63 \text{ lb per square foot of roof surface}$$
$$\text{Panel load} = (18.63)(150.93) = 2{,}810 \text{ lb}$$

The stresses due to wind from left and right are shown in Fig. 6-15a and b. These stresses may be obtained by multiplying those in Fig. 6-16a and d by 2.81. It should be noted that the hinged support is at the left end and the roller support is at the right end.

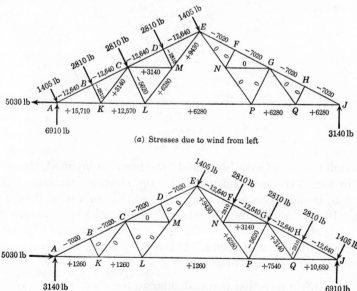

(a) Stresses due to wind from left

(b) Stresses due to wind from right

Fig. 6-15

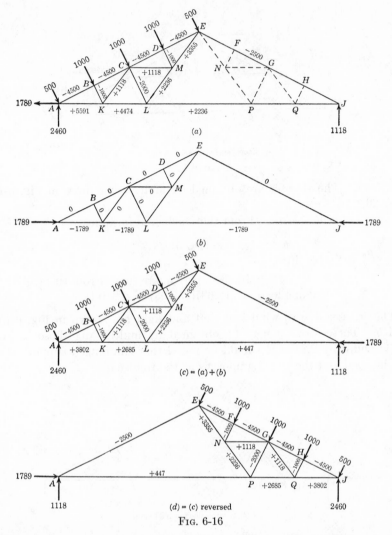

Fig. 6-16

Although the graphic method could have been readily used, the stresses shown in Fig. 6-16a were calculated by the algebraic method. In this connection, it should be noted that the members shown by dotted lines in Fig. 6-16a are not stressed because of wind from left. Because there are no loads acting at joints F, G, H, N, P, or Q, the triangle LEJ satisfies the requirements for truss action. It is also apparent that the stresses in FN and HQ must be zero to satisfy

$$\Sigma F \text{(in direction perpendicular to top chord)} = 0$$

at joints F and H. Then stresses in NG and GQ must be zero to satisfy

ΣF(in directions perpendicular to ENP or PQJ) = 0, respectively, at joints N and Q. Finally, similar considerations at joints G and P show that the stresses in GP and ENP must be zero. In the graphic solution points 7 to 13, inclusive, will coincide on the stress diagram.

By superimposing Fig. 6-16b on Fig. 6-16a, Fig. 6-16c is obtained. The stresses due to wind from left when the right support is hinged are shown in Fig. 6-16c, and the stresses due to wind from right when the left support is hinged are shown in Fig. 6-16d, which is Fig. 6-16c reversed.

(d) *Stresses due to dead load plus equivalent loading for snow and wind.*

Dead-load panel load = 2,640 lb
Panel load due to equivalent loading for snow and wind
$$= (20)(150.93) = 3,020 \text{ lb}$$
Total panel load due to dead and equivalent loads
$$= 2,640 + 3,020 = 5,660 \text{ lb}$$

The stresses due to dead load plus equivalent loading for snow and wind are shown in Fig. 6-17. These stresses are obtained by multiplying those in Fig. 6-12 by 5.66.

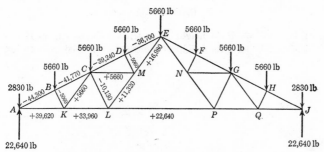

Stresses due to dead and equivalent superimposed loads

Fig. 6-17

(e) *Combination of stresses.* A comparison of Fig. 6-11 with Fig. 6-15 shows that wind from either side will not cause stress in any member opposite in sign to that of dead load. Load combination (2), therefore, is not critical. Only load combinations (1) and (3) need be considered. In Table 6-2, the maximum stress in each member, which is the larger of $D + S$ and $\frac{3}{4}(D + S/2 + W)$, is compared with the stress due to $D + E$, or dead plus equivalent loads. The $D + E$ stresses appear much larger than the maximum stresses as determined from the specified load combinations. However, the equivalent loading for both snow and wind, 20 psf in this case, is only 5 psf more than the snow load of 15 psf. Thus the effect of wind on design is rather small. This is to be expected because of the allowance of $33\frac{1}{3}$ per cent increase in unit working stresses when wind effect is included.

TABLE 6-2. COMBINATION OF STRESSES

Member	D	S	$W_L{}^*$	$W_R{}^*$	$D + S$	$D + \dfrac{S}{2} + W_L \text{ or } W_R$	$\dfrac{3}{4}\left(D + \dfrac{S}{2} + W_L \text{ or } W_R\right)$	Max stress	$D + E$
AB	−20,660	−17,690	−12,640	−12,640	−38,350	−42,150	−31,610	−38,350	−44,300
BC	−19,480	−16,680	−12,640	−12,640	−36,160	−40,460	−30,340	−36,160	−41,770
CD	−18,300	−15,670	−12,640	−12,640	−33,970	−38,780	−29,080	−33,970	−39,240
DE	−17,120	−14,650	−12,640	−12,640	−31,770	−37,080	−27,810	−31,770	−36,700
AK	+18,480	+15,820	+15,710	+10,680	+34,300	+42,100	+31,580	+34,300	+39,620
KL	+15,840	+13,560	+12,570	+ 7,540	+29,400	+35,190	+26,390	+29,400	+33,960
LP	+10,560	+ 9,040	+ 6,280	+ 1,260	+19,600	+21,360	+16,020	+19,600	+22,640
$BK\text{-}DM$	− 2,360	− 2,020	− 2,810	− 2,810	− 4,380	− 6,180	− 4,640	− 4,640	− 5,060
$CK\text{-}CM$	+ 2,640	+ 2,260	+ 3,140	+ 3,140	+ 4,910	+ 6,910	+ 5,180	+ 5,180	+ 5,660
CL	− 4,720	− 4,040	− 5,620	− 5,620	− 8,760	−12,360	− 9,270	− 9,270	−10,130
EM	+ 7,920	+ 6,780	+ 9,430	+ 9,430	+14,700	+20,740	+15,560	+15,560	+16,980
LM	+ 5,280	+ 4,520	+ 6,280	+ 6,280	+ 9,800	+13,820	+10,360	+10,360	+11,320

* The stresses recorded are the larger ones on either side of the center line since the truss will be symmetrically fabricated.

PROBLEMS

6-1. Rework Example 6-3 if the height of the truss is changed from 12 to 9 ft.

6-2. Rework Example 6-4 if the height of the truss is changed from 15 to 20 ft.

CHAPTER 7

ANALYSIS OF BUILDING BENTS

7-1. General Description. In Chap. 6 it was stated that roof trusses for buildings may rest on masonry walls or on columns along the sides of the building. A roof truss attached to its supporting columns is commonly called a bent. Two bents are shown in Fig. 7-1, one without and the other with *knee braces*. Before the action of the knee brace can be explained, however, it should be mentioned that the vertical faces of such buildings are usually covered with siding supported on *girts*, which are attached to the columns. The wind pressure on the vertical face of the building is therefore carried by the girts to the columns, just

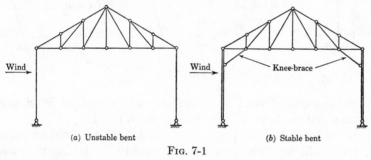

(a) Unstable bent   (b) Stable bent

Fig. 7-1

as the purlins bring their loads to the roof truss. There will be, then, horizontal forces acting on the windward column. Because it is assumed that the roof truss is pin-connected to the columns, the bent shown in Fig. 7-1a is unstable when wind is acting as indicated. However, if the knee braces are connected to the inner faces of the columns, as shown in Fig. 7-1b, the bent becomes stable, because the structure cannot collapse until the knee braces fail. The knee braces are therefore most essential in obtaining lateral stability when the structure resists inclined or horizontal loading. A bent as shown in Fig. 7-1b is often called a *knee-braced bent*.

Depending on the structural details used, the columns of a bent may be considered to be hinged, partially fixed, or fixed at the base. The stress analysis of a bent with hinged column bases will be considered first. For any kind of loading, there will be two unknown (horizontal and vertical) reaction components at each hinge, or a total of four unknowns. When the whole bent is taken as a free body, there are, however, only three independent equations of statics, and the structure is statically indeterminate. (Any structure whose reactions cannot be determined by the equations of statics alone is statically indeterminate. The treatment of statically indeterminate structures will be discussed in the latter part of this book.) By making a reasonable assumption about the values of the horizontal reactions, an "approximate" solution may be obtained by statics. When the bent is subjected to vertical loads only, it is usually assumed that the horizontal reactions at the hinges are zero; if inclined loads are present, the horizontal reactions are assumed to be equal. Once the reactions have been obtained, the analysis is straightforward. However, it should be noted that the two columns are three-force members, while all other members are two-force members. As

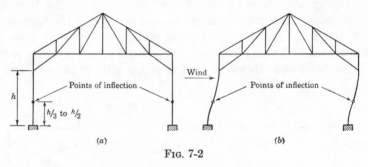

Fig. 7-2

far as a knee-braced bent is concerned, this statement is of utmost importance and needs to be kept constantly in mind.

When the lower ends of the columns of a bent are fixed or partially fixed, there will be three reaction components (a horizontal force, a vertical force, and a resisting moment) at each support, or a total of six unknowns. Three more conditions must therefore be provided or assumed in addition to the three equations of statics. The first assumption, which is the same as that for a bent with hinged supports, is that the horizontal reactions are zero for vertical loadings only and are equal for inclined loadings. The other two assumptions involve the arbitrary location of the points of inflection in the two columns, as shown in Fig. 7-2a. If the column bases are rigidly fixed, the points of inflection may be assumed at $h/2$ from the base wherein h is the distance from the foot of the knee brace to the base of the column. If the column bases are partially fixed (a condition comparable with the ordinary details of structural-steel columns), the points of inflection may be assumed at $h/3$ from the base. Because the points of inflection are equivalent to hinges, the analysis of the portion of the bent above the points of inflection will be identical with that of a bent with hinged supports. The deformation of the columns with fixed or partially fixed bases under the action of wind pressure is shown in Fig. 7-2b.

7-2. Methods of Analysis. In the analysis of a building bent, either algebraic or graphic methods may be used. A bent may be subjected to dead, snow, and wind loads as well as crane or other special loadings. Under dead and snow loads, the horizontal reactions at the bases of the columns are assumed to be zero. Thus the columns are subjected to direct axial compression only, the stresses in the knee braces are zero, and the stresses in the members of the truss proper are identical with those of a truss supported on masonry walls. Under wind loads, the columns are three-force members subjected to shear and bending moment, as well as direct stress. The knee braces and all other truss members are two-force members subjected to direct tension or compression. If the column bases are hinged and the horizontal reactions are assumed to be equal, all external reaction components may be found by taking the entire bent as a free body. If the column bases are partially or fully fixed, the horizontal and vertical forces at the assumed points of inflection are first found by taking the portion of bent above the points of inflection as a free body. If each column is taken as a free body, the stress in the knee brace and the horizontal and vertical reactions of the column at the end joint of the truss may be found. The shears and bending moments, as well as direct stresses in both columns, can then be determined. Finally, the stresses in all members of the truss may be found by either the algebraic or the graphic methods.

Example 7-1. On the basis of normal working stress for vertical loading, determine the maximum and minimum (if opposite in sign) stresses in all members of the knee-braced bent shown in Fig. 7-3. The data for the analysis are:

Bay length = 18 ft

The dead load consists of

Weight of roofing, rafters, and purlins
$$= 14 \text{ lb per square foot of roof surface}$$
Weight of bracing system = 1 lb per square foot of roof surface
Weight of truss = 0.4 + 0.04L
$$= 2.8 \text{ lb per square foot of horizontal surface}$$
No ceiling or other suspended loads.

The snow load is 15 lb per square foot of roof surface.

The wind load is 25 lb per square foot of vertical surface.

$$\text{Duchemin formula, } p_n = p \frac{2 \sin \theta}{1 + \sin^2 \theta}$$

Assume points of inflection to be at 7.5 ft above base of columns.

Structural details at column bases are such that the points of inflection may be taken at one-third of the distance from the column base to the foot of the knee brace. A $33\frac{1}{3}$ per cent increase in allowable working stress is permitted when wind effect is included.

Load combinations are

(1) dead + full snow on both sides
(2) dead + wind from either side
(3) dead + half snow on both sides + wind from either side

Equivalent loading for snow and wind: 20 lb per square foot of roof surface on the basis of normal working stress for vertical loading. (Compare maximum stresses due to load combinations stated above with stresses due to dead load plus the equivalent loading for snow and wind condition.)

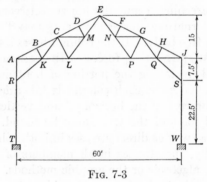

FIG. 7-3

SOLUTION. The dimensions of the Fink roof truss in Example 6-4 and also the vertical loadings on it are identical with those of the Fink truss of the knee-braced bent in this problem. When only vertical loads act on the bent, the horizontal reactions are assumed to be zero; thus the columns take direct stresses only and the dead-load stresses in the knee braces are zero. The stresses in the other members of the bent due to vertical loads are therefore identical with those of the truss supported on walls. Thus the values shown in columns (2), (3), and (10) of Table 7-1 are taken from Table 6-2 of Example 6-4.

The direct stresses, shears, and bending moments in the columns, and the direct stresses in all other members of the bent due to wind loads will now be determined. The inclined panel loads on the truss due to wind from the left are taken from Example 6-4 and are shown in Fig. 7-4. If the girts are spaced at 7.5 ft apart, the horizontal panel load on the column due to a wind pressure of 25 psf will be

(Bay length)(girt spacing)(wind pressure) = (18)(7.5)(25) = 3,375 lb

The horizontal panel loads due to wind are also shown in Fig. 7-4.

The complete stress analysis of the knee-braced bent due to wind loads is shown diagrammatically in Fig. 7-6. A brief description of the order of calculations follows.

1. Fig. 7-4: Determination of horizontal reactions. The wind load of 1,688 lb at the column base goes directly into the support; no transfer of any portion of it to the other support is made.

Total horizontal force (exclusive of wind load at column base)
$$= 1,688 + (3)(3,375) + (4)(2,810)(1/\sqrt{5}) = 16,840 \text{ lb}$$

Therefore

Horizontal reaction at the left support $= 1,688 + \frac{1}{2}(16,840)$
$$= (1,688 + 8,420) \text{ lb}$$
Horizontal reaction at the right support $= \frac{1}{2}(16,840) = 8,420 \text{ lb}$

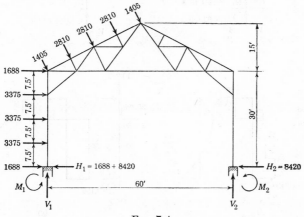

Fig. 7-4

2. Fig. 7-6a: Portion of bent above the points of inflection as free body. From $\Sigma(M$ about the left point of inflection$) = 0$,

$$V_2 = 6,925 \text{ lb}$$

From $\Sigma(M$ about the right point of inflection$) = 0$,

$$V_1 = 3,128 \text{ lb}$$

Check by $\Sigma F_y = 0$.

3. Fig. 7-6b and c: Lower portions of columns as free bodies. From $\Sigma M = 0$ (Fig. 7-6b),

$$M_1 = 63,150 \text{ ft-lb}$$

From $\Sigma M = 0$ (Fig. 7-6c),

$$M_2 = 63,150 \text{ ft-lb}$$

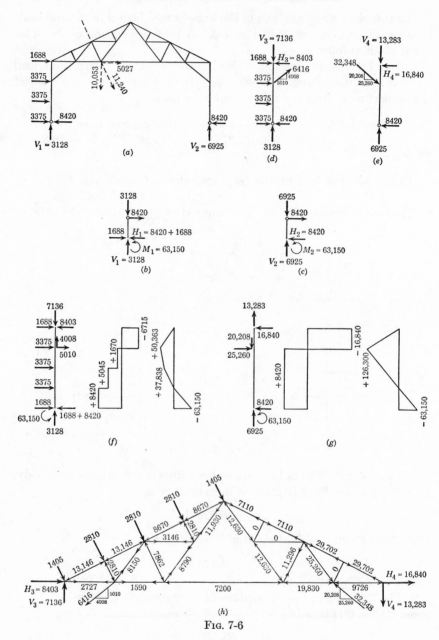

FIG. 7-6

4. Fig. 7-6d: Upper portion of left column as free body.

From ΣM about top = 0, H component of knee brace = 5,010 lb
Then V component of knee brace = 4,008 lb
and   Stress in knee brace = 6,416 lb
From $\Sigma H = 0$, H_3 = 8,403 lb
From $\Sigma V = 0$, V_3 = 7,136 lb
Check by $\Sigma(M$ about the point of inflection) = 0.

5. Fig. 7-6e: Upper portion of right column as free body.

From $\Sigma(M$ about top) = 0, H component of knee brace = 25,260 lb
Then V component of knee brace = 20,208 lb
and   Stress in knee brace = 32,348 lb
From $\Sigma H = 0$, H_4 = 16,840 lb
From $\Sigma V = 0$, V_4 = 13,283 lb
Check by $\Sigma(M$ about point of inflection) = 0.

6. Fig. 7-6f and g: Free-body, shear, and bending-moment diagrams of both columns. The sign convention for the shear and bending-moment diagrams is that of treating the columns as beams when viewed from the right side.

7. Fig. 7-6h: Stress analysis of truss. The methods of joints and sections are freely used. After all stresses have been determined, it will be advisable to check each joint by using the equations $\Sigma F_x = 0$ and $\Sigma F_y = 0$. A graphic solution for this part of the problem may be used to advantage.

Table 7-1 shows the various combinations of stresses. Values in columns 4 and 5 are taken from Fig. 7-6. Note that the W_R (wind from the right) stresses in the left half of the bent are equal to the W_L (wind from the left) stresses in the right half of the bent. The maximum stresses in column 7a are all of the same sign as those of dead-load stresses; naturally the $D + S/2 + W$ condition, not the $D + W$ condition, controls. In column 7b, only minimum stresses with signs opposite to those of maximum stresses are listed; these minimum stresses are, of course, due to the $D + W$ condition without the inclusion of $S/2$. The $D + S$ condition is entered in column 6. Maximum values shown in column 9 are the larger of the values listed in columns 6 and 8a. The minimum values in column 9 are identical with those in column 8b.

A comparison of columns 9 and 10 indicates that the use of equivalent vertical loading to replace both snow and wind effects is unsafe in the case of building bents. Column 9 shows that many members are subjected to stress reversals. They should be designed accordingly. As indicated in column 10, the use of an "equivalent" loading does not reveal the true nature of these reversals of stress.

TABLE 7-1. COMBINATION OF STRESSES

Member	D	S	W_L	W_R	D + S	(a) max $D + \dfrac{S}{2} + W_L$ or W_R	(b) min $D + W_L$ or W_R	(a) max ¾ of (7a)	(b) min ¾ of (7b)	Combined stress (a) max	Combined stress (b) min	D + E
(1)	(2)	(3)	(4)	(5)	(6)	(7)	(7)	(8)	(8)	(9)	(9)	(10)
AB	−20,660	−17,690	−13,150	+29,700	−38,350	−42,655	+ 9,040	−31,990	+ 6,780	−38,350	+ 6,780	−44,300
BC	−19,480	−16,680	−13,150	+29,700	−36,160	−40,970	+10,220	−30,730	+ 7,670	−36,160	+ 7,670	−41,770
CD	−18,300	−15,670	− 8,670	+ 7,110	−33,970	−34,805	....	−26,100	....	−33,970	....	−39,240
DE	−17,120	−14,650	− 8,670	+ 7,110	−31,770	−33,115	....	−24,840	....	−31,770	....	−36,700
AK	+18,480	+15,820	+ 2,730	− 9,730	+34,300	+29,120	....	+24,840	....	+34,300	....	+39,620
KL	+15,840	+13,560	+ 1,590	−19,830	+29,400	+24,210	− 3,990	+18,160	− 2,990	+29,400	− 2,990	+33,960
LP	+10,560	+ 9,040	− 7,200	− 7,200	+19,600	+ 7,880	....	+ 5,910	....	+19,600	....	+22,640
BK-DM	− 2,360	− 2,020	− 2,810	0	− 4,380	− 6,180	....	− 4,640	....	− 4,640	....	− 5,060
CK	+ 2,640	+ 2,260	+ 8,150	−25,260	+ 4,900	+11,920	−22,620	+ 8,940	−16,970	+ 8,940	−16,970	+ 5,660
CM	+ 2,640	+ 2,260	+ 3,150	0	+ 4,900	+ 6,920	....	+ 5,190	....	+ 5,190	....	+ 5,660
CL	− 4,720	− 4,040	− 7,860	+11,300	− 8,760	−14,600	+ 6,580	−10,950	+ 4,940	−10,950	+ 4,940	−10,130
EM	+ 7,920	+ 6,780	+11,930	−12,630	+14,700	+23,240	− 4,710	+17,430	− 3,530	+17,430	− 3,530	+16,980
LM	+ 5,280	+ 4,520	+ 8,790	−12,630	+ 9,800	+16,330	− 7,350	+12,250	− 5,510	+12,250	− 5,510	+11,320
RK	0	0	+ 6,420	−32,350	0	−32,350	+ 6,420	−24,260	+ 4,820	−24,260	+ 4,820	0
AR*	−10,560	− 9,040	− 7,140	+13,280	−19,600	−22,220	+ 2,720	−16,670	+ 2,040	−19,600	+ 2,040	−22,640
RT*	−10,560	− 9,040	− 3,130	− 6,930	−19,600	−22,010	....	−16,150	....	−19,600	....	−22,640

* Member is also subjected to shears and bending moments; only direct stresses are given in this table.

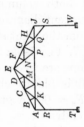

FIG. 7-5

PROBLEM

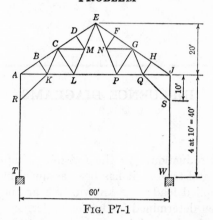

Fig. P7-1

7-1. Determine the maximum and minimum (if opposite in sign) stresses in all members of the knee-braced bent as shown on the basis of normal working stress for vertical loading. Assume points of inflection to be at 10 ft from base of columns. Other required data for the analysis are the same as given in Example 7-1.

CHAPTER 8

INFLUENCE DIAGRAMS

8-1. General Introduction. In the foregoing chapters involving the analysis of beams and trusses, it has been assumed in all cases that the position of the applied loads was known. Shear and bending-moment diagrams have been determined for beams under one given condition of loading. Methods for determining the stresses in the members of a truss under one given condition of loading have also been studied. Many structures, however, are subjected to the frequent passage in either direction of moving uniform or concentrated loads. A crane-runway girder in an industrial building is subjected to moving wheel loads. The beams which support conveyor systems transporting raw materials or finished products in factories are subjected to systems of moving loads. Girders or trusses in bridges carrying highway traffic are subjected to the moving wheel loads of heavy trucks or some equivalent loading system which simulates the streams of cars or trucks passing over the structure. Girders or trusses in bridges carrying railway traffic are subjected to the wheel loads of heavy locomotives followed by uniform train loads of indefinite length. These examples illustrate some types of moving loads for which structures must be designed.

In designing a beam or truss to carry moving loads, the position of these loads on the structure must be somehow determined so that the shear or bending moment at a section, or the stress in a member of the truss, will be the maximum which may ever happen. In this connection it must be emphasized that the position of loads which may cause maximum shear at a section will not necessarily cause maximum bending moment at this same section, or a condition of loading which causes maximum bending moment at one section may not cause maximum bending moment at some other section. When a maximum is being sought, whether it be shear, bending moment, or stress in a member, the first consideration is to determine the critical position of the moving loads. A study of influence diagrams will provide an understanding and, in many cases, the best solution for this problem.

8-2. Definition. Preliminary to the study of the effect of a system of moving loads which may consist of both concentrated and uniform loads, it will be desirable to consider first just one moving concentrated load. For instance, the effect of a single moving concentrated load on the shear at section C of simple beam AB (Fig. 8-1a) is to be found. As a matter of convenience, the magnitude of the moving concentrated load is assumed to be unity. When the unit load is on segment CB at a distance x_b from B (Fig. 8-1b), the left and right reactions are $R_A = x_b/L$ and $R_B = (L - x_b)/L$ and the shear at C is, by considering AC as a free body,

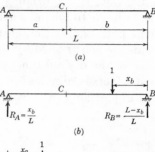

$$V_C = R_A = +\frac{x_b}{L}$$

or, by considering BC as a free body,

$$V_C = +1 - R_B = +1 - \frac{L - x_b}{L} = +\frac{x_b}{L}$$

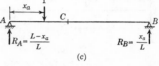

The shear at C, $V_C = +x_b/L$, due to a unit load at a distance of x_b from B, is plotted directly under the position of the unit load, as shown in Fig. 8-1d. When the unit load is at B, $V_C = 0$; when the unit load is at an infinitesimal distance to the right of C, $V_C = +b/L$. These values are plotted at B and C and the points B_1 and C_3 are connected by a straight line

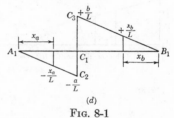

FIG. 8-1

(Fig. 8-1d). When the unit load is on segment AC at a distance x_a from A (Fig. 8-1c), the left and right reactions are $R_A = (L - x_a)/L$ and $R_B = x_a/L$. Considering AC as a free body, the shear at C is

$$V_C = R_A - 1 = \frac{L - x_a}{L} - 1 = -\frac{x_a}{L}$$

Or, with BC as a free body,

$$V_C = -R_B = -\frac{x_a}{L}$$

When the unit load is at A, $x_a = 0$, $V_C = 0$; when the unit load is at an infinitesimal distance to the left of C, $x_a = a$ and $V_C = -a/L$. These values are plotted at A and C and the straight line A_1C_2 is drawn as shown in Fig. 8-1d. The enclosed diagram $A_1C_2C_3B_1$ of Fig. 8-1d is called the influence diagram for shear at C. Thus the shear at C due to the unit load at any position on the span is equal to the ordinate on the influence diagram directly under the load.

From the preceding discussion, it is seen that an influence diagram is constructed to show the variation in the effect of a single moving unit load on some function at any section of a structure (such as shear at C in Fig. 8-1a). Influence diagrams may be constructed for a function such as reaction, shear, bending moment, deflection, or stress in a member. The ordinate to the influence diagram shows the desired function due to a single unit load at the position of the ordinate. Usually influence diagrams are constructed by first calculating ordinates which represent the value of the desired function due to a unit load as it moves across the span. These ordinates are calculated and plotted under the load at each critical position of the load, and then lines connecting the extremities of successive ordinates thus calculated enclose the influence diagram for the given function.

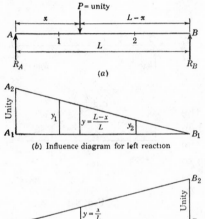

(a)

(b) Influence diagram for left reaction

(c) Influence diagram for right reaction

FIG. 8-2

It is well to note that a shear or bending-moment diagram shows the effect of *stationary* loads on shears or bending moments at *all* sections in a structure, while the shear or bending-moment influence diagram shows the effect of one *moving* unit load on the shear or bending moment at *the section* for which the influence diagram has been constructed.

8-3. Influence Diagrams for Reactions on a Beam. Let it be required to construct the influence diagrams for the left and right reactions of the simple beam AB as shown in Fig. 8-2a. A unit load is placed at a distance x from the left support. By taking moments about point B, the left reaction is found to be $R_A = (L - x)/L$, and by taking moments about point A, the right reaction is $R_B = x/L$.

In the influence diagram for the left reaction, the value of

$$R_A = y = \frac{L - x}{L}$$

is plotted directly under the unit load, which is now at x from point A. Since y is a linear function of x and its expression is applicable between $x = 0$ and $x = L$, it will only be necessary to compute the values of y at $x = 0$ and $x = L$. The influence diagram for R_A is then obtained by drawing a straight line connecting the upper extremities of the ordinates at points A and B. Thus,

When $x = 0$, $\qquad y = \dfrac{L - x}{L} = \dfrac{L - 0}{L} = 1$

When $x = L$, $\qquad y = \dfrac{L - x}{L} = \dfrac{L - L}{L} = 0$

The influence diagram for the left reaction is plotted as $A_1A_2B_1$ as in Fig. 8-2b. Similarly the influence diagram for the right reaction is $A_1B_1B_2$ as shown in Fig. 8-2c. At A the ordinate is $y = x/L = 0/L = 0$ and at B the ordinate is $y = x/L = L/L = 1$.

For concentrated loads P_1 and P_2 (not shown) at points 1 and 2 in Fig. 8-2a, the left reaction is

$$R_A = P_1y_1 + P_2y_2$$

This may be explained by the fact that, since y_1 is the left reaction due to a unit load at point 1, the left reaction due to P_1 at point 1 is P_1y_1. Similarly, the left reaction due to P_2 at point 2 is P_2y_2. The left reaction due to P_1 and P_2 is thus $R_A = P_1y_1 + P_2y_2$.

The above influence diagrams may also be used to calculate reactions due to uniform loading. Suppose a uniform load of intensity w lb per lin ft (not shown) is applied between points 1 and 2 on the beam shown in Fig. 8-2a. The reaction at A, due to a load $w\,dx$ is $dR_A = yw\,dx$, in which $y = (L - x)/L$. Thus,

$$R_A = \int_1^2 yw\,dx = w\int_1^2 y\,dx$$

It is noted that $\int_1^2 y\,dx$ is the area of the influence diagram between ordinates y_1 and y_2. Therefore the left reaction may be obtained if the area of the influence diagram covered by the uniform load is multiplied by the intensity of the uniform loading.

Example 8-1. Given a beam 25 ft long which overhangs the left support by 5 ft as shown in Fig. 8-3a. Construct the influence diagrams for R_A and R_B. Compute the maximum upward and downward (if any) reactions due to (a) a moving uniform live load of 400 lb per lin ft and (b) two concentrated loads of 10 kips each at 4 ft apart.

SOLUTION. First consider the influence diagram for R_A. When the unit load is at a distance x from B,

$$R_A = +\frac{x}{20} \qquad \text{for } 0 \leqq x \leqq 25$$

At point C: $\quad x = 25 \quad$ and $\quad R_A = +{}^{25}\!/_{20} = +1.25$
At point A: $\quad x = 20 \quad$ and $\quad R_A = +{}^{20}\!/_{20} = +1.00$
At point B: $\quad x = 0 \quad$ and $\quad R_A = +{}^{0}\!/_{20} = 0$

The influence diagram for the left reaction is plotted as $C_1C_2B_1$ as shown in Fig. 8-3b.

Next consider the influence diagram for R_B. When the unit load is between points A and B at a distance x from A,

$$R_B = +\frac{x}{20} \qquad \text{for } 0 \leqq x \leqq 20$$

At point A: $x = 0$ and $R_B = +\frac{0}{20} = 0$

At point B: $x = 20$ and $R_B = +\frac{20}{20} = +1.00$

When the unit load is between points C and A at a distance x from A,

$$R_B = -\frac{x}{20} \qquad \text{for } 0 \leqq x \leqq 5$$

At point A: $x = 0$ and $R_B = -\frac{0}{20} = 0$

At point C: $x = 5$ and $R_B = -\frac{5}{20} = -0.25$

The influence diagram for the right reaction is $C_1C_2A_1B_2B_1$ as shown in Fig. 8-3c.

To ascertain the effect of moving uniform or concentrated loads on the reactions, the loading positions for maxima must be determined. For

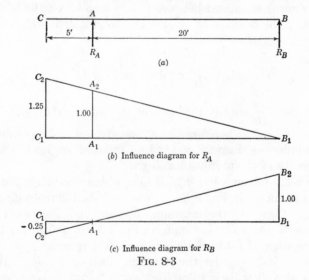

(a)

(b) Influence diagram for R_A

(c) Influence diagram for R_B

Fig. 8-3

simple systems of loadings as given in this example, such positions can be easily found by inspection of the influence diagrams. These positions are summarized in Fig. 8-4. Inasmuch as the entire influence diagram for the left reaction is above the base line, there is no possibility of a downward left reaction at any time. Thus, in Fig. 8-4b, no loads are shown. For a maximum upward reaction at the left support due to uniform load, the uniform load must cover the entire length. This reaction may be found by multiplying the intensity of the load and the entire area of the influence diagram. The two concentrated loads will produce a maximum left reaction when the loads are placed over the highest possible influence ordinates. These positions are shown in Fig. 8-4a. At the right support, however, the uniform load should cover only the 20-ft segment between the supports for maximum upward reaction and the 5-ft overhang for maximum downward reaction. The two concentrated loads are placed so that the positive or negative ordinates are, numerically and respectively, the largest possible (see Fig. 8-4c and d).

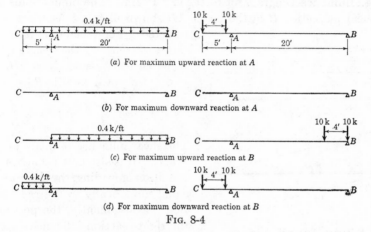

(a) For maximum upward reaction at A

(b) For maximum downward reaction at A

(c) For maximum upward reaction at B

(d) For maximum downward reaction at B

FIG. 8-4

The actual values of the maximums may be computed by considering the loadings shown in Fig. 8-4, or by use of the influence diagrams, viz., $R = w$ (area of influence diagram covered) or $R = \Sigma P y$. The beginner is advised to find all required maxima by both methods. For instance, the maximum downward reaction at the right support (Fig. 8-4d) due to the moving uniform load is

$$R_B = \frac{(0.4)(5)^2}{(2)(20)} = 0.25 \text{ kip downward}$$

By the use of the influence diagram,

$$R_B = (0.4)\frac{(0.25)(5)}{2} = 0.25 \text{ kip downward} \quad (check)$$

Due to the two moving concentrated loads,

$$R_B = \frac{(10)(1) + (10)(5)}{20} = 3 \text{ kips downward}$$

or $\quad R_B = (10)[0.25 + 0.25(\frac{1}{5})] = 3 \text{ kips downward} \quad (check)$

The required maxima are summarized in the following table. The reader may check each value by both methods as illustrated above.

Loading	Max R_A, kips		Max R_B, kips	
	(+) Upward	(−) Downward	(+) Upward	(−) Downward
Uniform load.......	+ 6.25	0	+ 4	−0.25
Concentrated loads..	+23	0	+18	−3

8-4. Influence Diagram for Shear in a Beam. The influence diagram for shear at section C in the beam AB shown in Fig. 8-5a will be constructed. For a load of unity at distance x from A, the reactions are $R_A = (L - x)/L$ and $R_B = x/L$. If $x < a$, the shear at C is $R_A - 1$; or

$$V_C = \frac{L - x}{L} - 1 = -\frac{x}{L}$$

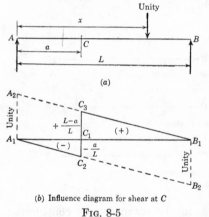

(a)

(b) Influence diagram for shear at C

Fig. 8-5

The shear influence ordinate at C is $-a/L$. It is also noted that, for this condition of loading, the shear at C is equal numerically to the right reaction. Consequently, the portion of the right-reaction influence diagram between A and C may be used as a shear influence diagram for C; however, to conform to usual sign conventions, this line is plotted as A_1C_2 with negative ordinates. If $x > a$, the shear at C is the left reaction and

$$V_C = \frac{L - x}{L}$$

Thus, for the portion of the beam between C and B, the left-reaction influence diagram is also the shear influence diagram. At C the shear influence ordinate is $(L - a)/L$ as shown in Fig. 8-5b.

The diagram $A_1C_2C_3B_1$ is the shear influence diagram for section C. It will be noted that the ordinate C_1C_3 is the positive shear at C when the unit load is applied at an infinitesimal distance to the right of C and the ordinate C_1C_2 is the negative shear at C when the unit load is applied at an infinitesimal distance to the left of C.

Example 8-2. A beam 50 ft long rests on its left support and extends 10 ft beyond the right support as shown in Fig. 8-6a. Construct the influence diagram for shear at a section midway between the supports.

Compute the numerical maximum shear at C due to (a) a moving uniform live load of 400 lb per lin ft and (b) two concentrated loads of 10 kips each at 4 ft apart.

SOLUTION. For loads between A and C, the shear at C is the left reaction *minus* the load, and for loads to the right of C, the shear at C is the left reaction.

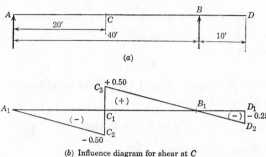

(a)

(b) Influence diagram for shear at C

FIG. 8-6

When the unit load is at a distance x from A and $x < 20$,

$$V_C = R_A - 1 = \frac{40 - x}{40} - 1 = -\frac{x}{40}$$

When $x > 20$,

$$V_C = R_A = \frac{40 - x}{40}$$

When $x > 40$,

$$V_C = R_A = -\frac{x - 40}{40}$$

Thus $V_C = 0$ at $x = 0$; $V_C = -0.50$ at $x = 20$; $V_C = +0.50$ at $x = 20$; $V_C = 0$ at $x = 40$; and $V_C = -0.25$ at $x = 50$. When values of V_C are plotted at points A, C, B, and D, the shear influence diagram $A_1C_2C_3B_1D_2D_1$ of Fig. 8-6b is obtained. It is noted that loads on segments AC and BD cause negative shears at C while positive shears are produced by loads on segment CB.

If it is assumed that the moving uniform load may be broken into segments of any length, numerically the maximum shear is the negative shear at C when the uniform load covers portions AC and BD of the span, as shown in Fig. 8-7a. This shear is (Fig. 8-7a)

$$V_C = R_A - (0.4)(20) = 5.5 - 8 = -2.5 \text{ kips}$$

From the influence diagram,

$$V_C = 0.4 \text{ (area of } A_1C_1C_2 + \text{ area of } B_1D_1D_2)$$
$$= 0.4 \left[-(\tfrac{1}{2})(0.50)(20) - \tfrac{1}{2}(0.25)(10) \right]$$
$$= -2.5 \text{ kips} \quad (check)$$

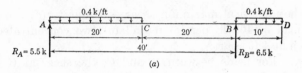

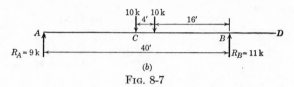

FIG. 8-7

The two concentrated loads should be placed as shown in Fig. 8-7b to cause maximum shear at C.

$$V_C = R_A = 9 \text{ kips}$$

Or, from the influence diagram,

$$V_C = \Sigma Py = 10(0.50) + 10(0.50)(^{16}\!/_{20}) = 9 \text{ kips} \qquad (check)$$

8-5. Influence Diagram for Bending Moment in a Beam. The influence diagram for bending moment at section C in the beam AB of Fig. 8-8a is shown in Fig. 8-8b. For a load of unity at distance of x from A in Fig. 8-8a, the reactions are $R_A = (L - x)/L$ and $R_B = x/L$. If $x < a$, the bending moment at C, by considering AC as the free body, is

$$M_C = R_A a - 1(a - x) = \frac{L - x}{L}(a) - 1(a - x)$$

$$= \frac{x}{L}(L - a)$$

The same expression can be obtained by considering CB as the free body. Thus,

$$M_C = R_B(L - a) = \frac{x}{L}(L - a)$$

This is the equation of line A_1C_2. Note that $M_C = 0$ when the unit load is at $x = 0$, and $M_C = (a/L)(L - a)$ when the unit load is at $x = a$. If $x > a$, by considering AC as the free body,

$$M_C = R_A a = \frac{a}{L}(L - x)$$

or, by considering CB as the free body,

$$M_C = R_B(L - a) - 1(x - a)$$

$$= \frac{x}{L}(L - a) - 1(x - a) = \frac{a}{L}(L - x)$$

This is the equation of line C_2B_1. Note that $M_C = (a/L)(L - a)$ when $x = a$, and $M_C = 0$ when $x = L$. It will be observed that A_1C_2 and B_1C_2 have a common ordinate $(a/L)(L - a)$ at C.

It happens that the influence diagram for bending moment at C in simple beam AB is identical with the bending moment diagram for a unit load at C. The interpretation of these two diagrams, however, is entirely different; the influence ordinate always gives the bending moment at C as the unit load moves across the span, while the bending-moment diagram shows the bending moments at various sections due to a fixed unit load at C. Thus, for convenience, the influence diagram for bending moment at C may be constructed by placing a load of unity at C (the critical section) and then drawing the bending-moment diagram for this load.

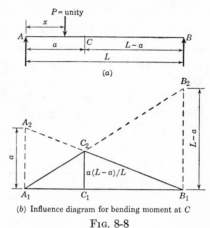

(a)

(b) Influence diagram for bending moment at C

Fig. 8-8

Example 8-3. A beam 33 ft long rests on the left support and overhangs the right support 6 ft, as shown in Fig. 8-9a. Construct the bending-moment influence diagram for a section C at 9 ft from the left support. Compute the maximum positive and negative bending moment

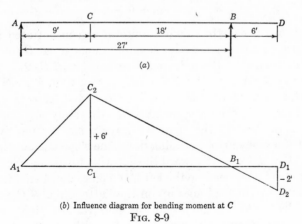

(a)

(b) Influence diagram for bending moment at C

Fig. 8-9

at C due to (a) a moving uniform live load of 400 lb per lin ft and (b) two concentrated loads of 10 kips each at 4 ft apart.

SOLUTION. This influence diagram may be constructed by placing unity at critical points A, C, B, and D and, in each instance, calculating the bending moment at C. The value of each bending moment will be plotted under the load as shown in Fig. 8-9b. Thus:

For unity at A: $M_C = 0$ and ordinate $A_1 = 0$
For unity at C: $M_C = +6$ ft and ordinate $C_1C_2 = +6$ ft
For unity at B: $M_C = 0$ and ordinate $B_1 = 0$
For unity at D: $M_C = -2$ ft and ordinate $D_1D_2 = -2$ ft

It is to be noted that, in the case of influence diagrams for reaction or shear, the influence ordinate is the ratio of the reaction or shear to the moving load and therefore is merely an abstract number without an attached dimensional unit. The bending-moment influence ordinate, however, is the ratio of the bending moment at the section to the moving load; or dimensionally speaking, this ratio is $FL/F = L$. Thus the ordinates C_1C_2 and D_1D_2 in Fig. 8-9 are $+6$ ft and -2 ft, respectively.

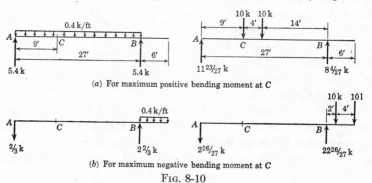

(a) For maximum positive bending moment at C

(b) For maximum negative bending moment at C

FIG. 8-10

The positions which the moving uniform or concentrated loads must take to cause positive or negative bending moments at C may be ascertained by inspection of the influence diagram $A_1C_2B_1D_2D_1$. These positions are shown in Fig. 8-10. Note that, for maximum positive bending moment at C, one of the two equal concentrated loads is placed at C and the other on the long segment CB. Certainly any movement of these two loads toward the right will give a smaller positive bending moment at C as they are descending down the slope C_2B_1 of the influence diagram. Any movement toward the left will also give a smaller positive bending moment at C because the left load comes down on a steeper slope C_2A_1, while the right load goes up on the flatter slope B_1C_2; or, the loss because of the left load is more than the gain from the right load. Thus any further movement of the two concentrated loads in either direction from the position shown in Fig. 8-10 will cause a decrease in the positive bending moment at C. Consequently the critical position for these loads has been determined. Had the system of concentrated loads been more complicated (more loads at varied spacings), it would have been difficult to determine by simple inspection the position which these loads should take on the span to cause a maximum positive bending moment at C. Problems of this nature will be rigorously treated in the next chapter.

The actual values of the maximum positive or negative bending moment at C may be found by applying statics to the free-body diagrams of Fig. 8-10 or by the use of the influence diagram. Computations by both methods are shown below.

For the uniform load,

Max $+M_C = (5.4)(9) - \frac{1}{2}(0.4)(9)^2 = 32.4$ kip-ft
or $\quad = (0.4)$ (area of $A_1C_2B_1$) $= (0.4)(\frac{1}{2})(6)(27) = 32.4$ kip-ft
Max $-M_C = -(0.267)(9) = -2.4$ kip-ft
or $\quad = (0.4)$ (area of $B_1D_1D_2$) $= (0.4)[-\frac{1}{2}(6)(2)] = -2.4$ kip-ft

For the two concentrated loads,

Max $+M_C = (^{320}\!/_{27})(9) = {}^{320}\!/_{3} = 106.67$ kip-ft
or $\quad = 10(6) + 10(6)(^{14}\!/_{18}) = 60 + 46.67 = 106.67$ kip-ft
Max $-M_C = -(^{80}\!/_{27})(9) = -{}^{80}\!/_{3} = -26.67$ kip-ft
or $\quad = -(10)(2) - 10(2)(\frac{2}{6}) = -20 - 6.67 = -26.67$ kip-ft

(MULLER - BRESSLAU)

8-6. Influence Diagram as a Deflection Diagram. Influence diagrams for reaction, shear, or bending moment as described in the preceding articles may be determined on the basis of the conception that influence diagrams are deflection diagrams.

1. To obtain the influence diagram for a reaction, remove the support giving resistance for this reaction and introduce a unit displacement in the direction of the reaction. The area enclosed between the original and the final positions of the beam is the required influence diagram. Thus, for the simple beam AB shown in Fig. 8-11a, the influence diagrams for R_A and R_B are given by the enclosed area between the original position (1) and the final position (2), as shown in Fig. 8-11b and c. Similarly, for the overhanging beam AB shown in Fig. 8-12a, the influence diagrams for R_A and R_B are shown in Fig. 8-12b and c. For the cantilever beam AB shown in Fig. 8-13a, the influence diagram for R_A is obtained by sliding the fixed support vertically upward a unit displacement as shown in Fig. 8-13b.

2. To obtain the influence diagram for shear at a section, cut the beam at the section and lift the cut end at the right a unit displacement relative to the cut end at the left, without introducing relative rotation at the section. Thus in Figs. 8-11d and 8-12d, A_1C_2 and C_3B_1 are parallel in order that there is no relative rotation at C. Similarly, the influence diagram for V_C in the cantilever beam AB is obtained by cutting the

beam at C and lifting the cut end at the right a unit distance, as shown in Fig. 8-13c. In this case, A_1C_1 must remain horizontal; therefore C_2B_2 must also be horizontal in order that there is no relative rotation at C.

3. To obtain the influence diagram for bending moment at a section, insert a hinge at the section so that there is no moment resistance at the section and introduce a unit relative rotation at the section. Thus, in Fig. 8-11e, if there were no relative rotation at C_2, A_1C_2 would have gone straight to B_2. C_2B_2 rotates around C_2 for 1 radian to position C_2B_1.

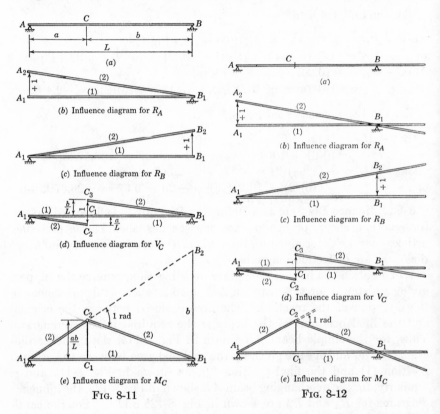

(a)

(b) Influence diagram for R_A

(c) Influence diagram for R_B

(d) Influence diagram for V_C

(e) Influence diagram for M_C

FIG. 8-11

(a)

(b) Influence diagram for R_A

(c) Influence diagram for R_B

(d) Influence diagram for V_C

(e) Influence diagram for M_C

FIG. 8-12

If all vertical displacements are small, $C_1B_1 = C_2B_1$, and $B_1B_2 = 1$ radian times C_1B_1. Thus $C_1C_2 = (B_1B_2/L)(A_1C_1) = ab/L$. It should be noted that, although 1 radian is defined to be about 57.3° in trigonometry, it must be regarded as a very small unit of measure so that the length B_1B_2 may be equal to that of an arc with radius equal to C_2B_1 or C_1B_1. Note also the influence diagrams for M_C in Fig. 8-12e and for M_A and M_C in Fig. 8-13d and e.

This method of obtaining influence diagrams by use of deflection diagrams has been demonstrated in a few cases, and the influence diagrams thus found are seen to be identical with those determined analytically. A formal proof of this principle now seems to be in order. If it can be proved that the ordinate y in Fig. 8-14c is equal to M_C in Fig. 8-14b, then by definition $A_1C_2B_1$ must be the influence diagram for M_C. The underlying principle for making this proof is this: If the force system acting on a structure is in equilibrium, and if this structure undergoes a change of shape or position without resulting internal stresses or strains, the total external work done by the elements (forces and moments) in the balanced force system must be zero, because, if there is any external work done, by the law of con-

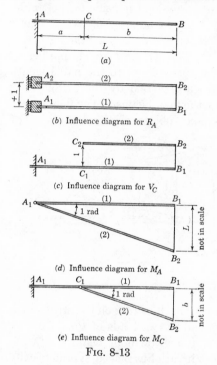

(a)

(b) Influence diagram for R_A

(c) Influence diagram for V_C

(d) Influence diagram for M_A

(e) Influence diagram for M_C

FIG. 8-13

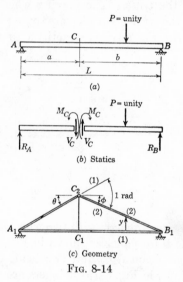

(a)

(b) Statics

(c) Geometry

FIG. 8-14

servation of energy, such work must be stored in the beam as internal elastic energy. Now, when the balanced force system $(R_A, R_B, P, V_C, V_C, M_C, M_C)$ in Fig. 8-14b goes through the change in shape from position 1 to position 2 as shown in Fig. 8-14c, the total external work W done is

$$W = R_A(0) + R_B(0) - Py + V_C(C_1C_2) - V_C(C_1C_2) + M_C(\theta) + M_C(\phi)$$
$$= -Py + M_C(\theta + \phi) = -Py + M_C(1 \text{ radian}) = -(1)(y) + M_C(1)$$
$$= 0$$

Thus

$$y = M_C$$

Note that the work done by the downward force P in going through the upward displacement y is negative. Similar procedures can be used to prove that influence diagrams for reaction and shear are deflection diagrams as described above in items (1) and (2).

The relationship between influence diagrams and deflection diagrams is important and can be used as an auxiliary method either to get a preliminary sketch of or to make a final visual check on any influence diagram.

8-7. Influence Diagrams for Simple Trusses. Usually two trusses are used to carry moving loads, with one truss on each side of the traffic. In Fig. 8-15 is shown a through truss bridge composed of two six-panel Pratt trusses joined together by floor beams and other bracing in the transverse direction (some not shown). The stringers are simple beams supported on the floor beams and with spans equal to the panel length of the truss. The moving wheel loads are carried by the bridge floor to the stringers, which are supported by the floor beams. The floor beams carry the stringer reactions to the panel points of the trusses. For

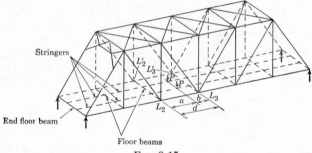

Fig. 8-15

instance, an axle with two wheel loads of P each, acting on the stringers in the third panel as shown in Fig. 8-15, causes two loads of Pb/d each on floor beam L_2L_2' and two loads of Pa/d each on floor beam L_3L_3'. As shown in Fig. 8-15, stringers are almost invariably placed symmetrically on the floor beams; consequently, in this case loads of Pb/d are transferred to joints L_2 and L_2'. Likewise loads of Pa/d are transferred to joints L_3 and L_3'.

This truss bridge is supported at the four corners as shown in Fig. 8-15 and requires the use of end floor beams to support the exterior ends of the stringers in the end panels of the bridge. Sometimes a bridge is designed so that both the trusses and the exterior ends of the stringers in the end panels rest directly on the abutments or piers. If such is the case, end floor beams are not required.

When structures, particularly railway and highway bridges, carry moving loads which may occupy any position on the span, it is necessary to determine the position of the load system causing maximum tensile and/or compressive stress in any one member of the truss. It is to be noted that the position which causes maximum stress in one member may not produce maximum stress in any other member. Obviously, the loading condition which causes maximum tensile stress in one member will not produce maximum compressive stress in this same member. Influence diagrams provide a convenient method for developing criteria which may be used to determine these critical loading conditions. Once the critical condition of loading has been determined, the methods explained in Chap. 5 or the influence diagram itself may be used to calculate the maximum stress in any member.

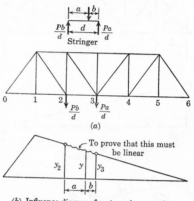

(a)

(b) Influence diagram for stress in a member

FIG. 8-16

8-8. Influence Diagram between Panel Points of a Truss. It will be proved that the influence diagram between panel points of a truss must be bounded by a straight line. Let Fig. 8-16b represent the influence diagram for the stress in some member of the bridge truss shown in Fig. 8-16a. Assume that a unit load at panel point 2 causes a stress equal to y_2 in this member and a unit load at panel point 3 causes a stress equal to y_3. If two wheel loads of P each act on symmetrical stringers in the third panel at distances a and b from the adjacent panel points, components Pb/d and Pa/d, respectively, will be transmitted to panel points 2 and 3 on each truss. Therefore, the stress y in the member is

$$y = \Sigma Py = \frac{Pb}{d}(y_2) + \frac{Pa}{d}(y_3) = \frac{Pb}{d}y_2 + \frac{P(d-b)}{d}y_3$$

$$= P\left[y_3 + (y_2 - y_3)\frac{b}{d}\right]$$

If a straight line is drawn connecting the upper ends of the ordinates y_2 and y_3, the expression within the brackets is seen to be the value of the ordinate under P in the influence diagram. This demonstration indicates that the influence diagram between panel points of a truss is always bounded by straight lines. This statement should be constantly kept in mind when influence diagrams for trusses are being determined because, once the influence ordinates at all the panel points have been calculated, the influence diagram is constructed by drawing straight lines through the extremities of these ordinates.

8-9. Influence Diagrams for Reactions on a Truss. The influence diagrams for reactions on a truss with end floor beams are different from those for a truss without end floor beams. When end floor beams are used, all the loads on the floor system are transmitted to the abutments or piers through the end pedestals. Thus, if end floor beams are used in the bridge shown in Fig. 8-17a, the reaction at the end pedestals is unity when a pair of unit loads is placed on symmetrical stringers at their junction with the end floor beam. When the unit load moves to the panel points b, c, d, e, f, or g, the end reaction decreases linearly from unity to zero. Thus, as shown in Fig. 8-17b, the influence diagram for the left reaction of a truss with end floor beams is the same as the reaction influence diagram for a simple beam.

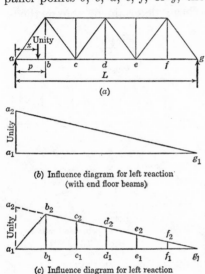

(a)

(b) Influence diagram for left reaction (with end floor beams)

(c) Influence diagram for left reaction (without end floor beams)

Fig. 8-17

When end floor beams are not used, the exterior ends of the end stringers rest directly on the abutments or piers. For the truss shown in Fig. 8-17a, when the moving unit load is to the right of panel point b, the left reaction is the same whether or not end floor beams are used. Thus the portion of the influence diagram between panel points b and g in Fig. 8-17c is identical with that drawn for this segment of the truss in Fig. 8-17b. When the unit load is on the end panel, however, say at a distance x $(x < p)$ from a, the left reaction of a truss is equal to that of a beam equal in length to that of the truss minus the stringer reaction at a, thus

$$R_1 = \frac{L - x}{L} - \frac{(p - x)}{p} = \frac{x}{p} - \frac{x}{L}$$

The left reaction may also be found by distributing the panel-point load at b to the left and right ends of the truss. The panel-point load at b is x/p, and the reaction at the left end is

$$R_1 = \frac{x}{p} \frac{(L - p)}{L} = \frac{x}{p} - \frac{x}{L}$$

When $x = 0$, the unit load acts at the end of the stringer and goes directly to the abutment or pier; therefore, its effect on the reaction at the left pedestal of the truss is zero. Thus the influence diagram for the left reaction of a truss without end floor beams is constructed as shown in Fig. 8-17c. It will be noted that the triangle $a_1 a_2 b_2$ is an influence diagram for the left reaction of a simple span p.

Example 8-4. The exterior ends of the end stringers of the bridge shown in Fig. 8-18a rest directly on abutments. Draw the influence diagram for the left reaction of one truss. Compute the maximum left reaction due to (*a*) a moving uniform load of 4 kips per lin ft on the bridge and (*b*) two moving axle loads of 60 kips each, at 8 ft on centers.

SOLUTION. The influence diagram for the left reaction is found to be as shown in Fig. 8-18b. When the unit load on the stringer is at panel point 0, all of it goes directly into the abutment and the left reaction on the truss is zero. The left reaction becomes $+\frac{5}{6}$, $+\frac{4}{6}$, $+\frac{3}{6}$, $+\frac{2}{6}$, $+\frac{1}{6}$, and zero as the unit load moves from panel points 1 to 6. Inasmuch as it has been proved that the influence curve between panel points is always linear, the correctness of the influence diagram shown in Fig. 8-18b is assured.

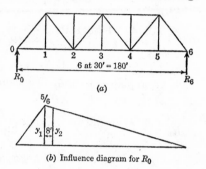

(a)

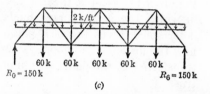

(b) Influence diagram for R_0

To cause a maximum left reaction on the truss, the moving uniform load of 4 kips per lin ft should cover the entire 180 ft of the bridge, or 2 kips per lin ft to each truss. The value of this reaction may be computed by several different methods. One procedure is to multiply the area of the

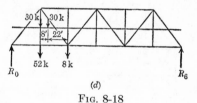

(c)

(d)

FIG. 8-18

influence diagram by the intensity of the uniform loading, or in this case by 2 kips per lin ft. Thus

$$R_0 = (2)[\tfrac{1}{2}(\tfrac{5}{6})(180)] = (2)(75) = 150 \text{ kips}$$

A second method is to subtract the exterior reaction on the end stringer from the left reaction of an equivalent simple beam 180 ft long. Thus

$$R_0 = \tfrac{1}{2}(2)(180) - \tfrac{1}{2}(2)(30) = 150 \text{ kips}$$

A third procedure is to determine the left reaction of the truss from the actual panel-point loads on the truss as shown in Fig. 8-18c; thus

$$R_0 = \tfrac{1}{6}(60)(5 + 4 + 3 + 2 + 1) = 150 \text{ kips}$$

From an inspection of the influence diagram, it is seen that the two moving concentrated loads should take positions in the second panel as shown in Fig. 8-18d.

(a) By the influence-diagram method,

$$R_0 = 30y_1 + 30y_2 = 30(\tfrac{5}{6}) + 30(\tfrac{5}{6})(^{142}\!/_{150}) = 48.67 \text{ kips}$$

(b) By the simple-beam method (truss replaced by equivalent simple beam),

$$R_0 = \frac{30(142) + 30(150)}{180} = 48.67 \text{ kips}$$

(c) By the panel-loads method (the loads are replaced by panel concentrations as shown in Fig. 8-18d),

$$R_0 = \frac{8(4) + 52(5)}{6} = 48.67 \text{ kips}$$

Although it may seem longer than the other procedures, the last method, in which the actual panel loads are determined, gives the best picture of how the loads on the floor system are transferred to the panel points of the truss.

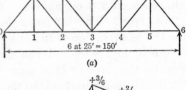

(a)

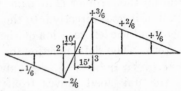

(b) Influence diagram for shear in panel 2-3

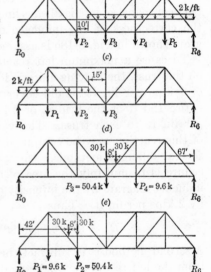

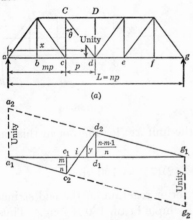

(b) Influence diagram for shear in panel cd

Fig. 8-19

Fig. 8-20

8-10. Influence Diagram for Shear in a Parallel-chord Truss. The influence diagram for shear in the panel cd of a parallel-chord truss ag (Fig. 8-19a) is shown in Fig. 8-19b. For loads at panel points to the left of panel cd, the shear in the panel is equal to the right reaction; therefore, the portion of the right-reaction influence diagram a_1c_2, plotted as shown, may be used for shear in panel cd. The ordinate c_1c_2 is $-m/n$.

For loads at panel points to the right of panel cd, the shear in the panel is equal to the left reaction; therefore, the portion of the left-reaction influence diagram g_1d_2, plotted as shown, may be used for shear in panel cd. The ordinate d_1d_2 is $+(n - m - 1)/n$. Joining c_2 and d_2 by a straight line will give the complete influence diagram for the shear in panel cd as $a_1c_2d_2g_1$.

Because c_1c_2i and d_1d_2i are similar triangles, the distances c_1i and id_1 in Fig. 8-19b may be found by simple proportion. Thus

$$c_1i = c_1d_1 \frac{c_1c_2}{c_1c_2 + d_1d_2} = p \frac{m/n}{(m/n) + [(n - m - 1)/n]} = \frac{m}{n - 1} p$$

and

$$id_1 = c_1d_1 \frac{d_1d_2}{c_1c_2 + d_1d_2} = p \frac{(n - m - 1)/n}{(m/n) + [(n - m - 1)/n]} = \frac{n - m - 1}{n - 1} p$$

At this time it will be interesting to note that the distances ig_1 and ia_1 are, respectively, n times the distances id_1 and ic_1. The proof follows:

$$\frac{ig_1}{id_1} = \frac{[(n - m - 1)/(n - 1)] p + (n - m - 1)p}{[(n - m - 1)/(n - 1)]p} = \frac{1/(n - 1) + 1}{1/(n - 1)} = n$$

and $\quad \dfrac{ia_1}{ic_1} = \dfrac{[m/(n - 1)]p + mp}{[m/(n - 1)]p} = \dfrac{1/(n - 1) + 1}{1/(n - 1)} = n$

The point i in the influence diagram for shear in panel cd is sometimes called the "load divide" because it is seen that, if a unit load is placed at i, the shear in the panel is zero. In other words, the point i divides the span a_1g_1 in the same ratio as it does the panel length c_1d_1 so that the left reaction at a is equal to the panel concentration at c. It is to be noted that loads to the right of point i produce positive shear (tension in member Cd) in the panel, while loads to the left of i produce negative shear (compression in member Cd) in the panel. Inasmuch as the stress in diagonal Cd equals the shear in the panel cd times sec θ, influence diagram $a_1c_2d_2g_1$ may be used as a stress influence diagram for member Cd.

It will be noted that the ordinates to diagram $a_1c_2d_2g_1$ may be constructed by placing unity first at c and then at d. With unity at c, the shear in panel cd is $-m/n$, which is plotted as c_1c_2. For unity at d, the shear in the panel is $+(n - m - 1)/n$, which is the value of ordinate d_1d_2.

Example 8-5. Draw the influence diagram for the shear in panel 2-3 of the parallel-chord truss shown in Fig. 8-20. Compute the maximum positive and negative shears in panel 2-3 due to (a) a moving uniform load of 2 kips per lin ft on each truss and (b) two moving concentrated loads of 30 kips each at 8 ft apart on each truss.

SOLUTION. The influence diagram for the shear in panel 2-3 as shown
in Fig. 8-20b can be obtained by placing the unit load at panel points
0, 1, 2, . . . , 6 in succession and computing the shear in the third panel.
The distances 2-i and i-3 are found by dividing the panel length 2-3, or
25 ft, into two parts in the ratio of $\frac{2}{6}$ to $\frac{3}{6}$, or 2 to 3; thus,

$$2\text{-}i = (25)\left(\frac{2}{2+3}\right) = 10 \text{ ft}$$

$$3\text{-}i = (25)\left(\frac{3}{2+3}\right) = 15 \text{ ft}$$

The critical positions of the uniform load to cause maximum positive
or negative shears in panel 2-3 are shown in Fig. 8-20c and d. These
values can be found by multiplying the area of the influence diagram
covered by the uniform load by the intensity of loading. Thus

$$\text{Max} +V_{2\text{-}3} = +(2)(\tfrac{1}{2})(\tfrac{3}{6})(90) = +45 \text{ kips}$$
$$\text{Max} -V_{2\text{-}3} = -(2)(\tfrac{1}{2})(\tfrac{2}{6})(60) = -20 \text{ kips}$$

These shear values can also be found from the free-body diagrams of the
truss as shown in Fig. 8-20c and d; thus

$$\text{Max} +V_{2\text{-}3} = R_0 - P_2$$
$$= \frac{2(90)^2}{(2)(150)} - \frac{2(15)^2}{2(25)} = 54 - 9 = +45 \text{ kips}$$

Note that R_0 is equal to the moment of P_2 to P_5, inclusive about the
right support divided by the span of the truss. Also the moment of
P_2 to P_5 about the right support is equal to that of the 90 ft of uniform
load about the right support. P_2 in Fig. 8-20c is the portion of the
uniform load on panel 2-3 transferred to panel point 2. Similarly,

$$\text{Max} -V_{2\text{-}3} = -(R_6 - P_3) = -\left[\frac{2(60)^2}{2(150)} - \frac{2(10)^2}{2(25)}\right] = -(24 - 4)$$
$$= -20 \text{ kips}$$

The critical positions for the two moving concentrated loads on the
span to cause maximum positive or negative shears in panel 2-3 are
shown in Fig. 8-20e and f. From the influence-diagram method,

$$\text{Max} +V_{2\text{-}3} = \Sigma Py = (30)(\tfrac{3}{6}) + 30(\tfrac{3}{6})(\tfrac{67}{75}) = +28.4 \text{ kips}$$
$$\text{Max} -V_{2\text{-}3} = \Sigma Py = -(30)(\tfrac{2}{6}) - 30(\tfrac{2}{6})(\tfrac{42}{50}) = -18.4 \text{ kips}$$

From the panel-loads method (Fig. 8-20e and f),

$$\text{Max} +V_{2\text{-}3} = R_0 = \frac{2P_4 + 3P_3}{6}$$
$$= \frac{2(9.6) + 3(50.4)}{6} = +28.4 \text{ kips}$$

$$\text{Max} \ -V_{2\text{-}3} = -R_6 = -\frac{P_1 + 2P_2}{6}$$

$$= -\frac{9.6 + 2(50.4)}{6} = -18.4 \ \text{kips}$$

8-11. Influence Diagram for Bending Moment at a Panel Point in the Loaded Chord of a Truss. The influence diagram for bending moment at panel point c of the truss in Fig. 8-21a is shown in Fig. 8-21b. This influence diagram is exactly the same as the influence diagram for bending moment at point c in a simple beam ag. For unity on the segment ac at a distance x from a, the bending moment at c is

$$M_c = R_2 s_2 = \frac{s_2 x}{L}$$

For unity on the segment cg, the

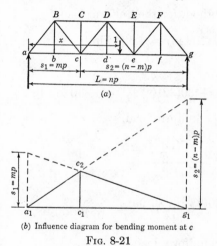

(a)

(b) Influence diagram for bending moment at c

Fig. 8-21

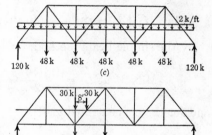

(a)

(b) Influence diagram for bending moment at L_2

(c)

(d)

Fig. 8-22

bending moment at c is

$$M_c = R_1 s_1 = \frac{s_1(L - x)}{L}$$

By substituting $x = s_1$ in either expression for M_c it is seen that the common ordinate $c_1 c_2$ is $s_1 s_2 / L$.

It just happens, as noted before in Art. 8-5, that the influence diagram $a_1 c_2 g_1$ may be obtained by constructing the bending-moment diagram for point c in beam ag. For unity at c, the left reaction is s_2 / L and the bending moment at c is $s_1 s_2 / L$.

Example 8-6. Draw the influence diagram for bending moment at L_2 in the loaded chord of the truss shown in Fig. 8-22a. Compute the maximum bending moment at L_2 due to (a) a moving uniform load of 2 kips per lin ft on each truss, and (b) two moving concentrated loads of 30 kips each at 8 ft on centers, on each truss.

SOLUTION.   The influence diagram for the bending moment at L_2 is shown in Fig. 8-22b.   The ordinate under L_2 is found by placing the unit load at L_2 and computing the bending moment at L_2; thus

$$\text{Influence ordinate at } L_2 = R_0(48) = (\tfrac{2}{3})(48) = 32 \text{ ft}$$

For maximum bending moment at L_2 it is seen that the moving uniform load covers the entire span as shown in Fig. 8-22c.   The two moving concentrated loads should be placed as shown in Fig. 8-22d to give the maximum bending moment at L_2.   As in previous illustrative problems, the value of the bending moment may be computed by the use of the influence-diagram, the equivalent simple-beam, or the panel-loads method.

Thus, by the influence-diagram method, for the uniform load,

$$\text{Max } M \text{ at } L_2 = (2)(\tfrac{1}{2})(32)(144) = 4{,}608 \text{ kip-ft}$$

and for the concentrated loads,

$$\text{Max } M \text{ at } L_2 = \Sigma Py = (30)(32) + 30(32)(\tfrac{88}{96}) = 1{,}840 \text{ kip-ft}$$

By the simple-beam method, for the uniform load,

$$\text{Max } M \text{ at } L_2 = \frac{(2)(144)}{2}(48) - \frac{2(48)^2}{2} = 4{,}608 \text{ kip-ft}$$

and for the concentrated loads,

$$\text{Max } M \text{ at } L_2 = 48R_0 = (48)\frac{30(88) + 30(96)}{144} = 1{,}840 \text{ kip-ft}$$

By the panel-loads method, for the uniform load,

$$\text{Max } M \text{ at } L_2 = [120(2) - 48(1)](24) = 4{,}608 \text{ kip-ft}$$

and for the concentrated loads,

$$\text{Max } M \text{ at } L_2 = 48R_0 = 48[\tfrac{1}{2}(10) + \tfrac{2}{3}(50)] = 1{,}840 \text{ kip-ft}$$

It will be noted that the maximum compressive stress in members U_1U_2 or U_2U_3 of this truss due to any moving load system is equal to the maximum bending moment at L_2 divided by the height of the truss.

8-12. Influence Diagram for Pier or Floor-beam Reaction.   The common support for the two simple structures shown in Fig. 8-23a must take the right reaction of span L_1 and the left reaction of span L_2.   The reaction influence diagrams are constructed as shown in Fig. 8-23b.   It will be noted that this diagram is similar to the bending-moment influence diagram for point e in a simple beam with span $L_1 + L_2$ as shown in Fig. 8-23c. The ordinates in Fig. 8-23c when multiplied by $(L_1 + L_2)/L_1L_2$ become numerically equal to those in Fig. 8-23b. Thus, to cause a maximum pier reaction at e, the critical position for a system of moving loads is the same as that for maximum bending moment at point e in a simple beam with span $(L_1 + L_2)$. In fact, it is seen that the value of this reaction is equal to $(L_1 + L_2)/L_1L_2$ times the value of the bending moment at e.

In Fig. 8-24a, the floor beam at panel point b must support the right reaction from the stringer in panel ab and the left reaction from the stringer in panel bc. The reaction influence diagram for both stringers as shown in Fig. 24b is therefore the influence diagram for the floor-beam

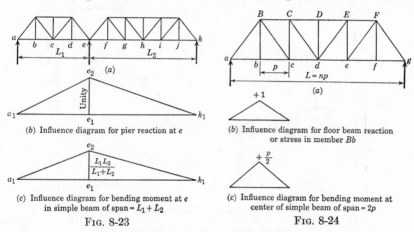

(b) Influence diagram for pier reaction at e

(c) Influence diagram for bending moment at e in simple beam of span $= L_1 + L_2$

FIG. 8-23

(b) Influence diagram for floor beam reaction or stress in member Bb

(c) Influence diagram for bending moment at center of simple beam of span $= 2p$

FIG. 8-24

reaction, and, in this case, also for the stress in member Bb. This diagram is similar to the bending-moment influence diagram for the mid-point of a simple beam with a span equal to two times the panel length, or $2p$, as shown in Fig. 8-24c. The ordinates in Fig. 8-24c are $p/2$ times those in Fig. 8-24b. Thus the maximum floor-beam reaction or the maximum stress in member Bb due to any system of moving loads is $2/p$ times the maximum bending moment at the mid-point of a simple beam with span $2p$.

On occasions the conversion of one influence diagram to another as illustrated above is a useful device.

8-13. Influence Diagram for Bending Moment at a Panel Point in the Unloaded Chord of a Truss. The influence diagram for bending moment at panel point C in the unloaded chord of the truss shown in Fig. 8-25a is drawn in Fig. 8-25b.

As has been previously shown, for a load of unity at a distance x

from a on segment ac, the bending moment at C is

$$M_C = R_2 s_2 = \frac{x}{L} s_2$$

and for unity on segment dg, the bending moment at C is

$$M_C = R_1 s_1 = \frac{(L - x)}{L} s_1$$

Thus, if the unit load is outside the panel cd, the bending moment at C is the same as though C were on the loaded chord. In Fig. 8-25b, $a_1 C_3 g_1$

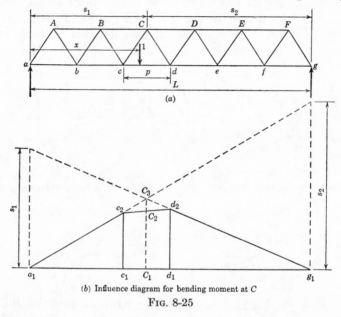

(a)

(b) Influence diagram for bending moment at C

Fig. 8-25

is the influence diagram for bending moment at C if C were on the loaded chord. The ordinates $c_1 c_2$, $d_1 d_2$, and $C_1 C_3$ are

$$c_1 c_2 = R_2 s_2 = \frac{ac}{L} s_2$$

$$d_1 d_2 = R_1 s_1 = \frac{dg}{L} s_1$$

and

$$C_1 C_3 = \frac{s_1 s_2}{L}$$

In Art. 8-8 it has been shown that the influence diagram between adjacent panel points of a truss is composed of straight-line segments; consequently diagram $a_1 c_2 d_2 g_1$ is the influence diagram for bending moment at panel point C. It is to be noted that, had point C been vertically above point c, the influence diagrams for bending moment at C or c would be identical.

In an actual problem, it is suggested that ordinates c_1c_2, d_1d_2, and C_1C_3 as shown in Fig. 8-25b be all computed and their correctness be checked by verifying that $a_1c_2C_3$ and $g_1d_2C_3$ are straight lines. The usable part of the influence diagram is, of course, only $a_1c_2d_2g_1$.

Example 8-7. Draw the influence diagram for bending moment at panel point U_3 in the unloaded chord of the Warren truss shown in Fig. 8-26a. Compute the maximum bending moment at U_3 due to (a) a moving uniform load of 2 kips per lin ft on each truss, and (b) two moving concentrated loads of 30 kips each at 8 ft on centers on each truss.

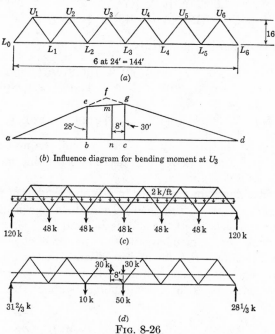

(a)

(b) Influence diagram for bending moment at U_3

(c)

(d)

Fig. 8-26

SOLUTION. The influence diagram for bending moment at U_3 is shown in Fig. 8-26b. The ordinate be is the bending moment at U_3 due to unity at L_2; thus

$$be = R_6(84) = \tfrac{1}{3}(84) = 28 \text{ ft}$$

Similarly, cg is the bending moment at U_3 due to unity at L_3, or

$$cg = R_0(60) = \tfrac{1}{2}(60) = 30 \text{ ft}$$

The ordinate at point f in Fig. 8-26b is the bending moment at a point 60 ft from the left end of a 144-ft simple beam, or

$$\text{Ordinate at point } f = \frac{(60)(84)}{144} = 35 \text{ ft}$$

Verification that aef and fgd are straight lines may now be made.

From an inspection of the influence diagram, it is seen that for maximum bending moment at U_3 the critical loading conditions for the moving uniform load and the system of concentrated loads, respectively, are as shown in Fig. 8-26c and d. The maximum bending moment at U_3 may be conveniently found by the influence-diagram method or the panel-loads (Fig. 8-26c and d) method. Due to the uniform load,

$$\text{Max } M \text{ at } U_3 = (2)(\text{area of } aegda)$$
$$= (2)[\tfrac{1}{2}(28)(48) + \tfrac{1}{2}(28 + 30)(24) + \tfrac{1}{2}(30)(72)]$$
$$= 4{,}896 \text{ kip-ft}$$

or $\text{Max } M \text{ at } U_3 = (120)(60) - 48(12 + 36) = 4{,}896 \text{ kip-ft}$

Due to the two concentrated loads,

$$\text{Max } M \text{ at } U_3 = \Sigma Py = (50)(30) + 10(28) = 1{,}780 \text{ kip-ft}$$
or $= (30)(30) + 30[30 - \tfrac{8}{24}(2)] = 1{,}780 \text{ kip-ft}$
or
$$\text{Max } M \text{ at } U_3 = R_0(60) - 10(12) = (31\tfrac{2}{3})(60) - 120 = 1{,}780 \text{ kip-ft}$$

It is to be noted that, for any live-load system, the maximum tensile stress in member L_2L_3 is equal to the maximum bending moment at U_3 divided by the height of the truss.

8-14. Influence Diagram for Stress in a Web Member of a Truss with Inclined Chords. A truss with inclined chords is shown in Fig. 8-27a. Point O is the intersection of the chords BC and bc. The stress in web member Bc equals the moment about O of the forces on either side of a section through members BC, Bc, and bc, divided by the arm t.

For unity on segment ab, at a distance x from a, the compressive stress in Bc is conveniently found by taking moments about O with the right side of the section as a free body; thus

$$Bc = -R_2 \frac{e + L}{t} = -\frac{x}{L} \frac{e + L}{t}$$

If the length ab is substituted for x in the above equation, ordinate b_1b_2 in the influence diagram for stress in member Bc is found to be

$$b_1b_2 = -\frac{ab}{L} \frac{e + L}{t}$$

If b_2a_1 is prolonged to o_2 directly under O, ordinate o_1o_2 is found to be

$$o_1o_2 = \frac{e}{L} \frac{e + L}{t}$$

For unity at a distance x from a and on segment cg, the tensile stress in Bc may be found by taking moments about O with the left side of the section as a free body; thus

$$Bc = +R_1\frac{e}{t} = +\frac{L-x}{L}\frac{e}{t}$$

If the length ac is substituted for x in the above equation, ordinate c_1c_2 is found to be

$$c_1c_2 = +\frac{cg}{L}\frac{e}{t}$$

Now points o_2, c_2, and g_1 can be proved to lie in the same straight line by showing that

$$\frac{c_1c_2}{o_1o_2} = \frac{c_1g_1}{o_1g_1}$$

Because the influence diagrams between adjacent panel points must be bounded by straight lines, it is seen that $a_1b_2c_2g_1$ is the influence diagram

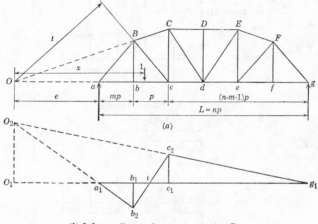

(a)

(b) Influence diagram for stress in member Bc

FIG. 8-27

for the stress in member Bc. In a problem like this, the critical ordinates b_1b_2 and c_1c_2 are first calculated by placing unity at panel points b and c, and a verification of their correctness can be made by showing that b_2a_1 and g_1c_2, when both prolonged, intersect under the moment center O. Point i may be located by dividing the panel length bc in the ratio of b_1b_2 and c_1c_2.

Example 8-8. Draw the influence diagram for stress in member U_1L_2 of the truss shown in Fig. 8-28a. Compute the maximum tensile and compressive stresses in member U_1L_2 due to (a) a moving uniform load of 2 kips per lin ft on each truss, and (b) two moving concentrated loads of 30 kips each at 8 ft on centers on each truss.

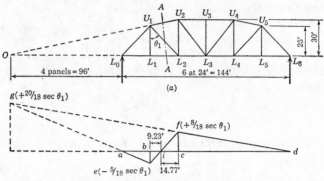

(a)

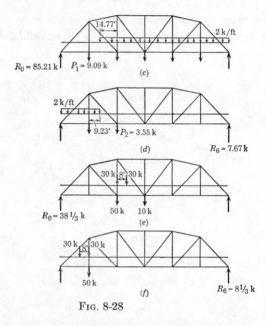

(b) Influence diagram for stress in member U_1L_2

(c)

(d)

(e)

(f)

FIG. 8-28

SOLUTION. The influence diagram for the stress in member U_1L_2 is shown in Fig. 8-28b. The ordinate be is the compressive stress in member U_1L_2 found by placing unity at joint L_1 and taking moments about point O with the right side of section AA as a free body.

$$be = -R_6 \frac{10 \text{ panels}}{6 \text{ panels}} (\sec \theta_1)$$
$$= -\tfrac{1}{6}(\tfrac{10}{6}) \sec \theta_1 = -\tfrac{5}{18} \sec \theta_1$$

The minus sign for be means compressive stress. The ordinate cf is the tensile stress in member U_1L_2 found by placing unity at joint L_2 and taking moments about point O with the left side of section AA as a free body.

$$cf = +R_0 \frac{4 \text{ panels}}{6 \text{ panels}} (\sec \theta_1)$$

$$= +\tfrac{2}{3}(\tfrac{4}{6}) \sec \theta_1 = +\tfrac{6}{18} \sec \theta_1$$

The positive sign for cf means tensile stress. The calculated values of ordinates be and cf should be verified by showing that point g, the intersection of the prolongations of ea and df, is directly under point O.

The distances bi and ci in the influence diagram may be found by dividing the panel length bc in the ratio of be to cf or 5 to 8. Thus

$$bi = (24) \frac{5}{5+8} = 9.23 \text{ ft}$$

$$ci = (24) \frac{8}{5+8} = 14.77 \text{ ft}$$

The critical conditions for the moving uniform load, to cause maximum tensile and compressive stresses in member U_1L_2, are shown in Fig. 8-28c and d, respectively. Thus

$$\text{Max tension in } U_1L_2 = +(2)(\text{area of } ifd)$$
$$= +(2)(\tfrac{1}{2})(\tfrac{8}{18} \sec \theta_1)(110.77)$$
$$= +49.23 \sec \theta_1 \text{ kips}$$
$$\text{Max compression in } U_1L_2 = -(2)(\text{area of } aei)$$
$$= -(2)(\tfrac{1}{2})(\tfrac{5}{18} \sec \theta_1)(33.23)$$
$$= -9.23 \sec \theta_1 \text{ kips}$$

Or, from Fig. 8-28c and d,

$$\text{Max tension in } U_1L_2 = \frac{R_0(4) - P_1(5)}{6} \sec \theta_1$$
$$= \frac{4(85.21) - 5(9.09)}{6} \sec \theta_1$$
$$= +49.23 \sec \theta_1 \text{ kips}$$
$$\text{Max compression in } U_1L_2 = -\frac{R_6(10) - P_2(6)}{6} \sec \theta_1$$
$$= -\frac{(7.67)(10) - (3.55)(6)}{6} \sec \theta_1$$
$$= -9.23 \sec \theta_1 \text{ kips}$$

The positions which the two moving concentrated loads should take for maximum tensile and compressive stresses in member U_1L_2 are shown in Fig. 8-28e and f. Thus

$$\text{Max tension in } U_1L_2 = \Sigma Py = +30(\tfrac{8}{18} \sec \theta_1)(1 + \tfrac{88}{96})$$
$$= +25.55 \sec \theta_1 \text{ kips}$$
$$\text{Max compression in } U_1L_2 = \Sigma Py = -30(\tfrac{5}{18} \sec \theta_1)(1 + \tfrac{16}{24})$$
$$= -13.88 \sec \theta_1 \text{ kips}$$

Or, from Fig. 8-28e and f,

$$\text{Max tension in } U_1L_2 = +\frac{R_0(4)}{6}\sec\theta_1$$

$$= +\frac{(38\frac{1}{3})(4)}{6}\sec\theta_1 = +25.55\sec\theta_1 \text{ kips}$$

$$\text{Max compression in } U_1L_2 = -\frac{R_6(10)}{6}\sec\theta_1$$

$$= -\frac{(8\frac{1}{3})(10)}{6}\sec\theta_1$$

$$= -13.88\sec\theta_1 \text{ kips}$$

PROBLEMS

8-1. Given a simple beam 30 ft long, construct influence diagrams and compute the maximum values due to a moving uniform load of 1 kip per ft and a movable concentrated load of 30 kips for (a) the left reaction, (b) the shear and bending moment at a section 10 ft from the left end, and (c) the shear and bending moment at the mid-point of the beam.

8-2. Given a simple beam 40 ft long, construct influence diagrams and compute the maximum values due to a moving uniform load of 0.8 kip per ft for (a) the left reaction and shear at sections 5 ft, 10 ft, 15 ft, and 20 ft from the left end, (b) the bending moment at sections 5 ft, 10 ft, 15 ft, and 20 ft from the left end.

8-3. A cantilever beam 15 ft long is fixed at the right end. Construct shear and bending-moment influence diagrams for sections 5 ft, 10 ft, and 15 ft from the free end. Calculate maximum shears and bending moments at these sections in this beam due to a moving uniform load of 0.6 kip per ft and a movable concentrated load of 10 kips.

8-4. A beam 40 ft long is supported at the left end and 30 ft from the left end. Construct influence diagrams and compute the maximum values due to a dead load of 0.5 kip per ft, a moving live load of 0.8 kip per ft, and a movable concentrated load of 12 kips for (a) the reactions, (b) the shear at sections 10 ft and 20 ft from the left end, (c) the bending moment at sections 10 ft, 20 ft, and 30 ft from the left end.

8-5. A beam 65 ft long is supported at 10 ft from the left end and at 15 ft from the right end. Construct influence diagrams and compute the maximum values due to a dead load of 0.8 kip per ft, a moving live load of 1.2 kips per ft, and two movable concentrated loads of 10 kips each spaced 10 ft apart, for (a) the reactions, (b) the shear at the left support, at 10 ft and 20 ft from the left support, and at the right support, (c) the bending moment at the left support, at 10 ft and 20 ft from the left support, and at the right support.

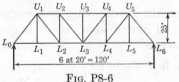

FIG. P8-6

8-6. Given the Pratt truss as shown, construct influence diagrams and compute the maximum and minimum (if any) values due to a moving uniform load of 1.2 kips per ft and a movable concentrated load of 20 kips for (a) the shear in panels L_0L_1, L_1L_2, and L_2L_3; (b) the bending moment at panel points L_1, L_2, and L_3; (c) the stress in member U_1L_1.

8-7. Given the Warren truss as shown, construct influence diagrams and compute the maximum and minimum (if any) values due to a moving uniform load of 0.64 kip per ft and a movable concentrated load of 18 kips for (a) the shear in panels L_0L_1, L_1L_2, and L_2L_3; (b) the bending moment at panel points L_1 and L_2; (c) the bending moment at panel points U_1, U_2, and U_3.

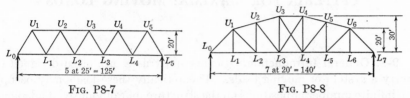

FIG. P8-7   FIG. P8-8

8-8. Given the Parker truss as shown, construct influence diagrams and compute the maximum and minimum (if any) values due to a dead load of 0.5 kip per ft, a moving live load of 1.5 kips per ft, and a movable concentrated load of 20 kips for (a) the stress in members U_1U_2 and L_2L_3, (b) the stress in member U_1L_1, (c) the stress in member U_2L_2 and U_2L_3.

CHAPTER 9

CRITERIA FOR MAXIMA: MOVING LOADS

9-1. General Introduction. Beams and trusses are often designed to carry a system of moving loads. Theoretically these loads may occupy an infinite number of positions on the structure, but obviously, if adequate provision is made for the maximum condition of loading, the designer need not be concerned about the other and lesser conditions of loading on the structure. Thus when a system of moving loads such as a uniform load or a series of concentrated loads, or both, passes over a beam or a bridge floor supported by trusses, the effect on the shear or bending moment at a section of the beam, or the stress in a member of the truss is different for various positions of the loading. Before its maximum effect on a specific function such as, for instance, the bending moment at a section, can be calculated, the critical *position* of the load system on the structure must be determined. In general these critical positions are different for maximum effects on different functions.

In the preceding chapter, influence diagrams for various functions in beams and trusses have been studied. As illustrated in Chap. 8, the critical position of a uniform load or one or two concentrated loads can usually be determined by simple inspection of the appropriate influence diagrams. For more complicated conditions of loading, however, it is difficult to find the critical positions by ordinary inspection or by other simple cut-and-try methods. In this chapter, definite criteria for determining the positions of loading necessary to cause maximum effects will be developed.

9-2. Maximum Reactions and Shears in Simple Beams: Uniform Loads. A glance at the reaction influence diagrams in Fig. 8-2 indicates at once that a uniformly distributed load produces a maximum left or right reaction when the beam is fully loaded, i.e., when the uniformly distributed load covers the full length of the beam.

The maximum positive and negative shears at some section C, at distance x from A in the simple beam AB (Fig. 9-1a), due to the passage of a uniformly distributed load, will be determined. The influence diagram for shear at section C is constructed as shown in Fig. 9-1b. From this influence diagram it is seen that segment CB should be covered with uniform load to cause the maximum positive shear at C, and segment AC should be covered to cause maximum negative shear at C. Thus, from Fig. 9-1c,

$$\text{Max positive shear at } C = +R_1 = +p \frac{(L-x)^2}{2L}$$

and, from Fig. 9-1d,

$$\text{Max negative shear at } C = -R_2 = -\frac{px^2}{2L}$$

In the course of design of a beam or girder to carry a uniform live load, the variation in the maximum positive or negative shears throughout the length of the beam may be required. For instance, in a built-up steel girder, the rivet pitch at a section depends on the maximum combined (dead plus live plus impact) shear at the section. If the beam in Fig. 9-2a is subjected to a uniform live load of p per linear foot, the

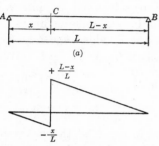

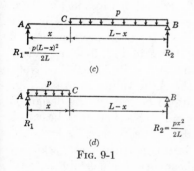

(b) Influence diagram for shear at C

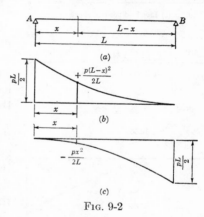

FIG. 9-1 FIG. 9-2

maximum positive and negative shear curves due to live load are shown, respectively, in Fig. 9-2b and c. Both curves are parabolic.

Usually only the maximum numerical shear, regardless of sign, is needed in the design. Thus, for the beam in Fig. 9-2a, the maximum numerical value of the shear in the left half of the beam is the sum of the dead-load positive shear plus the maximum positive shear due to live load and impact; the maximum numerical value of the shear in the right half of the beam is the sum of the dead-load negative shear plus the maximum negative shear due to live load and impact. Of course, the maximum value of the shear at a section in the left half of the beam is numerically equal to that at a similar section in the right half of the beam.

Example 9-1.   Determine the maximum combined shears at 6-ft intervals in a 48-ft simple beam subjected to a dead load of 500 lb per lin ft, a live load of 1,200 lb per lin ft, and an impact equal to 20 per cent of the live load.

SOLUTION.   The complete solution is shown diagrammatically in Fig. 9-3.   The shear diagram due to dead load is shown in Fig. 9-3b.   Note

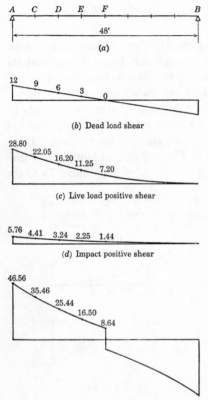

(a)

(b) Dead load shear

(c) Live load positive shear

(d) Impact positive shear

(e) Combined shear diagram

FIG. 9-3

that the end shear due to dead load is $\frac{1}{2}(0.5)(48) = 12$ kips.   The curve for maximum positive shear due to live load is shown in Fig. 9-3c.   This curve is drawn after live-load shears have been calculated at intervals along the beam.   For instance, the value of the live-load shear at section D (36 ft from the right end) is

$$V_D \text{ due to } LL = + \frac{(1.2)(36)^2}{2(48)}$$
$$= +16.20 \text{ kips}$$

As shown in Fig. 9-3d, the maximum positive shear values due to impact are 20 per cent of those in Fig. 9-3c. The combined shear values in the left half of the beam are the sums of those shown in Fig. 9-3b, c, and d, and are shown in Fig. 9-3e.   Note that the shear values shown in Fig. 9-3e are to be used in the design of the beam.   This curve may be called the "design shear curve."   Also note that, in this particular case, the design shear curve is almost linear.   To save time, some designers simply compute the shear ordinates at the ends and center of the beam (46.56 and 8.64 in Fig. 9-3e) and connect the upper extremities of these ordinates by straight lines.   When this is done, the slight errors at the intervening sections are on the side of safety.

9-3. Maximum Reactions and Shears in Simple Beams : Concentrated Loads.   A simple beam AB carrying a moving system of five concentrated loads $P_1, P_2, \ldots, P_5$ at spacings b_1, b_2, b_3, b_4 is shown in Fig. 9-4. These five loads may take any position on the span.   It is required to determine the critical loading position to cause the maximum numerical shear at section C.   By inspection of the influence diagram for shear at section C of Fig. 9-4d, it is obvious that the maximum positive shear at C will be numerically larger than the maximum negative shear at C; this will always be true as long as section C is to the left of the mid-span.

If the loads shown in Fig. 9-4a come upon the span from the right toward the left, certainly most of these loads should climb up the influence line B_1C_3 as high as possible, in order to give the maximum positive shear at C. It is to be expected that the position of the loads shown in Fig. 9-4b (P_1 at C) may produce the maximum positive shear at C. With P_1 at C, the shear at C is equal to R_1. By bringing P_2 to C (Fig. 9-4c), the shear at C is equal to $R_1 - P_1$. The loading condition with P_2 at C, however, may or may not cause a larger shear at C than the loading condition with P_1 at C. This depends on whether the gain in the value of R_1 is greater than the decrease in shear due to P_1.

Let $G = P_1 + P_2 + P_3 + P_4 + P_5$. The left reaction in Fig. 9-4c is larger than that in Fig. 9-4b by the amount Gb_1/L because, in finding the value of R_1 by taking moments about B, the moment arm of all the loads is increased by b_1 and the increase in reaction due to the forward movement of b_1 is $\Sigma P(b_1)/L$ or Gb_1/L. Thus in determining whether P_1 at C or P_2 at C will give the larger shear at C it is only necessary to compare the *gain* (Gb_1/L) with the *loss* (P_1). If the gain is smaller than the loss, then P_1 at C causes maximum shear at C. If the gain is larger than the loss, then P_2 at C causes a larger shear at C than does P_1 at C. If the latter happens, it will then be necessary to find out whether P_3 will have to be moved forward to C. Theoretically this comparison between two successive loads should be repeated until the forward load at the section controls, but very rarely will it be necessary to bring up the third or fourth load.

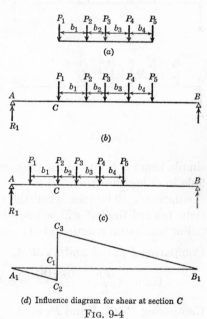

(d) Influence diagram for shear at section C

Fig. 9-4

If the traffic shown in Fig. 9-5a moves on the span from the left toward the right, one of the loading conditions shown in Fig. 9-5b or c may cause the larger positive shear at C. The loading conditions with P_5 at C (Fig. 9-5b) and then P_4 at C (Fig. 9-5c) should first be compared. If P_5 at C controls, stop with P_5 at C. If P_4 at C controls, then compare the effect of P_4 at C with that of P_3 at C.

Referring to Fig. 9-4, suppose that it is now required to determine whether P_1 at A or P_2 at A will cause the larger left reaction. By bringing P_2 forward to A, the gain in the reaction is $(P_2 + P_3 + P_4 + P_5)b_1/L$ exclusive of P_1b/L, because P_1 moves off the span. The loss in the left reaction is, of course, P_1. Let G' be the sum of all loads on the span exclusive of the forward load.

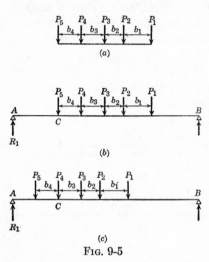

(a)

(b)

(c)

Fig. 9-5

Then if the gain $(G'b_1/L)$ is larger than the loss (P_1), P_2 at A will give the larger R_1; otherwise P_1 at A will cause a larger reaction. If the former is true, the conditions with P_2 at A and then with P_3 at A should be compared. The gain now becomes $G'b_2/L$ in which $G' = P_3 + P_4 + P_5$ and the loss is P_2.

The reader will probably be wise to check through the example below and then read the preceding discussion again.

Example 9-2. Determine the maximum left reaction and the maximum numerical shear at section C in the simple beam AB due to the passage of the six wheel loads shown in Fig. 9-6a in either direction.

SOLUTION. The case when the traffic comes upon the span from the right toward the left will be considered first.

For maximum reaction at A,

Comparing P_1 at A and P_2 at A,

$$\text{Gain} = \frac{G'b}{L} = \frac{(63)(8)}{60} = 8.4 \text{ kips} \qquad \text{Loss} = P_1 = 6 \text{ kips}$$

Comparing P_2 at A and P_3 at A,

$$\text{Gain} = \frac{G'b}{L} = \frac{(54)(5)}{60} = 4.5 \text{ kips} \qquad \text{Loss} = P_2 = 9 \text{ kips}$$

Thus, as shown in Fig. 9-6b, the reaction R_A is a maximum with P_2 at A.

$$\text{Max } R_A = 9 + \frac{12(38 + 44 + 49) + 18(55)}{60} = 51.7 \text{ kips}$$

For maximum shear at C,

Comparing P_1 at C and P_2 at C,

$$\text{Gain} = \frac{Gb}{L} = \frac{(69)(8)}{60} = 9.2 \text{ kips} \qquad \text{Loss} = P_1 = 6 \text{ kips}$$

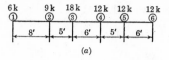

(a)

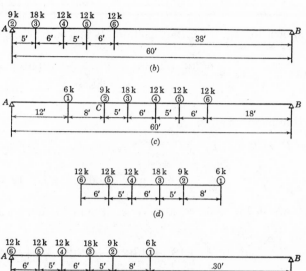

(b)

(c)

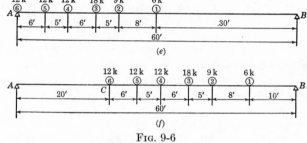

(d)

(e)

(f)

FIG. 9-6

Comparing P_2 at C and P_3 at C,

$$\text{Gain} = \frac{Gb}{L} = \frac{(69)(5)}{60} = 5.75 \text{ kips} \qquad \text{Loss} = P_2 = 9 \text{ kips}$$

Thus, as shown in Fig. 9-6c, the maximum shear at C occurs when P_2 is at C.

$$\text{Max } V_C = R_A - P_1$$
$$= \frac{12(18 + 24 + 29) + 18(35) + 9(40) + 6(48)}{60} - 6$$
$$= 29.5 \text{ kips}$$

When the traffic comes upon the span from the left toward the right, the six concentrated loads will move on the span in the order shown in Fig. 9-6d.

For maximum reaction at A,

Comparing P_6 at A and P_5 at A,

$$\text{Gain} = \frac{G'b}{L} = \frac{(57)(6)}{60} = 5.7 \text{ kips} \qquad \text{Loss} = P_6 = 12 \text{ kips}$$

Thus, as shown in Fig. 9-6e, the reaction R_A is a maximum with P_6 at A.

$$\text{Max } R_A = 12 + \frac{6(30) + 9(38) + 18(43) + 12(49 + 54)}{60}$$

$$= 54.2 \text{ kips}$$

For maximum shear at C,

Comparing P_6 at C and P_5 at C,

$$\text{Gain} = \frac{Gb}{L} = \frac{(69)(6)}{60} = 6.9 \text{ kips} \qquad \text{Loss} = P_6 = 12 \text{ kips}$$

Thus, as shown in Fig. 9-6f, P_6 at C causes the maximum shear at C.

$$\text{Max } V_C = R_A$$

$$= \frac{6(10) + 9(18) + 18(23) + 12(29 + 34 + 40)}{60}$$

$$= 31.2 \text{ kips}$$

From the above computations it is seen that the larger maximum left reaction and maximum shear at C are both caused by traffic moving from the left toward the right. Thus, the final results are

$$\text{Max } R_A = 54.2 \text{ kips}$$
$$\text{Max } V_C = 31.2 \text{ kips}$$

9-4. Maximum Bending Moment at a Point in a Simple Beam: Uniform Load. It is obvious that the maximum bending moment at any section in a simple beam due to a moving uniform live load occurs when the beam is fully loaded. This is evident from a consideration of the influence diagram for bending moment shown in Fig. 8-8. The beam must be fully covered by the moving uniform load to produce maximum bending moment at any point between the supports.

9-5. Maximum Bending Moment at a Point in a Simple Beam: Concentrated Loads. Let it be required to determine the position which a series of concentrated loads must take on the simple beam AB (Fig. 9-7a) so that the bending moment at C will be a maximum.

Before a direct attack is made on this problem, some preliminary considerations will be discussed. First, if the line segment in Fig. 9-8b is a portion of an influence diagram, the value of the function when the structure is loaded with P_1, P_2, P_3, P_4 as shown in Fig. 9-8a is ΣPy or $(P_1y_1 + P_2y_2 + P_3y_3 + P_4y_4)$. If y_G is the influence ordinate under the resultant $G = \Sigma P$ of these forces, it can be shown that

$$Gy_G = P_1y_1 + P_2y_2 + P_3y_3 + P_4y_4$$

PROOF. The proof follows:

$$Ga_G = P_1a_1 + P_2a_2 + P_3a_3 + P_4a_4 \qquad \text{(principle of moments)}$$

but $\quad a_G = \dfrac{y_G}{m} \qquad a_1 = \dfrac{y_1}{m} \qquad a_2 = \dfrac{y_2}{m} \qquad$ etc.   (by geometry, Fig. 9-8b)

Substituting, $\quad G\left(\dfrac{y_G}{m}\right) = P_1\left(\dfrac{y_1}{m}\right) + P_2\left(\dfrac{y_2}{m}\right) + P_3\left(\dfrac{y_3}{m}\right) + P_4\left(\dfrac{y_4}{m}\right)$

or $\qquad\qquad\qquad Gy_G = P_1y_1 + P_2y_2 + P_3y_3 + P_4y_4$

The influence diagram for bending moment at C in a simple beam AB is shown in Fig. 9-7b. The ordinate at C_2 is ab/L. The slope of line A_1C_2 is b/L (vertical) on 1 (horizontal) and the slope of line B_1C_2 is a/L on 1.

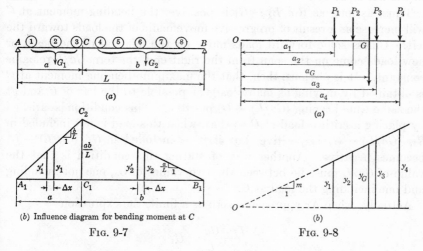

(b) Influence diagram for bending moment at C

FIG. 9-7

FIG. 9-8

When a series of concentrated loads takes a position as shown in Fig. 9-7a on the span, with G_1 to the left of C and G_2 to the right of C, the bending moment at C is

$$M_C = G_1y_1 + G_2y_2$$

If these loads move a small distance Δx to the left, the bending moment at C is

$$M'_C = G_1y'_1 + G_2y'_2$$

As a result of this small movement of Δx toward the left, the increase in the bending moment at C is

$$\begin{aligned}
\Delta M &= M_C' - M_C \\
&= G_2(y_2' - y_2) - G_1(y_1 - y_1') \\
&= G_2 \frac{a}{L} \Delta x - G_1 \frac{b}{L} \Delta x
\end{aligned}$$

By dividing every term in the above equation by Δx, the rate of increase of the bending moment at C with respect to the movement Δx is

$$\begin{aligned}
\frac{\Delta M}{\Delta x} &= G_2 \frac{a}{L} - G_1 \frac{b}{L} \\
&= G_2 \frac{a}{L} - G_1 \left(\frac{L - a}{L} \right) \\
&= (G_1 + G_2) \frac{a}{L} - G_1 = G \frac{a}{L} - G_1
\end{aligned}$$

in which G is the total load on the span and G_1 is the portion of the load to the left of C.

Thus, as long as $(Ga/L) - G_1$ is positive, the bending moment at C will increase as a result of progressive movements of the loads toward the left. Unless some forward loads move off the left end of the beam or new loads come on the span from the right end, the term Ga/L remains constant. It is obvious, then, that the maximum bending moment at C is obtained by moving as many loads as possible to the left of C and at the same time keeping $(Ga/L) - G_1$ positive. This condition is satisfied by placing a critical load at C so that, when this load is not included in G_1, $(Ga/L) - G_1$ is positive, but if it is included in G_1, $(Ga/L) - G_1$ becomes negative. Another way of stating this condition is that the value of Ga/L must lie between the two values of G_1, one not including and one including the load at C.

Actually, when Δx approaches zero as a limit, the expression

$$\frac{\Delta M}{\Delta x} = \frac{Ga}{L} - G_1$$

becomes

$$\frac{dM}{dx} = \frac{Ga}{L} - G_1$$

In calculus, it has been demonstrated that, when $dM/dx = 0$, or when there is no change in the bending moment at C by an infinitesimal movement dx, the condition for maximum or minimum is obtained. A *maximum* condition exists when dM/dx passes through zero from positive to negative; viz., when the load at C is considered at an infinitesimal distance to the right of C, $dM/dx = (Ga/L) - G_1$ is positive, but when this load is considered at an infinitesimal distance to the left of C,

$$\frac{dM}{dx} = \frac{Ga}{L} - G_1$$

becomes negative.

Summarizing, the criterion for maximum bending moment at C, distant a and b, respectively, from the left and right ends of a simple beam AB with span equal to L, is that the critical load should be placed at C so that the value of Ga/L lies between the two values of G_1, one

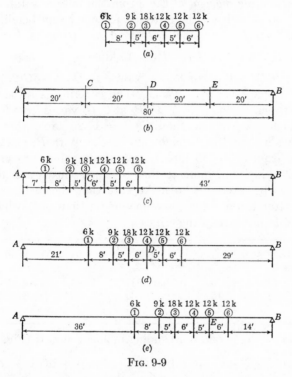

Fig. 9-9

including and the other not including the load at C (G is the total load on the span and G_1 is the load on segment AC).

Sometimes several different loads at C will satisfy the criterion. In this event, the bending moment at C must be calculated for each condition of loading that satisfies the criterion. A comparison of calculated results will determine the greatest bending moment.

Example 9-3. Determine the maximum bending moments at sections C, D, and E of the simple beam AB (Fig. 9-9b) due to the passage of the series of concentrated loads as shown in Fig. 9-9a.

SOLUTION. Traffic may come on the span from either direction but, if maximum bending moments at corresponding sections on either side of the center line are computed, it is necessary to assume traffic in one direction only (usually right to left). A little reflection will show that the maximum bending moment at C due to loads from the left is identical with that at E due to loads from the right, and the maximum bending moment at E due to loads from the left is identical with that at C due to loads from the right.

In view of the above discussion, only traffic from the right toward the left will be considered in the subsequent computations.

Maximum bending moment at C. From the following tabulation it is seen that P_3 at C satisfies the criterion for maximum bending moment at C. With P_3 at C, the value of G_1 (load on AC) varies from

$$P_1 + P_2 = 15 \text{ kips}$$

to $P_1 + P_2 + P_3 = 33$ kips; the value of $Ga/L = G(20)/80 = \frac{1}{4}G$ (G = load on span) is $69/4 = 17.25$. The criterion is satisfied because 17.25 lies between 15 and 33. With P_2 at C, G_1 varies from 6 to 15 kips, both of which are smaller than $Ga/L = 17.25$ kips, thus indicating that more loads should be brought onto AC. With P_4 at C, the value of G_1 varies from 33 to 45 kips. Both these values are greater than $Ga/L = 17.25$ kips, and indicate that too many loads are now on AC. It will always be advisable to make sure that the loads immediately before and after the one (or sometimes more than one) which satisfies the criterion do not satisfy the criterion.

Load at C	G_1	$\frac{1}{4}G$	G_1	Yes or no
P_2	6	$\frac{1}{4}(69) = 17.25$	15	No
P_3	15	17.25	33	Yes
P_4	33	17.25	45	No

With P_3 at C, the maximum bending moment at C is (Fig. 9-9c)

$$\text{Max } M_C = \frac{12(43 + 49 + 54) + 18(60) + 9(65) + 6(73)}{80} \quad (20)$$

$$- [9(5) + 6(13)]$$

$$= \frac{3,855}{4} - 123 = 840.75 \text{ kip-ft}$$

Maximum bending moment at D

Load at D	G_1	$\frac{1}{2}G$	G_1	Yes or no
P_3	15	$\frac{1}{2}(69) = 34.5$	33	No
P_4	33	34.5	45	Yes
P_5	45	34.5	57	No

With P_4 at D, the maximum bending moment at D is (Fig. 9-9d)

$$\text{Max } M_D = \frac{12(29 + 35 + 40) + 18(46) + 9(51) + 6(59)}{80} \quad (40)$$

$$- [(18)(6) + 9(11) + 6(19)]$$

$$= \frac{2,889}{2} - 321 = 1,123.5 \text{ kip-ft}$$

Maximum bending moment at E

Load at E	G_1	$\frac{3}{4}G$	G_1	Yes or no
P_4	33	$\frac{3}{4}(69) = 51.75$	45	No
P_5	45	51.75	57	Yes
P_6	57	51.75	69	No

With P_5 at E, the maximum bending moment at E is (Fig. 9-9e)

$$\text{Max } M_E = \frac{12(14 + 20 + 25) + 18(31) + 9(36) + 6(44)}{80} \quad (60)$$

$$- [12(5) + 18(11) + 9(16) + 6(24)]$$

$$= \frac{3}{4}(1,854) - 546 = 844.5 \text{ kip-ft}$$

Thus the maximum bending moment due to the passage of this series of concentrated loads in either direction is 844.5 kip-ft at C or E and 1,123.5 kip-ft at D.

Example 9-4.   Determine the maximum bending moment at section C of the simple beam AB (Fig. 9-10b) due to the passage of a uniform load which is 12 ft in length as shown in Fig. 9-10a.

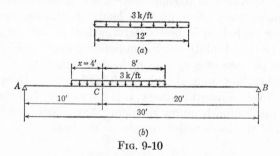

(a)

(b)

Fig. 9-10

SOLUTION.   If the uniform load is of indefinite length, the whole span must be loaded to produce maximum bending moment at C, or, for that matter, at any other section in the beam.   In this problem, however, the uniform load has a definite length of 12 ft.   The criterion $dM/dx = (Ga/L) - G_1 = 0$ will be used to determine the position of this uniform load on the span so that the bending moment at C may be maximum. Let x be the length of uniform load on segment AC, then $G_1 = 3x$ kips, $G = 36$ kips, and $a/L = {}^{10}\!/_{30} = \frac{1}{3}$.

$$\frac{Ga}{L} - G_1 = 0 \qquad 36(\tfrac{1}{3}) - 3x = 0 \qquad x = 4 \text{ ft}$$

Thus the uniform load should take the position shown in Fig. 9-10b.

$$\text{Max } M_C = R_A(10) - \frac{3(4)^2}{2} = \frac{(36)(18)}{30}(10) - 24 = 192 \text{ kip-ft}$$

9-6. Absolute Maximum Bending Moment in a Simple Beam: Concentrated Loads. Absolute maximum bending moment in a beam is defined as the largest bending moment which may ever occur in the beam due to the passage of a series of loads. The significance of this moment is apparent, as the strength of the beam must be proportioned accordingly. It is to be noted that the location of the section at which the absolute maximum bending moment occurs is not yet known. Inasmuch as the

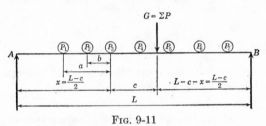

Fig. 9-11

bending-moment diagram of a beam subjected to concentrated loads only is composed of broken straight-line segments, the absolute maximum bending moment must occur under a load. If the maximum bending moment under each load is first determined, then the largest of these maxima will be the absolute maximum bending moment in the beam.

During the passage of a system of loads across a beam as shown in Fig. 9-11, the bending moment under each load varies with every position of the load. In the present instance, it is required to find the position of P_3 for the maximum bending moment that occurs under P_3. Let $G = $ the sum of the loads which remain on the beam, then

$$R_A = G\,\frac{(L - c - x)}{L}$$

and the bending moment under P_3 is

$$
\begin{aligned}
M &= R_A x - (P_1 a + P_2 b) \\
  &= G\,\frac{(L - c - x)}{L}\,x - (P_1 a + P_2 b)
\end{aligned}
$$

Differentiating to determine the condition for maximum bending moment,

$$\frac{dM}{dx} = \frac{G}{L}(L - 2x - c) = 0$$

and
$$x = \frac{L - c}{2}$$

In the above equation $x = (L - c)/2$ is the distance from A to P_3; the distance from G to B (Fig. 9-11) is $(L - c - x)$ which is also $(L - c)/2$. Thus for the bending moment at P_3 to be a maximum, P_3 and G must be equidistant from the left and right ends of the beam. In other words, the maximum bending moment under any concentrated load occurs when this concentrated load and the center of gravity of all loads on the span are at equal distances from the center line of the beam.

It should be noted that, during the passage of a system of concentrated loads, a maximum bending moment occurs under each load when the above condition is met. Several calculations may be necessary to determine the greatest of these maximum bending moments. Usually, however, it will be found that the absolute maximum bending moment, or the greatest of these maximum bending moments, occurs under the load which is nearest to the center of gravity of the system.

If two unequal loads at a fixed distance apart move across a beam, the maximum bending moment occurs under the heavier load.

If two equal loads of P at a distance of a apart move across a simple beam, maximum bending moment occurs under either load when it is at a distance $x = L/2 - a/4$ from either end of the beam. The maximum bending moment can be found to be

$$M = \frac{P}{2L}\left(L - \frac{a}{2}\right)^2$$

If the spacing of the loads is greater than $0.586L$, maximum bending moment occurs under one load when it is at mid-span. This maximum spacing may be obtained by equating the single-load bending moment to the maximum value for two equal loads. Thus

$$\frac{PL}{4} = \frac{P}{2L}\left(L - \frac{a}{2}\right)^2$$

Solving,
$$a = 0.586L$$

The maximum bending moment caused by three equal loads, spaced a distance a apart, occurs at the middle and is $M = (P/4)(3L - 4a)$. If a is greater than $0.450L$, the maximum bending moment occurs with two loads on the beam.

Example 9-5.   Determine the absolute maximum bending moment in the simple beam AB (Fig. 9-12b) due to the passage of the series of concentrated loads shown in Fig. 9-12a.

SOLUTION.   Let $\bar{x}$ be the distance between P_6 and the center of gravity G of the six loads.   Taking moments about P_6,

$$69\bar{x} = 12(6 + 11) + 18(17) + 9(22) + 6(30)$$
$$\bar{x} = 12.870 \text{ ft}$$

Thus wheel P_4 is nearest to G and the distance between P_4 and G is 1.870 ft. P_4 and G are placed at equal distances from the center or the

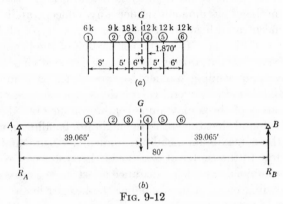

(a)

(b)

FIG. 9-12

ends of the beam as shown in Fig. 9-12b.   With the loads in this position, the bending moment under P_4 is

$$\text{Max } M \text{ at } P_4 = R_A(40.935) - (\text{moments of } P_1, P_2, \text{ and } P_3 \text{ about } P_4)$$
$$= \frac{69(40.935)}{80} (40.935) - [18(6) + 9(11) + 6(19)]$$
$$= 1{,}445.3 - 321 = 1{,}124.3 \text{ kip-ft}$$

This moment may also be found by using the right free body; thus

$$\text{Max } M \text{ at } P_4 = R_B(39.065) - (\text{moments of } P_5 \text{ and } P_6 \text{ about } P_4)$$
$$= \frac{69(39.065)}{80} (39.065) - 12(5 + 11)$$
$$= 1{,}316.2 - 192 = 1{,}124.2 \text{ kip-ft} \qquad (check)$$

In Example 9-3, it was found that these same loads produced a maximum bending moment of 1,123.5 kip-ft at the middle of an 80-ft beam with P_4 at the middle.   With P_4 placed at 0.935 ft from the center, the bending moment at P_4 becomes a trifle larger, or 1,124.3 kip-ft.   Usually the absolute maximum bending moment in a beam occurs under the load, which, if placed at the mid-point of the beam, will cause the maximum bending moment at the mid-section.   It is seen that the difference between the absolute maximum bending moment and the maximum bending moment at the center is, in this case, very small. This difference, however, may be quite large in cases when there are fewer than three or four concentrated loads in the system, or when the spacings between loads are relatively large in comparison with the span.

9-7. Maximum Reactions on Trusses. As explained in Art. 8-9, the influence diagram for the left reaction of a simple truss with end floor beams, wherein all loads on the structure are transferred to the abutments through the pedestals, is the same as that for a simple beam and is shown in Fig. 9-13b. The criteria previously explained in Art. 9-3 may be used, but in general, the maximum reaction will occur when the span is loaded as much as possible and heavy loads are near the support.

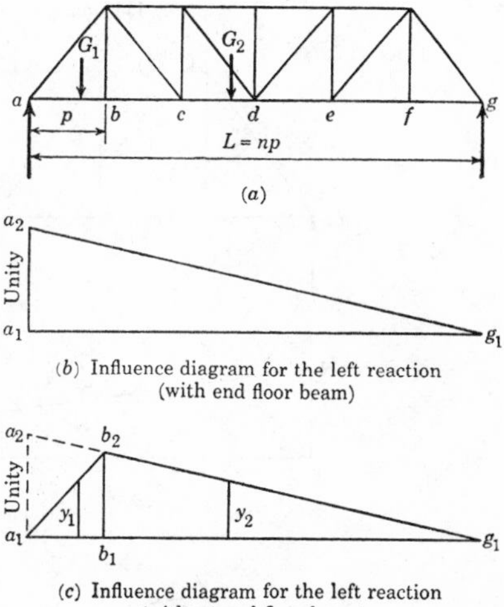

(a)

The influence diagram shown in Fig. 9-13c is for a truss with end panel stringers resting directly on the abutment. This influence diagram is similar to that for bending moment at b of a simple beam with span = ag. The criterion for maximum is

(b) Influence diagram for the left reaction (with end floor beam)

(c) Influence diagram for the left reaction (without end floor beam)

FIG. 9-13

$$\frac{G_1}{p} = \frac{G}{L} \quad \text{or} \quad G_1 = \frac{G}{n}$$

This means that the maximum end reaction on a truss having stringers which rest on the abutment occurs when the average load on the end panel equals the average load on the entire span.

Example 9-6. Compute the maximum left reaction for the simple truss of Fig. 9-14b due to the passage of the system of loads as shown in Fig. 9-14a, (a) when end floor beams are used, (b) when the exterior ends of the end stringers rest directly on the abutments.

SOLUTION. When end floor beams are used, the influence diagram for R_0 of the truss is identical with that of the left reaction of a 108-ft simple beam (Fig. 9-14c). Thus the maximum simple-beam reaction is also the maximum reaction on the truss. For the maximum simple-beam reaction,

Comparing P_1 at L_0 and P_2 at L_0,

$$\text{Gain} = \frac{G'b}{L} = \frac{139.6(5)}{108} + \frac{6(2.5)}{108} = 6.60 \text{ kips}$$

$$\text{Loss} = P_1 = 5 \text{ kips}$$

Comparing P_2 at L_0 and P_3 at L_0,

$$\text{Gain} = \frac{G'b}{L} = \frac{135.6(5)}{108} + \frac{6(2.5)}{108} = 6.42 \text{ kips}$$

$$\text{Loss} = P_2 = 10 \text{ kips}$$

From the above calculations, it is seen that, when P_2 is at L_0, reaction R_0 is a maximum. Note that, in the comparison of P_1 and P_2 at L_0,

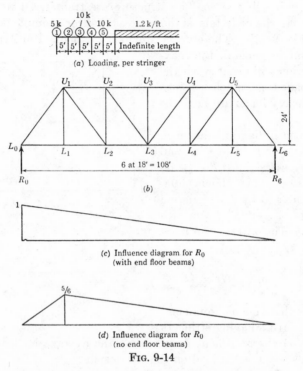

(a) Loading, per stringer

(b)

(c) Influence diagram for R_0
(with end floor beams)

(d) Influence diagram for R_0
(no end floor beams)

Fig. 9-14

the load on the span exclusive of P_1 when P_1 is at L_0 is equal to

$$40 + 1.2(83) = 139.6 \text{ kips}$$

In the forward movement of 5 ft, this load of 139.6 kips climbs up the influence line for a full distance of 5 ft; thus the gain in reaction is $(139.6)(5)/108 = 6.46$ kips, but, in the meantime 5 ft more of the uniform load, or $(1.2)(5) = 6$ kips come onto the span. The center of gravity of this 5-ft uniform load climbs up the span only 2.5 ft; thus the gain in reaction is only $6(2.5)/108 = 0.14$ kip. The total gain is therefore $6.46 + 0.14 = 6.60$ kips. With P_2 at L_0,

$$\text{Max } R_0 = \frac{\frac{1}{2}(1.2)(88)^2 + 10(93 + 98 + 103 + 108)}{108} = 80.24 \text{ kips}$$

When the exterior ends of the end stringers rest directly on the abutments, the influence diagram for R_0 of the truss is shown in Fig. 9-14d. This influence diagram is similar to that for the bending moment at the one-sixth point of a simple beam. The criterion is therefore $G_1 = G/6$.

Load at L_1	G_1	$\dfrac{G}{6}$	G_1	Yes or no
P_2	5	$^{129}\!/_6 = 21.5$	15	No
P_3	15	$^{135}\!/_6 = 22.5$	25	Yes
P_4	25	$^{141}\!/_6 = 23.5$	35	No

With P_3 at L_1, the reaction R_0 on the truss may be found by subtracting the reaction at the exterior end of the end stringer from the simple-beam reaction at L_0.

$$R_0 = \frac{\frac{1}{2}(1.2)(7.5)^2 + 10(80 + 85 + 90 + 95) + 5(100)}{108} - \frac{10(5) + 5(10)}{18}$$

$$= \frac{7{,}375}{108} - \frac{100}{18} = 68.29 - 5.55 = 62.74 \text{ kips}$$

9-8. Maximum Shear in a Panel of a Parallel-chord Truss.

The influence diagram for shear in panel cd of the parallel-chord truss ag of Fig. 9-15a is shown in Fig. 9-15b. The slope of g_1d_2 is a_1a_2 over a_1g_1, or 1 on L. The slope of c_2d_2 is $(c_1c_2 + d_1d_2)$ over p, or $[m/n + (n - m - 1)/n]$ over p, or $(n - 1)$ on L.

Now it is desired to determine the condition of loading which causes maximum positive shear in panel cd. Let G_2 equal the resultant of the loads on panel cd (although G_2 may fall on either segment c_1i or id_1 of the influence diagram, in most cases it will be within segment id_1). Let G_3 equal the resultant of the loads on segment dg. Generally there will be no load on segment ac because such loading will cause negative shear in panel cd (see next paragraph). As shown by the ordinates in Fig. 9-15b, the positive shear in panel cd is

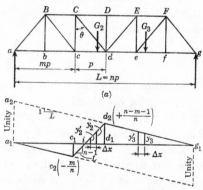

(a)

(b) Influence diagram for shear in panel cd

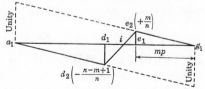

(c) Influence diagram for shear in panel de

FIG. 9-15

$$V_{cd} = G_2 y_2 + G_3 y_3$$

If the system of loads moves a small distance Δx toward the left, the positive shear in panel cd becomes

$$V'_{cd} = G_2 y'_2 + G_3 y'_3$$

The change or increase in shear is

$$\Delta V = V'_{cd} - V_{cd} = G_3(y'_3 - y_3) - G_2(y_2 - y'_2)$$

but

$$y'_3 - y_3 = \frac{1}{L} \Delta x \qquad \text{and} \qquad y_2 - y'_2 = \frac{n-1}{L} \Delta x$$

Thus

$$\Delta V = \frac{G_3}{L} \Delta x - G_2 \frac{n-1}{L} \Delta x$$

or, by letting $G = G_2 + G_3$,

$$\frac{\Delta V}{\Delta x} = \frac{G - G_2}{L} - \frac{G_2(n-1)}{L} = \frac{G - G_2 n}{np} = \frac{1}{p}\left(\frac{G}{n} - G_2\right)$$

A maximum is attained if $\Delta V/\Delta x$ changes from positive through zero to negative. This will be possible if a load is placed at panel point d, so that, if it is not considered in G_2, $(G/n) - G_2$ is positive, and if it is, $(G/n) - G_2$ is negative. The criterion for maximum positive shear in panel cd is therefore

$$\frac{G}{n} = G_2$$

or, in other words, the value of G/n falls within the two values of G_2.

In fact, for the usual case this criterion may be developed by referring to Art. 8-10. In this article it was proved that, in the triangle $ig_1 d_2$ in Fig. 9-15b, the ratio of id_1 to ig_1 is $1/n$. Inasmuch as the portion of the influence diagram which should be loaded for positive shear in panel cd is similar to the influence diagram for bending moment at point d_1 in a simple beam of span equal to ig_1, the criterion becomes

$$G_2 = G \frac{id_1}{ig_1} = \frac{G}{n}$$

in which G_2 is the load on id_1 and G is the load on both id_1 and $d_1 g_1$. It will be noted that this derivation by similarity excludes the possibility of any load on segment $c_1 i$ while the preceding derivation does not dictate this requirement. As pointed out before, except in unusual cases there is generally no load on segment $c_1 i$. In any event, the criterion $G_2 = G/n$ is applicable whenever G_2 is the load within panel cd and G is the total load on the span.

It may be pointed out that on occasion the criterion $G_2 = G/n$ can be satisfied only when some loads extend into segment ac. In such a case, G is the sum of G_1, G_2, and G_3, where G_1 is the load on segment ac, G_2 is the load within panel cd, and G_3 is the load on segment dg. The derivation follows the usual pattern, except that in Fig. 9-15a and b, there should be added a load G_1 on segment ac and ordinates y_1 and y_1' on portion a_1c_2 of the influence diagram.

For maximum negative shear in panel cd, portion a_1ic_2 of the influence diagram in Fig. 9-15b should be loaded when the traffic comes from the left toward the right. However, a comparison of the influence diagrams for shear in panels cd and de as shown in Fig. 9-15b and c will show that triangle a_1ic_2 of Fig. 9-15b is identical with triangle ig_1e_2 of Fig. 9-15c except they are on opposite sides of the base line. Therefore the maximum negative shear in any panel of a truss is numerically equal to the maximum positive shear in an opposite-handed panel on the other side of the center line of the truss. Once the maximum positive shear in every panel of a truss between the left and right supports has been found, the maximum positive and negative shears in every panel are known.

In Fig. 9-15a, the maximum tensile and compressive stresses in diagonal Cd in the parallel-chord truss ag equal the maximum positive and negative shears, respectively, in the panel Cd multiplied by the secant of the angle the diagonal makes with the vertical.

Example 9-7. Compute the maximum positive and negative shears in panels 0-1, 1-2, and 2-3 of the truss of Fig. 9-16b due to the passage of the system of loads shown in Fig. 9-16a.

SOLUTION. The maximum positive shears in all panels due to the given loading will first be computed.

Maximum positive shear in panel 0-1

Load at L_1	G_2	$\dfrac{G}{6}$	G_2	Yes or no
P_2	5	$129\!\!\!/_6 = 21.5$	15	No
P_3	15	$135\!\!\!/_6 = 22.5$	25	Yes
P_4	25	$141\!\!\!/_6 = 23.5$	35	No

With P_3 at L_1, the shear in panel 0-1 may be found by subtracting the panel load at L_0 from the left reaction on the truss (see Fig. 9-16c).

$$V_{0-1} = \frac{\frac{1}{2}(1.2)(75)^2 + 10(80 + 85 + 90 + 95) + 5(100)}{108} - \frac{10(5) + 5(10)}{18}$$

$$= 68.29 - 5.55 = 62.74 \text{ kips}$$

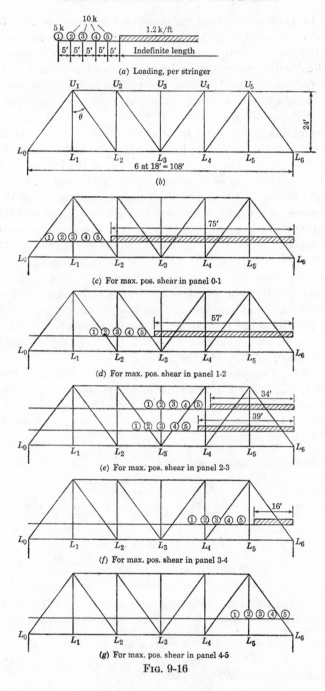

(a) Loading, per stringer

(b)

(c) For max. pos. shear in panel 0-1

(d) For max. pos. shear in panel 1-2

(e) For max. pos. shear in panel 2-3

(f) For max. pos. shear in panel 3-4

(g) For max. pos. shear in panel 4-5

FIG. 9-16

Maximum positive shear in panel 1-2

Load at L_2	G_2	$\dfrac{G}{6}$	G_2	Yes or no
P_2	5	$\dfrac{107.4}{6} = 17.9$	15	No
P_3	15	$\dfrac{113.4}{6} = 18.9$	25	Yes
P_4	25	$\dfrac{119.4}{6} = 19.9$	35	No

With P_3 at L_2, the shear in panel 1-2 may be found by subtracting the panel load at L_1 from the left reaction on the truss (see Fig. 9-16d).

$$V_{1\text{-}2} = \frac{\frac{1}{2}(1.2)(57)^2 + 10(62 + 67 + 72 + 77) + 5(82)}{108} - \frac{10(5) + 5(10)}{18}$$
$$= 47.59 - 5.55 = 42.04 \text{ kips}$$

Maximum positive shear in panel 2-3

Load at L_3	G_2	$\dfrac{G}{6}$	G_2	Yes or no
P_1	0	$\dfrac{79.8}{6} = 13.3$	5	No
P_2	5	$\dfrac{85.8}{6} = 14.3$	15	Yes
P_3	15	$\dfrac{91.8}{6} = 15.3$	25	Yes
P_4	25	$\dfrac{97.8}{6} = 16.3$	35	No

The criterion is satisfied by two conditions of loading: P_2 at L_3 or P_3 at L_3. To determine the maximum shear in panel 2-3, it will be necessary to compare the computed values of the shear in the panel for both conditions of loading. These positions are shown in Fig. 9-16e.

With P_2 at L_3,

$$V_{2\text{-}3} = \frac{\frac{1}{2}(1.2)(34)^2 + 10(39 + 44 + 49 + 54) + 5(59)}{108} - \frac{5(5)}{18}$$
$$= 26.38 - 1.39 = 24.99 \text{ kips}$$

With P_3 at L_3,

$$V_{2\text{-}3} = \frac{\frac{1}{2}(1.2)(39)^2 + 10(44 + 49 + 54 + 59) + 5(64)}{108} - \frac{10(5) + 5(10)}{18}$$
$$= 30.49 - 5.55 = 24.94 \text{ kips}$$

Thus the maximum positive shear of 24.99 kips occurs in panel 2-3 when P_2 is at L_3.

Maximum positive shear in panel 3-4

Load at L_4	G_2	$\dfrac{G}{6}$	G_2	Yes or no
P_1	0	$\dfrac{58.2}{6} = 9.7$	5	No
P_2	5	$\dfrac{64.2}{6} = 10.7$	15	Yes
P_3	15	$\dfrac{70.2}{6} = 11.7$	25	No

With P_2 at L_4, the shear in panel 3-4 may be found by subtracting the panel load at L_3 from the left reaction on the truss (see Fig. 9-16f).

$$V_{3\text{-}4} = \frac{\frac{1}{2}(1.2)(16)^2 + 10(21 + 26 + 31 + 36) + 5(41)}{108} - \frac{5(5)}{18}$$

$$= 13.88 - 1.39 = 12.49 \text{ kips}$$

Maximum positive shear in panel 4-5

Load at L_5	G_2	$\dfrac{G}{6}$	G_2	Yes or no
P_1	0	$3\frac{5}{6} = 5.83$	5	No
P_2	5	$4\frac{5}{6} = 7.50$	15	Yes
P_3	15	$\dfrac{48.6}{6} = 8.10$	25	No

With P_2 at L_5, the shear in panel 4-5 is, from Fig. 9-16g,

$$V_{4\text{-}5} = \frac{10(3 + 8 + 13 + 18) + 5(23)}{108} - \frac{5(5)}{18} = 4.95 - 1.39 = 3.56 \text{ kips}$$

The shear in panel 5-6 can only be negative; therefore the maximum positive shear in this panel due to the given system of loads is zero.

Thus the maximum positive shear in panel 0-1 is 62.74 kips, or the maximum compression in member L_0U_1 is 62.74 sec θ = 78.42 kips. The maximum negative shear in panel 0-1 is zero; therefore tension never occurs in diagonal L_0U_1.

The maximum positive shear in panel 1-2 is 42.04 kips; so the maximum tension in member U_1L_2 is 42.04 sec θ = 52.55 kips. The maximum negative shear in panel 1-2 is 3.56 kips; so the maximum compression in member U_1L_2 is 3.56 sec θ = 4.45 kips.

The maximum positive shear in panel 2-3 is 24.99 kips; so the maximum compression in member U_2L_2 is 24.99 kips and the maximum tension in diagonal U_2L_3 is 24.99 sec θ = 31.24 kips. The maximum negative shear in panel 2-3 is 12.49 kips; so the maximum tension in member U_2L_2 is 12.49 kips, and the maximum compression in member U_2L_3 is 12.49 sec θ = 15.61 kips.

9-9. Maximum Bending Moment at a Panel Point in the Loaded Chord of a Truss. The influence diagram for bending moment at c of the parallel-chord truss ag (Fig. 9-17a) is shown in Fig. 9-17b. Let G_1 equal the resultant of the loads to the left of panel point c and G_2 the resultant of the loads on segment cg. Inasmuch as this influence diagram is

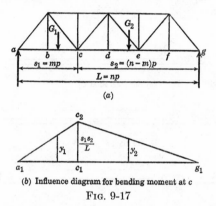

(a)

(b) Influence diagram for bending moment at c

FIG. 9-17

identical with that for bending moment at c of a simple beam ag, the criterion for maximum is

$$\frac{Gs_1}{L} = G_1$$

in which $G = G_1 + G_2$. It is thus seen that the maximum bending moment at a panel point in the loaded chord of a truss occurs when the average load to the left of the panel point equals the average load on the entire span.

Example 9-8. Compute the maximum bending moments at panel points L_1, L_2, and L_3 of the truss shown in Fig. 9-18b due to the passage of the system of loads shown in Fig. 9-18a.

SOLUTION. From the discussion in Art. 9-5 it will be recalled that the maximum bending moment at a point in the left half of a beam due to traffic from the left is equal to that at a corresponding point in the right half due to traffic from the right. For instance, in the present problem, the maximum bending moment at panel point L_2 due to traffic from the left is equal to that at L_4 due to traffic from the right. Thus, in order to find the maximum bending moments at points L_1, L_2, and L_3 due to traffic in either direction, it will only be necessary to find the maximum bending moments at all five panel points due to traffic from the right only.

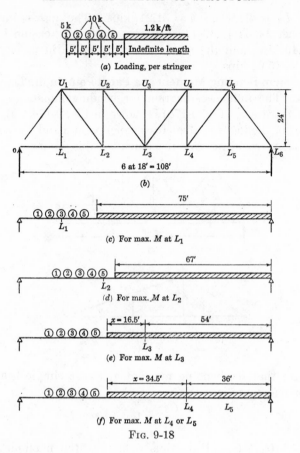

(a) Loading, per stringer

(b)

(c) For max. M at L_1

(d) For max. M at L_2

(e) For max. M at L_3

(f) For max. M at L_4 or L_5

FIG. 9-18

Maximum bending moment at L_1

Load at L_1	G_1	$\dfrac{G}{6}$	G_1	Yes or no
P_2	5	$129\%{6} = 21.5$	15	No
P_3	15	$135\%{6} = 22.5$	25	Yes
P_4	25	$141\%{6} = 23.5$	35	No

With P_3 at L_1 (Fig. 9-18c),

$$M \text{ at } L_1 = \frac{\frac{1}{2}(1.2)(75)^2 + 10(80 + 85 + 90 + 95) + 5(100)}{6}$$
$$- [10(5) + 5(10)]$$

$$= \frac{7{,}375}{6} - 100 = 1{,}229.2 - 100 = 1{,}129.2 \text{ kip-ft}$$

Maximum bending moment at L_2

Load at L_2	G_1	$\dfrac{G}{3}$	G_1	Yes or no
P_4	25	$\dfrac{119.4}{3} = 39.8$	35	No
P_5	35	$\dfrac{125.4}{3} = 41.8$	45	Yes

With P_5 at L_2 (Fig. 9-18d),

$$M \text{ at } L_2 = \frac{\frac{1}{2}(1.2)(67)^2 + 10(72 + 77 + 82 + 87) + 5(92)}{3}$$

$$- [10(5 + 10 + 15) + 5(20)]$$

$$= \frac{6{,}333.4}{3} - 400 = 1{,}711.1 \text{ kip-ft}$$

Maximum bending moment at L_3. Let x equal the length of uniform load which passes to the left of panel point L_3 (Fig. 9-18e). Then $G_1 = 45 + 1.2x$ and $G = 45 + 1.2(x + 54)$. Equating $G_1 = \frac{1}{2}G$,

$$45 + 1.2x = \frac{1}{2}[45 + 1.2(x + 54)]$$

Solving for x,

$$x = 16.5 \text{ ft}$$

For this loading condition (Fig. 9-18e),

$$M \text{ at } L_3 = \frac{\frac{1}{2}(1.2)(70.5)^2 + 10(75.5 + 80.5 + 85.5 + 90.5) + 5(95.5)}{2}$$

$$- [\frac{1}{2}(1.2)(16.5)^2 + 10(21.5 + 26.5 + 31.5 + 36.5) + 5(41.5)]$$

$$= \frac{6{,}779.65}{2} - 1{,}530.85 = 1{,}859.0 \text{ kip-ft}$$

Maximum bending moment at L_4. Let x equal the length of uniform load which passes to the left of panel point L_4 (Fig. 9-18f). Then $G_1 = 45 + 1.2x$ and $G = 45 + 1.2(x + 36)$. Equating $G_1 = \frac{2}{3}G$,

$$45 + 1.2x = \frac{2}{3}[45 + 1.2(x + 36)]$$

Solving for x,

$$x = 34.5 \text{ ft}$$

This condition of loading is shown in Fig. 9-18f. A comparison of Fig. 9-18f and Fig. 9-18e will show that the two conditions of loading are identical. This must be true because the average load on the span, $[(1.2)(70.5) + 45]/108 = 1.2$ kips per lin ft, is equal to the average load on either L_0L_3 or L_0L_4. Thus Fig. 9-18e or f must also be the critical loading position for maximum bending moment at L_5.

$$
\begin{aligned}
\text{Max } M \text{ at } L_4 &= \tfrac{2}{3}(6{,}779.65) - [\tfrac{1}{2}(1.2)(34.5)^2 \\
&\quad + 10(39.5 + 44.5 + 49.5 + 54.5) + 5(59.5)] \\
&= 4{,}519.8 - 2{,}891.7 = 1{,}628.1 \text{ kip-ft}
\end{aligned}
$$

Maximum bending moment at L_5. Referring to either Fig. 9-18e or f,

$$
\begin{aligned}
\text{Max } M \text{ at } L_5 &= \tfrac{5}{6}(6{,}779.65) - [\tfrac{1}{2}(1.2)(52.5)^2 \\
&\quad + 10(57.5 + 62.5 + 67.5 + 72.5) + 5(77.5)] \\
&= 5{,}649.7 - 4{,}641.2 = 1{,}008.5 \text{ kip-ft}
\end{aligned}
$$

From the above computations it is seen that the maximum bending moments at L_4 and L_5 are somewhat smaller, respectively, than those at L_2 and L_1. This is understandable because the average load per linear foot in the range of the concentrated loads is larger than the intensity of the subsequent uniform load.

The maximum bending moment at L_1, which in this case is also the maximum bending moment at U_1, is the larger of 1,129.2 or 1,008.5 kip-ft. The maximum tension in members L_0L_1 or L_1L_2 is therefore

$$
\frac{1{,}129.2}{24} = 47.05 \text{ kips}
$$

The maximum bending moment at L_2 or U_2 is the larger of 1,711.1 or 1,628.1 kip-ft. The maximum compression in member U_1U_2 or the maximum tension in member L_2L_3 is therefore 1,711.1/24 = 71.30 kips.

The maximum bending moment at L_3 is 1,859.0 kip-ft. The maximum compression in member U_2U_3 is therefore 1,859.0/24 = 77.42 kips.

9-10. Maximum Bending Moment at a Panel Point in the Unloaded Chord of a Truss. Let it be required to derive the criterion for maximum bending moment at panel point C in the unloaded chord of the truss shown in Fig. 9-19a. In this derivation, the panel point C need not be at halfway between panel points c and d, but it is at a horizontal distance b to the right of panel point c. The influence diagram for bending moment at C is shown in Fig. 9-19b. The ordinate c_1c_2 is the bending moment at C with unity at c; thus

$$
c_1c_2 = R_2s_2 = \frac{m}{n}s_2
$$

The ordinate d_1d_2 is the bending moment at C with unity at d; thus

$$
d_1d_2 = R_1s_1 = \frac{n - m - 1}{n}s_1
$$

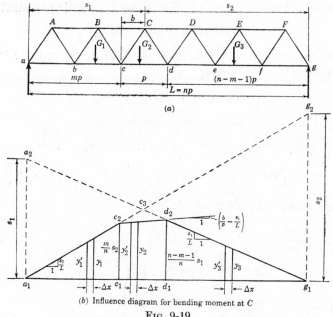

(a)

(b) Influence diagram for bending moment at C

Fig. 9-19

The slope of a_1c_2 is c_1c_2 divided by a_1c_1, or

$$\text{Slope of } a_1c_2 = \frac{c_1c_2}{a_1c_1} = \frac{(m/n)s_2}{mp} = \frac{s_2}{L}$$

The slope of g_1d_2 is d_1d_2 divided by g_1d_1, or

$$\text{Slope of } g_1d_2 = \frac{d_1d_2}{g_1d_1} = \frac{[(n-m-1)/n]s_1}{(n-m-1)p} = \frac{s_1}{L}$$

The slope of c_2d_2 is $(d_1d_2 - c_1c_2)$ divided by c_1d_1, or

$$\text{Slope of } c_2d_2 = \frac{d_1d_2 - c_1c_2}{c_1d_1} = \frac{n-m-1}{n}\frac{s_1}{p} - \frac{m}{n}\frac{s_2}{p}$$

$$= \frac{1}{L}(s_1n - s_1m - s_1 - s_2m)$$

$$= \frac{1}{L}[s_1n - s_1m - s_1 - (np - s_1)m]$$

$$= \frac{1}{L}(s_1n - pmn - s_1)$$

$$= \frac{1}{L}[n(s_1 - mp) - s_1] = \frac{1}{L}(nb - s_1) = \frac{b}{p} - \frac{s_1}{L}$$

Now if G_1, G_2, and G_3 are, respectively, the resultant loads on segments

ac, cd, and dg of the truss shown in Fig. 9-19a, the bending moment at C is

$$M_C = G_3 y_3 + G_2 y_2 + G_1 y_1$$

If the system of loads moves a small distance Δx toward the left, the bending moment at C becomes

$$M'_C = G_3 y'_3 + G_2 y'_2 + G_1 y'_1$$

The change in M_C due to this movement of Δx is

$$M_C = M'_C - M_C = G_3(y'_3 - y) - G_2(y_2 - y'_2) - G_1(y_1 - y'_1)$$

but $\quad y'_3 - y = \dfrac{s_1}{L} \Delta x \qquad y_2 - y'_2 = \left(\dfrac{b}{p} - \dfrac{s_1}{L} \right) \Delta x \qquad y_1 - y'_1 = \dfrac{s_2}{L} \Delta x$

Substituting and solving for $\Delta M/\Delta x$,

$$
\begin{aligned}
\frac{\Delta M}{\Delta x} &= G_3 \frac{s_1}{L} - G_2 \left(\frac{b}{p} - \frac{s_1}{L} \right) - G_1 \frac{s_2}{L} \\
&= (G - G_1 - G_2) \frac{s_1}{L} - G_2 \left(\frac{b}{p} - \frac{s_1}{L} \right) - G_1 \frac{s_2}{L} \\
&= G \frac{s_1}{L} - G_1 \frac{s_1 + s_2}{L} - G_2 \frac{b}{p} \\
&= \frac{G s_1}{L} - \left(G_1 + G_2 \frac{b}{p} \right)
\end{aligned}
$$

As explained before, the loading condition for maximum bending moment at C is attained when $(G s_1/L) - [G_1 + G_2(b/p)]$ changes from positive to negative. This will occur when a load is at either panel point c or d. When a load is placed at panel point c, $\Delta M/\Delta x$ may be positive if the load is included in G_2, but $\Delta M/\Delta x$ is negative if this load is included in G_1. The criterion may also be satisfied by placing a load at panel point d, the condition being that, if this load is not included in G_2, $\Delta M/\Delta x$ is positive, and if it is included in G_2, $\Delta M/\Delta x$ becomes negative. Usually, then, there are two loading conditions (one with a load at c, the other with a load at d) which satisfy the criterion; the larger of the two results thus obtained is the required maximum.

In most cases the panel point in the unloaded chord is halfway between the adjacent panel points in the loaded chord, thus making b/p equal to $\frac{1}{2}$. The criterion for maximum bending moment then becomes

$$\frac{dM}{dx} = \frac{G s_1}{L} - (G_1 + \tfrac{1}{2} G_2) = 0$$

If the panel point in the unloaded chord is directly above a panel point in the loaded chord, the maximum bending moment at the former is identical with that at the latter.

Example 9-9.   Compute the maximum bending moment at panel points U_1, U_2, and U_3 of the truss of Fig. 9-20b due to the passage of the system of loads shown in Fig. 9-20a.

SOLUTION.   The maximum bending moment at each of the five panel points in the unloaded chord due to traffic from the right will be determined.   Then the larger of the maximum bending moments at U_1 or U_5 will be the maximum at either U_1 or U_5, and the larger of those at U_2 or U_4 will be the maximum at either U_2 or U_4.

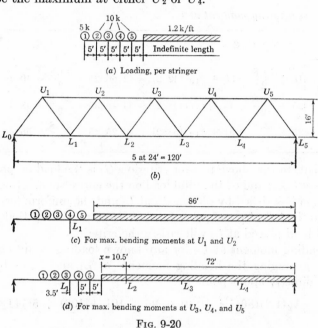

(a) Loading, per stringer

(b)

(c) For max. bending moments at U_1 and U_2

(d) For max. bending moments at U_3, U_4, and U_5

FIG. 9-20

Maximum bending moment at U_1.   The influence diagram for bending moment at U_1 is similar to that for bending moment at L_1.   Both are in the form of a triangle with a peak ordinate of 9.6 ft at L_1 in the first case and of 19.2 ft in the latter.   Therefore the maximum bending moment at U_1 is one-half of that at L_1 with a load at L_1.   The criterion becomes G_1 on L_0L_1 equals $G/5$.

Load at L_1	G on L_0L_1	$G/5$	G on L_0L_1	Yes or no
P_3	15	$\dfrac{142.2}{5} = 28.44$	25	No
P_4	25	$\dfrac{148.2}{5} = 29.64$	35	Yes
P_5	35	$\dfrac{154.2}{5} = 30.84$	45	No

With P_4 at L_1 (Fig. 9-20c),

$$\text{Max } M \text{ at } U_1 = \frac{1}{2}\left[\frac{\frac{1}{2}(1.2)(86)^2 + 10(91 + 96 + 101 + 106) + 5(111)}{5}\right.$$
$$\left. - 10(5 + 10) - 5(15)\right]$$
$$= \frac{1}{2}\left(\frac{8,932.6}{5} - 225\right) = \frac{1}{2}(1,786.52 - 225) = 780.8 \text{ kip-ft}$$

Maximum bending moment at U_2

Load at L_1	$G_1 + G_2/2$	$\frac{3}{10}G$	$G_1 + G_2/2$	Yes or no
P_3	$15 + \dfrac{40.8}{2} = 35.4$	$\frac{3}{10}(142.2) = 42.66$	$25 + \dfrac{30.8}{2} = 40.4$	No
P_4	$25 + \dfrac{36.8}{2} = 43.4$	$\frac{3}{10}(148.2) = 44.46$	$35 + \dfrac{26.8}{2} = 48.4$	Yes
P_5	$35 + \dfrac{32.8}{2} = 51.4$	$\frac{3}{10}(154.2) = 46.26$	$45 + \dfrac{22.8}{2} = 56.4$	No

Note that, in the above test for criterion, G_1 is the load on panel L_0L_1; G_2, on panel L_1L_2; and G, the total load on the entire span. Inasmuch as the criterion is satisfied by placing P_4 at L_1 and the uniform load already extends to the left of panel point L_2, there is no possibility that any concentrated load placed at L_2 will satisfy the criterion.

The bending moment at U_2 for any known loading condition may be found by averaging the bending moments at L_1 and L_2 or by taking moments directly about point U_2.

$$M \text{ at } L_1 = \frac{\frac{1}{2}(1.2)(86)^2 + 10(91 + 96 + 101 + 106) + 5(111)}{5}$$
$$- [10(5 + 10) + 5(15)]$$
$$= \frac{8,932.6}{5} - 225 = 1,561.5 \text{ kip-ft}$$

$$M \text{ at } L_2 = \frac{(8,932.6)(2)}{5}$$
$$- [\frac{1}{2}(1.2)(14)^2 + 10(19 + 24 + 29 + 34) + 5(39)]$$
$$= 3,573.0 - 1,372.6 = 2,200.4 \text{ kip-ft}$$

$M \text{ at } U_2 = \frac{1}{2}(M \text{ at } L_1 + M \text{ at } L_2) = \frac{1}{2}(1,561.5 + 2,200.4)$
$$= 1,881.0 \text{ kip-ft}$$

Taking moments directly about U_2,

$$M \text{ at } U_2 = \frac{3}{10}(8,932.6) - \left[\frac{\frac{1}{2}(1.2)(14)^2 + 10(19 + 24)}{24}\right.(12)$$
$$\left. + 10(17 + 22) + 5(27)\right]$$
$$= 2,679.8 - 798.8 = 1,881.0 \text{ kip-ft}$$

It should be remembered that the above loads are actually applied only at the panel points of the truss.

Maximum bending moment at U_3. Let x equal the length of the uniform load which passes to the left of L_2 (Fig. 9-20d). Then,

$$G_1 = 45 + 1.2x \qquad G_2 = 28.8 \qquad G = 45 + 1.2(x + 72)$$

Remembering that $G_1 + \frac{1}{2}G_2 = \frac{1}{2}G$,

$$(45 + 1.2x) + \frac{1}{2}(28.8) = \frac{1}{2}[45 + 1.2(x + 72)]$$

Solving, $x = 10.5$ ft

For this condition of loading (Fig. 9-20d),

$$\text{Max } M \text{ at } U_3 = \frac{1}{2}[\frac{1}{2}(1.2)(82.5)^2 + 10(87.5 + 92.5 + 97.5 + 102.5)$$
$$+5(107.5)] - [(14.4)(12) + (12.6)(17.25)$$
$$+ 10(27.5 + 32.5 + 37.5 + 42.5) + 5(47.5)]$$
$$= \frac{8,421.25}{2} - 2,027.6 = 2,183.0 \text{ kip-ft}$$

Maximum bending moment at U_4. It is apparent that the condition of loading shown in Fig. 9-20d will satisfy the criterion $G_1 + \frac{1}{2}G_2 = \frac{7}{10}G$. A check, however, will be made.

$$G_1 = 45 + 1.2(34.5) = 86.4 \qquad G_2 = 28.8 \qquad G = 45 + 1.2(82.5) = 144$$
$$84.6 + \frac{1}{2}(28.8) = \frac{7}{10}(144)$$
$$100.8 = 100.8$$
$$\text{Max } M \text{ at } U_4 = \frac{7}{10}(8,421.25) - [(14.4)(12) + (1.2)(34.5)(29.25)$$
$$+ 10(51.5 + 56.5 + 61.5 + 66.5) + 5(71.5)]$$
$$= 5,894.9 - 4,101.2 = 1,793.7 \text{ kip-ft}$$

Maximum bending moment at U_5. The maximum bending moment at U_5 is one-half of that at L_4. The criterion becomes G_1 on L_0L_4 equals $4G/5$. For the condition of loading shown in Fig. 9-20d,

$$G_1 \text{ on } L_0L_4 = 45 + 1.2(58.5) = 115.2 \qquad G = 144$$

Thus, $G_1 = \frac{4}{5}G$

$$\text{Max } M \text{ at } U_5 = \frac{1}{2}(M \text{ at } L_4)$$
$$= \frac{1}{2}\{\frac{4}{5}(8,421.25) - [\frac{1}{2}(1.2)(58.5)^2$$
$$+ 10(63.5 + 68.5 + 73.5 + 78.5) + 5(83.5)]\}$$
$$= \frac{1}{2}(6,737.0 - 5,310.8) = 713.1 \text{ kip-ft}$$

Because traffic moves in either direction, the maximum bending moment at U_1 is therefore 780.8 kip-ft (the larger of the computed values 780.8 or 713.1), that at U_2 is 1,881.0 kip-ft (the larger of 1,881.0 or 1,793.7), and that at U_3 is 2,183.0 kip-ft.

9-11. Maximum Stress in a Web Member of a Truss with Inclined Chords. The influence diagram for the stress in member Bc of the truss with inclined chords (Fig. 9-21a) is shown in Fig. 9-21b. Due to a moving load system the stress in member Bc is tensile when portion ig_1 of the span is loaded, and it is compressive when portion ia_1 is loaded. Because triangle ic_2g_1 is similar to the influence diagram for bending moment at section c_1 of a simple beam with span equal to ig_1, the criterion

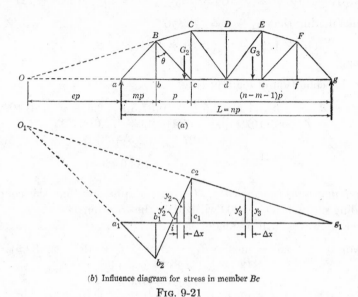

(b) Influence diagram for stress in member Bc

Fig. 9-21

for maximum tension in the member, due to a series of concentrated loads coming on the span from the right, becomes

$$G_2 \text{ (on } ic_1 \text{ or } b_1c_1) = G\frac{ic_1}{ig_1}$$

Likewise the criterion for maximum compression in the member, due to a series of concentrated loads coming on the span from the left, is

$$G_2 \text{ (on } ib_1 \text{ or } c_1b_1) = G\frac{ib_1}{ia_1}$$

In a numerical problem it will be best to compute first the ordinates b_1b_2 and c_1c_2 in the influence diagram by placing unity at panel points b and c, respectively. The influence diagram thus obtained, such as $a_1b_2c_2g_1$ in Fig. 9-21b, should be checked in that the prolongations of b_2a_1 and g_1c_2 should intersect at point O_1 directly under the moment center O. The distances b_1i and ic_1 are then computed by dividing the panel length into segments proportional to the ratio of b_1b_2 to c_1c_2.

Finally, the criteria may be numerically established by using the relationships described in the preceding paragraph.

For completeness in treatment, the criterion for maximum stress in a web member of a truss with inclined chords will be derived for the general case. Referring to Fig. 9-21b,

$$c_1c_2 = R_a \frac{e}{e+m+1} \sec \theta = \frac{n-m-1}{n} \frac{e}{e+m+1} \sec \theta$$

$$b_1b_2 = R_g \frac{e+n}{e+m+1} \sec \theta = \frac{m}{n} \frac{e+n}{e+m+1} \sec \theta$$

$$\text{Slope of } g_1c_2 = \frac{c_1c_2}{c_1g_1} = \frac{e}{L(e+m+1)} \sec \theta$$

$$\text{Slope of } a_1b_2 = \frac{b_1b_2}{a_1b_1} = \frac{e+n}{L(e+m+1)} \sec \theta$$

$$\text{Slope of } b_2c_2 = \frac{b_1b_2 + c_1c_2}{b_1c_1} = \frac{mn+en-e}{L(e+m+1)} \sec \theta$$

Let G_2 equal the total load on panel bc and G_3 the total load on the segment cg. The tensile stress in member Bc is

$$T = G_3 y_3 + G_2 y_2$$

If the system of loads moves a small distance Δx to the left, the tensile stress in Bc becomes

$$T' = G_3 y_3' + G_2 y_2'$$

The increase in tension is

$$\Delta T = T' - T = G_3(y_3' - y_3) - G_2(y_2 - y_2')$$

$$= G_3 \frac{e \sec \theta}{L(e+m+1)} \Delta x - G_2 \frac{(mn+en-e) \sec \theta}{L(e+m+1)} \Delta x$$

from which

$$\frac{\Delta T}{\Delta x} = \frac{\sec \theta}{L(e+m+1)} [G_3 e - G_2(mn+en-e)]$$

$$= \frac{\sec \theta}{L(e+m+1)} [(G - G_2)e - G_2(mn+en-e)]$$

$$= \frac{n(m+e) \sec \theta}{L(e+m+1)} \left(\frac{G}{n} \frac{e}{e+m} - G_2 \right)$$

Therefore, to obtain the maximum tension in member Bc, the load placed at panel point c must yield a value of $(G/n)[e/(e+m)]$ which is between the two values of G_2, when the load is assumed to be off or on the panel bc. Note that G_2 is the total load on panel bc and G is the total load on the span.

Similarly, the criterion for maximum compression in member Bc will be derived by letting G_2 equal the total load on panel cb and G_3 equal the total load on segment ab. If the system of loads moves a small distance Δx to the right, the increase in compression in member Bc is

$$\Delta C = G_3(\text{slope of } a_1b_2)\Delta x - G_2(\text{slope of } b_2c_2)\Delta x$$

$$= G_3 \frac{(e+n)\sec\theta}{L(e+m+1)}\Delta x - G_2\frac{(mn+en-e)\sec\theta}{L(e+m+1)}\Delta x$$

from which $\quad \dfrac{\Delta C}{\Delta x} = \dfrac{\sec\theta}{L(e+m+1)}[G_3(e+n) - G_2(mn+en-e)]$

$$= \frac{\sec\theta}{L(e+m+1)}[(G-G_2)(e+n) - G_2(mn+en-e)]$$

$$= \frac{n(e+m+1)\sec\theta}{L(e+m+1)}\left(\frac{G}{n}\frac{e+n}{e+m+1} - G_2\right)$$

$$= \frac{n\sec\theta}{L}\left(\frac{G}{n}\frac{e+n}{e+m+1} - G_2\right)$$

Thus, to obtain the maximum compression in member Bc, a load must be placed at panel point b so that the value of $(G/n)[(e+n)/(e+m+1)]$ lies between the two values of G_2. As in the preceding case, G is the total load on the span and the two values of G_2 are the load in panel bc without and then with the load at point b.

Although general expressions for the criteria have been derived, a direct procedure based on the use of the influence diagram is usually preferable.

Example 9-10. Compute the maximum tension and compression in member U_1L_2 of the truss in Fig. 9-22b due to the passage of the system of loads shown in Fig. 9-22a.

SOLUTION. The moment center O (Fig. 9-22b) is found to be at four panels to the left of L_0. The influence diagram for the stress in member U_1L_2 is shown in Fig. 9-22c. The ordinate at L_2 is the stress in U_1L_2 due to unity at L_2; or

$$\text{Ordinate at } L_2 = R_0\frac{4}{4+2}\sec\theta$$

$$= (\tfrac{4}{6})(\tfrac{4}{6})\sec\theta = \tfrac{8}{18}\sec\theta \text{ tension}$$

The ordinate at L_1 is the stress in U_1L_2 due to unity at L_1; or

$$\text{Ordinate at } L_1 = R_6\frac{4+6}{4+2}\sec\theta$$

$$= (\tfrac{1}{6})(\tfrac{10}{6})\sec\theta = \tfrac{5}{18}\sec\theta \text{ compression}$$

The correctness of the influence diagram may be verified by extending the two line segments to an intersection under O and with an ordinate of $\tfrac{20}{18}\sec\theta$ at this point.

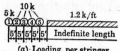

(a) Loading, per stringer

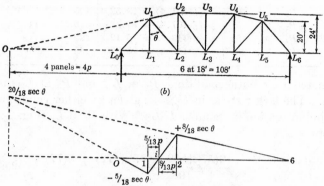

(b)

(c) Influence diagram for stress in member $U_1 L_2$

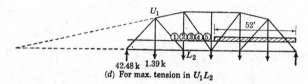

(d) For max. tension in $U_1 L_2$

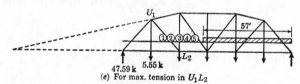

(e) For max. tension in $U_1 L_2$

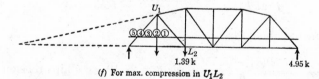

(f) For max. compression in $U_1 L_2$

FIG. 9-22

Maximum tension in $U_1 L_2$. In Fig. 9-22c, the ratio of 1-i to i-2 is 5 to 8; so i-2 = $\frac{8}{13}$ of the panel length. The criterion for maximum tension in $U_1 L_2$ is

$$G_2 = G \frac{i\text{-}2}{i\text{-}6} = G \frac{\frac{8}{13}}{\frac{8}{13} + 4} = G \frac{8}{8 + 52} = \frac{2}{15} G$$

This same result may be obtained by substituting $n = 6$, $e = 4$, and $m = 1$ in $G_2 = (G/n)[e/(e + m)]$.

Load at L_2	G_2	$\frac{2}{15}G$	G_2	Yes or no
P_1	0	$\frac{2}{15}(101.4) = 13.52$	5	No
P_2	5	$\frac{2}{15}(107.4) = 14.32$	15	Yes
P_3	15	$\frac{2}{15}(113.4) = 15.12$	25	Yes
P_4	25	$\frac{2}{15}(119.4) = 15.92$	35	No

It is noted that both conditions P_2 at L_2, and P_3 at L_2 satisfy the criterion. The larger stress in U_1L_2 as given by either condition will be the maximum tension in member U_1L_2. With P_2 at L_2 (Fig. 9-22d), the left reaction is

$$R_0 = \frac{\frac{1}{2}(1.2)(52)^2 + 10(57 + 62 + 67 + 72) + 5(77)}{108} = 42.48 \text{ kips}$$

The panel load at L_1 is

$$P \text{ at } L_1 = \frac{(5)(5)}{18} = 1.39 \text{ kips}$$

Taking moments about O,

$$U_1L_2 = \frac{(42.48)(4) - (1.39)(5)}{6} \sec \theta = 27.16 \sec \theta \text{ kips tension}$$

With P_3 at L_2 (Fig. 9-22e),

$$R_0 = \frac{\frac{1}{2}(1.2)(57)^2 + 10(62 + 67 + 72 + 77) + 5(82)}{108} = 47.59 \text{ kips}$$

$$P \text{ at } L_1 = \frac{10(5) + 5(10)}{18} = 5.55 \text{ kips}$$

Taking moments about O,

$$U_1L_2 = \frac{47.59(4) - 5.55(5)}{6} \sec \theta = 27.10 \sec \theta \text{ kips tension}$$

Thus the maximum tension in U_1L_2 is 27.16 $\sec \theta$ kips.

Maximum compression in U_1L_2. In Fig. 9-22c, the ratio of 1-i to i-2 is 5 to 8; so 1-i = $\frac{5}{13}$ of the panel length. The criterion for maximum compression in U_1L_2 is

$$G_2 = G\frac{i\text{-}1}{i\text{-}0} = G\frac{\frac{5}{13}}{\frac{5}{13} + 1} = G\frac{5}{5 + 13} = \frac{5}{18}G$$

The same result may be obtained by substituting $n = 6$, $e = 4$, and $m = 1$ in $G_2 = (G/n)[(e + n)/(e + m + 1)]$

Load at L_1	G_2	$\frac{5}{18}G$		G_2	Yes or no
P_1	0	$\frac{5}{18}(35)$	$= 9.72$	5	No
P_2	5	$\frac{5}{18}(45)$	$= 12.5$	15	Yes
P_3	15	$\frac{5}{18}(48.6)$	$= 13.5$	25	No

With the loads moving on the span from the left and P_2 at L_1 (Fig. 9-22f),

$$R_6 = \frac{10(3 + 8 + 13 + 18) + 5(23)}{108} = 4.95 \text{ kips}$$

$$\text{Panel load at } L_2 = \frac{5(5)}{18} = 1.39 \text{ kips}$$

Taking moments about O;

$$U_1L_2 = \frac{4.95(10) - 1.39(6)}{6} \sec \theta = 6.86 \sec \theta \text{ kips compression}$$

Thus the maximum compression in U_1L_2 is 6.86 sec θ kips.

PROBLEMS

9-1. Determine the maximum combined shears at 10-ft intervals in a 60-ft simple beam subjected to a dead load of 0.8 kip per lin ft, a live load of 1.6 kips per lin ft, and an impact equal to 20 per cent of the live load.

9-2. A simple beam 40 ft long carries moving loads of 5 kips, 10 kips, and 10 kips spaced 5 ft apart. Calculate the maximum left reaction, and the maximum shear and bending moment at a section 10 ft from the left end.

9-3. A beam 50 ft long is supported at 10 ft from the left end and at the right end. It carries moving loads of 5 kips, 10 kips, and 10 kips spaced 5 ft apart. Calculate the maximum left reaction, and the maximum bending moment at the left support.

9-4. A simple beam 15 ft long carries two moving concentrated loads of 12 kips each spaced 10 ft apart. Calculate the maximum shear and the absolute maximum bending moment in the beam due to these loads.

9-5. A simple beam 20 ft long carries two moving concentrated loads of 10 kips each spaced 4 ft apart. Calculate the maximum shear and the absolute maximum bending moment in the beam due to these loads.

9-6. A simple beam 28 ft long carries two moving concentrated loads of 10 kips and 5 kips spaced 9 ft on centers. Calculate the maximum shear and the absolute maximum bending moment in the beam due to these loads.

9-7. A simple beam 24 ft long carries moving loads of 10 kips, 20 kips, and 15 kips spaced 4 ft apart. Find the maximum shear and the absolute maximum bending moment caused by this system of loading.

9-8. A simple beam 24 ft long carries a system of loads spaced 5 ft on centers. The loads are 20 kips each. Calculate the maximum end shear and the maximum shear at a section 4 ft from the left end. Also calculate the absolute maximum bending moment.

9-9. A simple beam 30 ft long carries a moving uniform load of 2 kips per ft and two moving concentrated loads of 8 kips each spaced 10 ft on centers. Calculate the maximum shear and the absolute maximum bending moment in the beam due to these loads.

9-10. A simple beam 48 ft long carries a system of moving loads as shown. Calculate (a) the maximum left reaction and the maximum shear at sections 12 ft and 24 ft from the left end, (b) the maximum bending moment at sections 12 ft and 24 ft from the left end, and (c) the absolute maximum bending moment in the beam.

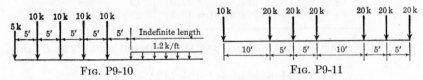

FIG. P9-10   FIG. P9-11

9-11. A simple beam 60 ft long carries a system of moving loads as shown. Calculate (a) the maximum left reaction and the maximum shear at sections 15 ft and 30 ft from the left end, (b) the maximum bending moment at sections 15 ft and 30 ft from the left end, and (c) the absolute maximum bending moment in the beam.

9-12. A simple beam 50 ft long carries a system of moving loads as shown. Calculate (a) the maximum bending moment at sections 20 ft and 25 ft from the left end, and (b) the absolute maximum bending moment in the beam.

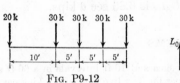
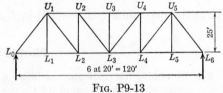

FIG. P9-12   FIG. P9-13

9-13. Given the Pratt truss as shown, calculate the maximum and minimum (if any) values due to the system of moving *wheel* loads as given in Prob. 9-10 for (a) the shear in panels L_0L_1, L_1L_2, and L_2L_3; (b) the bending moment at panel points L_1, L_2, and L_3; (c) the stress in member U_1L_1.

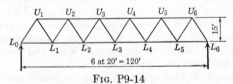

FIG. P9-14

9-14. Given the Warren truss as shown, calculate the maximum bending moments at panel points U_1, U_2, and U_3 due to the system of moving *wheel* loads as given in Prob. 9-10.

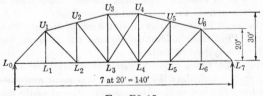

FIG. P9-15

9-15. Given the Parker truss as shown, calculate the maximum and minimum (if any) values due to the system of moving *wheel* loads as given in Prob. 9-10 for (a) the stress in members U_1U_2 and L_2L_3, (b) the stress in member U_1L_1, (c) the stress in members U_2L_2 and U_2L_3.

CHAPTER 10

ANALYSIS OF HIGHWAY AND RAILWAY BRIDGES

10-1. General Description. Although the floor slab in highway bridges is usually a reinforced-concrete slab, superstructures of bridges which carry highway or railway traffic are often built of steel. In recent years there has been a marked increase in the number of rigid-frame highway bridges built of reinforced concrete. The discussion in the present chapter will be limited to steel-girder or truss bridges carrying highway or railway traffic.

Bridge superstructures may make use of floor beams, but when the span is short and loading is light, a highway-bridge floor slab may rest on several joists or stringers running in the direction of the traffic and supported directly on the end piers. In the case of deck railway bridges, the ties supporting the rails commonly rest on the top of two parallel built-up plate girders. In the analysis of the highway stringers or railway plate girders as described above, it is necessary to determine the variation in the shear and bending moment along the span due to dead load, live load, and impact.

For longer spans and heavier loadings, or for other reasons, it may be uneconomical or infeasible to support the stringers or plate girders directly on the end piers. The span is then divided into several panels of equal length. The two or more stringers supporting the highway-bridge slab or the two stringers supporting the open or solid deck of a single-track railway bridge, now are simple beams with spans equal to the panel length and supported on floor beams in the transverse direction. The floor beams are in turn supported at the panel points of the two main girders on each side of the traffic. In the analysis of the main girders, the object is to determine the maximum combined shears in the panels and maximum combined bending moments at the panel points. When trusses are used instead of plate girders, the object will then be to determine the maximum and minimum combined stresses in all members of either truss.

A *deck-truss* bridge is one in which the floor beams as described in the preceding paragraph are connected to the panel points at the top chords of the main trusses; a *pony-truss* bridge is one in which the floor beams are connected to the panel points at the lower chords of the two main trusses, which are so low that overhead lateral bracing becomes impracticable; a *through-truss* bridge is one in which the floor beams are connected to the panel points at the lower chords of the main trusses, with complete upper and lower lateral, portal, and sway bracing. A typical four-panel through-truss bridge with stringers, end floor beams, floor beams, main trusses, upper lateral bracing, lower lateral bracing, portal bracing, and sway bracing is shown in Fig. 10-1.

The more commonly used types of trusses are shown in Fig. 10-2. The diagonals of the Pratt truss serve mainly in tension; while those of the Howe truss are usually in compression. The Howe truss is generally built of timber, with vertical members of steel. The Parker truss is really a curved-chord Pratt truss; the decreasing height of the truss toward the ends tends to equalize the required chord areas. The Baltimore and Pennsylvania trusses are, respectively, horizontal-chord and

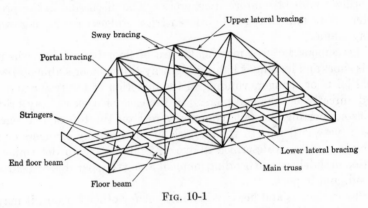

FIG. 10-1

curved-chord Pratt trusses with subdivided panels, with the main diagonals running across two subpanels. Subverticals, subties, or substruts are shown in Fig. 10-2e and f; the subties or substruts may be used in either the Baltimore or the Pennsylvania truss. The K truss, like the Baltimore and Pennsylvania trusses, may be used to provide an appropriate panel length (20 to 25 ft) and a height consistent with the length of span (one-sixth to one-eighth of the span).

10-2. Dead Load. The dead load carried by any structure is the weight of the structure itself. The dead load on a bridge includes (1) the weight of the floor system, and (2) the weight of the main girders or trusses together with the bracing system. The floor system is usually designed first; and its weight is therefore known prior to the analysis of the main girders or trusses. The weight of the main girders or trusses together with lateral bracing, however, must be assumed in the analysis and then reviewed after the design has been made. A fairly good estimate of the weight of the structure can usually be made by comparison with existing bridges or by use of an appropriate empirical formula.

An ordinary 6-in. reinforced-concrete slab in the floor system of a highway bridge may weigh 75 psf while the stringers and floor beams may weigh from 12 to 20 psf of roadway surface. The weight of an open-floor railway track, including rails, ties, and fastenings, may be approximately 500 lb per lin ft.

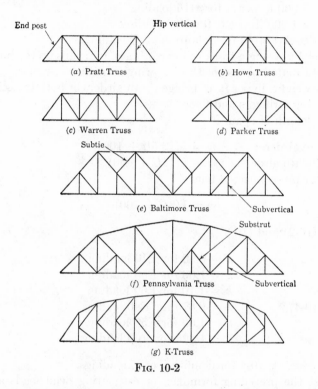

(a) Pratt Truss

(b) Howe Truss

(c) Warren Truss

(d) Parker Truss

(e) Baltimore Truss

(f) Pennsylvania Truss

(g) K-Truss

Fig. 10-2

For estimating the weight of highway girders or trusses, the following formula, which is a variation of the old Johnson, Bryan, Turneaure formula,[1] is suggested here. Again, any weight formula must be used with discretion; the actual weight of the structure as designed should be compared with the assumed weight used in the analysis. It may then be necessary to revise the design.

$$w = \frac{L\sqrt{p}}{20} + \frac{L(b - 16)}{10} + 50 \tag{10-1}$$

[1] Hool and Kinne, "Steel and Timber Structures," 2d ed., p. 359, McGraw-Hill Book Company, Inc., New York, 1942.

where w = weight, lb per ft, of each girder or truss, including floor beams, but not stringers and slab

L = length of span, ft

b = width of roadway, ft

p = live load per lin ft of each girder or truss

For two-lane bridges, use

p = 600 lb per ft for H10 loading

p = 900 lb per ft for H15 loading

p = 1,200 lb per ft for H20 loading

Note: See Art. 10-3 for definition of H loadings.

The weight of single-track railway bridges, for spans up to about 300 ft, may be estimated by use of the following formulas,[1] where

w = weight, lb per ft, of bridge (both girders or both trusses) including stringers and floor beams

L = length of span, ft

Deck plate girders: $w = k(12.5L + 100)$ (10-2)

Through plate girders: $w = k(14L + 450)$ (10-3)

Riveted or pin-connected trusses:

$$w = k(8L + 700) \qquad\qquad (10\text{-}4)$$

In Eqs. (10-2) and (10-3),

k = 0.90 for E40 loading

k = 1.00 for E50 loading

k = 1.10 for E60 loading

In Eq. (10-4), k = 0.875 for E40 loading

k = 1.00 for E50 loading

k = 1.125 for E60 loading

Note: See Art. 10-4 for definition of E loadings.

At best the preceding formulas for estimating dead loads are rough approximations of the weight of various types of bridge structures. They should not be considered as entirely reliable for design. The type of truss, the amount of live load and impact, and especially the allowable working stresses to be used in the design of the structure are all factors influencing the weight of the structure. Perhaps the best estimate of dead load can be obtained by comparison with similar existing structures.

Example 10-1. Estimate the dead load on a 120-ft highway girder bridge with floor beams spaced 20 ft apart. The width of roadway is 20 ft and live load is H20 loading. Draw shear and bending-moment diagrams due to the estimated dead load for the main girder.

[1] *Ibid.*, p. 290.

SOLUTION. The dead load per foot of girder is estimated as follows:
Assume a 6-in. floor slab which weighs 75 psf

$$(\tfrac{1}{2})(75)(\text{width of roadway}) = (\tfrac{1}{2})(75)(20) = 750 \text{ lb per ft}$$

Assume weight of stringers at 6 psf

$$(\tfrac{1}{2})(6)(\text{width of roadway}) = (\tfrac{1}{2})(6)(20) = 60 \text{ lb per ft}$$

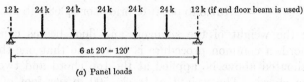

(a) Panel loads

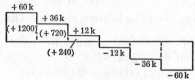

(b) Shear diagram

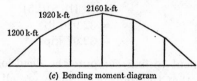

(c) Bending moment diagram

FIG. 10-3

Estimate weight of main girder including floor beams (Eq. 10-1)

$$w = \frac{L\sqrt{p}}{20} + \frac{L(b-16)}{10} + 50$$

$$= \frac{120\sqrt{1,200}}{20} + \frac{120(20-16)}{10} + 50 = 306 \text{ lb per ft}$$

Total dead load per ft of girder

$$750 + 60 + 306 = 1,116, \text{ or say, } 1,200 \text{ lb per ft}$$

In practice the slab, stringers, and floor beams are designed before-hand; thus their weight (with allowance for sidewalks, curbs, handrails, etc.) can be determined prior to the analysis of the girder.

Dead panel load = (1.2 kips per ft)(panel length in ft)
= (1.2)(20) = 24 kips

The panel loads and shear and bending-moment diagrams due to dead load for the main girder are shown in Fig. 10-3.

Example 10-2.   Estimate the dead load on a single-track, 150-ft through-truss railway bridge with floor beams spaced 25 ft apart.   The truss is a six-panel Parker truss as shown in Fig. 10-4.   Assume Cooper's E60 loading.   Determine the dead-load stresses in all members.

SOLUTION.   Assuming the use of Eq. (10-4), the weight of the trusses, bracing systems, stringers, and floor beams per foot of *bridge* is

$$w = k(8L + 700) = 1.125(8 \times 150 + 700)$$
$$= 2,137.5 \text{ lb per ft of bridge or } 1,068.8 \text{ lb per ft of truss}$$

Because the weight of the stringers and floor beams comes to the bottom chord, a common procedure is to assume that one-third of the weight estimated above be applied at the top chord and two-thirds at the bottom chord.   The weight of track, 500 lb per foot of bridge or 250 lb per foot of truss, also acts on the bottom chord.   Thus

$$\text{Top panel load} = (\tfrac{1}{3})(1,068.8)(25) = 8,907 \text{ lb} = 8.91 \text{ kips}$$
$$\text{Bottom panel load} = [(\tfrac{2}{3})(1,068.8) + 250](25)$$
$$= 24,063 \text{ lb} = 24.06 \text{ kips}$$
$$\text{Total panel load (top and bottom)} = (1,068.8 + 250)(25)$$
$$= (1,318.8)(25)$$
$$= 32,970 \text{ lb}$$
$$= 32.97 \text{ kips}$$

It is usually more convenient to first compute the dead-load stresses in all members by assuming that the total panel loads are all applied at the bottom chord, and then modify the stresses in the vertical members by adding a compressive stress equal to the amount of the top panel load. An inspection of Fig. 10-5 will show how this can be done.

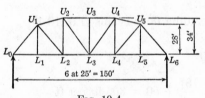

FIG. 10-4

The dead panel loads and the corresponding stresses in all members are shown in Fig. 10-5a, which shows the summation of the stresses shown in Fig. 10-5b and c.

The stresses shown in Fig. 10-5b are usually found by the algebraic method.   The suggested procedure is (1) to determine the amount and kind of stress in each member (also the horizontal and vertical components of the stress if the member is inclined) independently by the method of sections, and (2) to review the equilibrium of each joint and see that the two equations of equilibrium $\Sigma F_x = 0$ and $\Sigma F_y = 0$ are satisfied at each joint.   In this way, the correctness of the solution can be demonstrated.

The lengths of all inclined members are computed and entered on Fig. 10-6.   The inclined top chord member $U_1 U_2$ is extended to intersect

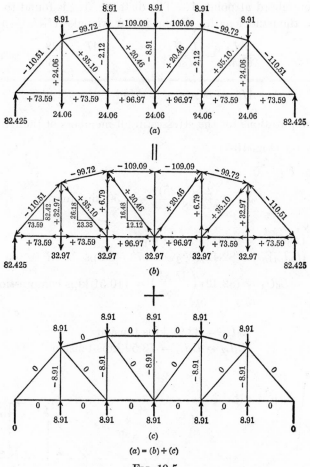

$(a) = (b) + (c)$

FIG. 10-5

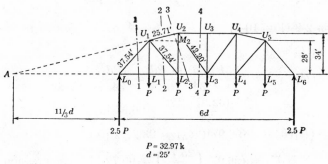

$P = 32.97 \text{ k}$
$d = 25'$

FIG. 10-6

the bottom chord at point A. The distance AL_0 is found to be $11\tfrac{1}{3}d$ in which d is the panel length. From similar triangles $U_1U_2M_2$ and AU_1L_1,

$$\frac{AL_1}{U_1M_2} = \frac{L_1U_1}{M_2U_2} \quad \text{or} \quad \frac{AL_1}{d} = \frac{28}{6}$$

Solving, $\qquad\qquad AL_1 = {}^{28}\!/_6\,d = 14\tfrac{2}{3}d$

and $\qquad\qquad AL_0 = AL_1 - d = 14\tfrac{2}{3}d - d = 11\tfrac{1}{3}d$

The computations for the stresses in all members of the truss follow.

Section 1-1 (Fig. 10-7)

ΣM about $U_1 = 0$:

$$L_0L_1 = \frac{(2.5P)(25)}{28} = 73.59 \text{ kips tension}$$

$\Sigma F_y = 0$:

$$(L_0U_1)_V = 2.5P = 82.42 \text{ kips}$$
$$(L_0U_1)_H = (82.42)({}^{25}\!/_{28}) = 73.59 \text{ kips}$$
$$L_0U_1 = (82.42)\left(\frac{37.54}{28}\right) = 110.51 \text{ kips compression}$$

Joint L_1

$\Sigma F_y = 0$: $\qquad L_1U_1 = 32.97 \text{ kips tension}$

$\Sigma F_x = 0$: $\qquad L_1L_2 = L_0L_1 = 73.59 \text{ kips tension}$

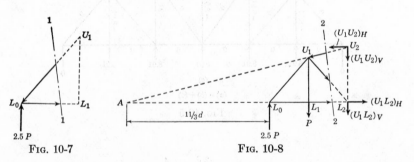

Fig. 10-7   Fig. 10-8

Section 2-2 (Fig. 10-8)

ΣM about $L_2 = 0$:

$$(U_1U_2)_H = \frac{(2.5P)(2d) - Pd}{34} = \frac{4Pd}{34}$$

$$= \frac{(4)(32.97)(25)}{34} = 96.97 \text{ kips}$$

$$(U_1U_2)_V = (96.97)({}^6\!/_{25}) = 23.27 \text{ kips}$$

$$U_1U_2 = (96.97)\left(\frac{25.71}{25}\right) = 99.72 \text{ kips compression}$$

ΣM about $A = 0$:

$$(U_1L_2)_V = \frac{(2.5P)(11\frac{1}{3}d) - P(14\frac{2}{3}d)}{17\frac{1}{3}d}$$

$$= \frac{13.5}{17} P = 26.18 \text{ kips}$$

$$(U_1L_2)_H = (26.18)(25\frac{7}{28}) = 23.38 \text{ kips}$$

$$U_1L_2 = (26.18)\left(\frac{37.54}{28}\right) = 35.10 \text{ kips tension}$$

Section 3-3 (Fig. 10-9)

ΣM about $A = 0$:

$$L_2U_2 = \frac{(2.5P)(11\frac{1}{3}d) - P(14\frac{2}{3}d) - P(17\frac{1}{3}d)}{17\frac{1}{3}d}$$

$$= -\frac{3.5P}{17} = -6.79 \text{ kips or } 6.79 \text{ kips tension}$$

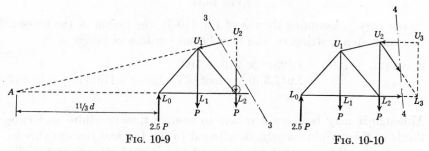

FIG. 10-9   FIG. 10-10

Section 4-4 (Fig. 10-10)

ΣM about $L_3 = 0$:

$$U_2U_3 = \frac{(2.5P)(3d) - (P)(2d) - Pd}{34}$$

$$= \frac{(4.5)(32.97)(25)}{34} = 109.09 \text{ kips compression}$$

ΣM about $U_2 = 0$:

$$L_2L_3 = \frac{(2.5P)(2d) - Pd}{34}$$

$$= \frac{(4)(32.97)(25)}{34} = 96.97 \text{ kips tension}$$

$\Sigma F_y = 0$:

$$(U_2L_3)_V = 2.5P - P - P = (0.5)(32.97) = 16.48 \text{ kips}$$
$$(U_2L_3)_H = (16.48)(25\frac{7}{34}) = 12.12 \text{ kips}$$

$$U_2L_3 = (16.48)\left(\frac{42.20}{34}\right) = 20.46 \text{ kips tension}$$

Joint U_3
$\Sigma F_y = 0$: $$U_3L_3 = 0$$

The stresses, with their horizontal and vertical components in the case of inclined members, are entered on Fig. 10-5b. The equilibrium of each joint can then be checked.

Example 10-3. Estimate the dead load on a single-track 240-ft through-truss railway bridge with floor beams spaced 20 ft apart. The truss is of the Baltimore type shown in Fig. 10-11. The live load is Cooper's E60 loading. Determine the dead-load stresses in all members.

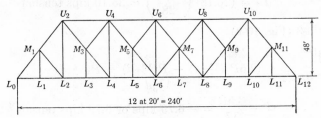

FIG. 10-11

SOLUTION. Assuming the use of Eq. (10-4), the weight of the trusses, bracing systems, stringers, and floor beams per foot of bridge is

$$w = k(8L + 700) = 1.125(8 \times 240 + 700)$$
$$= 2{,}947.5 \text{ lb per foot of bridge or } 1{,}473.8 \text{ lb per foot of truss}$$

Although it may be more accurate to assume that one-third and two-thirds of this weight may be distributed to the top and bottom chords, respectively, the fact that the live-load and impact stresses (especially in long spans) are relatively much larger than the dead-load stresses probably makes this refinement unnecessary. In this problem all dead load will be assumed to act on the lower chord. By assuming the weight of track to be 250 lb per foot of truss, the total panel load will be

$$(1{,}473.8 + 250)(20) = 34{,}476 \text{ lb or } 34.48 \text{ kips}$$

The dead-load stresses shown in Fig. 10-12 have been found by the algebraic method. As in the preceding problem, it is advisable to determine the stress in each member independently, and then check the equilibrium of each joint by inspection. The procedure is outlined below, but detailed computations are not shown.

1. *Members* L_1M_1, L_3M_3, *and* L_5M_5. For example, from $\Sigma F_y = 0$ at joint L_1,

$$L_1M_1 = 34.48 \text{ kips tension}$$

2. *Members* L_2M_1, M_3U_4, *and* M_5U_6. For example, from ΣM about $L_0 = 0$ with joint M_1 as the free body (Fig. 10-13a),

$$(L_2M_1)_V = \tfrac{1}{2}P$$

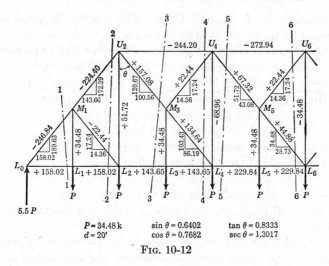

$$P = 34.48 \text{ k} \qquad \sin\theta = 0.6402 \qquad \tan\theta = 0.8333$$
$$d = 20' \qquad \cos\theta = 0.7682 \qquad \sec\theta = 1.3017$$

Fig. 10-12

and from ΣM about $U_2 = 0$ with joint M_3 as the free body (Fig. 10-13b),

$$(M_3U_4)_V = \tfrac{1}{2}P$$

3. *Members* L_0M_1, M_1U_2, U_2M_3, M_3L_4, U_4M_5, M_5L_6. By cutting sections 1-1 through 6-6, respectively (Fig. 10-12), the vertical component of the stress in each of these members can be found from $\Sigma F_y = 0$.

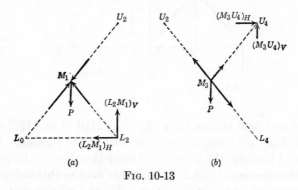

Fig. 10-13

4. *Members* $L_0L_1 = L_1L_2$, $L_3L_3 = L_3L_4$, *and* $L_4L_5 = L_5L_6$. The stresses in these members can be found from ΣM about $M_1 = 0$, section 2-2; ΣM about $U_2 = 0$, section 3-3; and ΣM about $U_4 = 0$, section 5-5.

5. *Members* U_2U_4 *and* U_4U_6. The stresses in these members can be found from ΣM about $L_4 = 0$, section 3-3; and ΣM about $L_6 = 0$, section 5-5. It should be noted that, in taking moments about L_4, section 3-3, the panel load at joint L_3 is outside of the free body (Fig. 10-14). This also happens with ΣM about $L_6 = 0$, section 5-5.

6. *Members U_2L_2, U_4L_4, and U_6L_6.* The stresses in these members can be found by using $\Sigma F_y = 0$ at joints L_2, U_4, and U_6.

10-3. Live Load on Highway Bridges. The live load to be used in the design of highway bridges is given in "Standard Specifications for Highway Bridges," 6th edition, 1953, published by the American Association of State Highway Officials. As noted in Art. 3.2.5 of these Specifications, there are five typical loadings for highway bridges:

1. H20-44 standard truck or lane loading
2. H15-44 standard truck or lane loading
3. H10-44 standard truck or lane loading
4. H20-S16-44 standard truck or lane loading
5. H15-S12-44 standard truck or lane loading

These standard truck or lane loadings are shown diagrammatically in Fig. 10-15.

The standard *truck* and *lane* loadings under the same designation are approximately equivalent loadings; however, the one which causes the

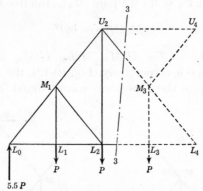

Fig. 10-14

larger stresses should always be used. The H10-44 and H15-44 loadings are 50 and 75 per cent, respectively, of the H20-44 loading. In the case of the H20-44 loading, the number 20 after H indicates the gross weight in tons of the standard truck. Twenty per cent of this weight, 4 tons or 8 kips, is assumed to be on the front axle; and 80 per cent, 16 tons or 32 kips, is assumed on the rear axle. The front and rear axles are 14 ft apart. The manner in which the total weight of the standard truck and load is assumed to be distributed to the four wheels is shown in Fig. 10-16. The number 44 after each loading designation refers to the 1944 edition of the specifications. The H15-S12-44 loading is 75 per cent of the H20-S16-44 loading. The number after H indicates the gross weight in tons of the tractor truck, which is identical with the standard truck of the same weight; and the number after S indicates the gross weight

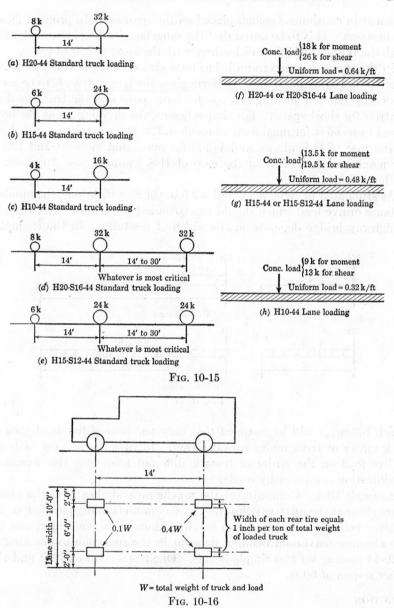

FIG. 10-15

Width of each rear tire equals
1 inch per ton of total weight
of loaded truck

W = total weight of truck and load

FIG. 10-16

in tons of the single axle of the semitrailer. The variable spacing of 14 to 30 ft is intended to approximate closely the tractor trailers now in common use and to provide a more satisfactory loading for continuous spans. The lane loading consists of a uniform load of indefinite length and a floating concentrated load (or two concentrated loads for negative moment in continuous spans) placed on the span so as to produce maximum stress. It is to be noted that the same lane loading is equivalent to both the H and the H-S truck loadings with the same weight of truck.

In the preceding paragraph it has been stated that the truck and lane loadings are equivalent, but whichever gives the larger stress is to be used. Generally the lane loading controls for long spans and the truck loading controls for short spans. For simple beams, the dividing point has been found to be 56 ft for maximum moment and 33 ft for maximum end shear in the case of H loadings, and 140 ft for maximum moment and 120 ft for maximum end shear in the case of H-S loadings (see Appendix A of the Specifications).

As stipulated in Arts. 3.2.6 and 3.2.9 of the Specifications, the number of lanes of live load which should be attributed to each girder or truss of a highway bridge depends on the width of roadway. In the examples

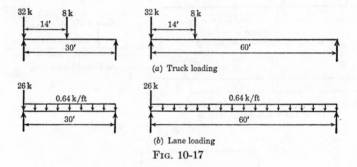

(a) Truck loading

(b) Lane loading

FIG. 10-17

which follow, it will be assumed that only one lane of live load goes to each girder or truss under consideration. In case the number of lanes of live load on the girder or truss is different from one, the necessary modification can be easily made.

Example 10-4. Compute (a) the maximum end shear, (b) the maximum shear at the quarter point, (c) the maximum bending moment at the quarter point, (d) the maximum bending moment at the center, and (e) the absolute maximum bending moment in the span due to one lane of H20-44 loading on two simple beams. One has a span of 30 ft and the other a span of 60 ft.

SOLUTION

(a) *Maximum end shear* (Fig. 10-17)

30-ft span:

$$\text{Truck loading, } V_e = 32 + (8)(^{16}\!/_{30}) = 36.27 \text{ kips } (\textit{controls})$$
$$\text{Lane loading, } V_e = 26 + (\tfrac{1}{2})(0.64)(30) = 35.6 \text{ kips}$$

60-ft span:

Truck loading, $V_e = 32 + (8)(^{46}\!/_{60}) = 38.13$ kips

Lane loading, $V_e = 26 + (\frac{1}{2})(0.64)(60) = 45.2$ kips (*controls*)

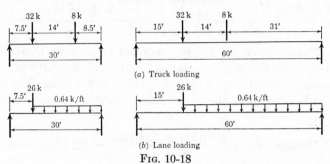

(*a*) Truck loading

(*b*) Lane loading

FIG. 10-18

(*b*) *Maximum shear at the quarter point* (Fig. 10-18)

30-ft span:

Truck loading, $V_{\frac{1}{4}} = (32)\left(\dfrac{22.5}{30}\right) + (8)\left(\dfrac{8.5}{30}\right) = 26.27$ kips (*controls*)

Lane loading, $V_{\frac{1}{4}} = (26)(\frac{3}{4}) + \dfrac{(0.64)(22.5)^2}{(2)(30)} = 24.9$ kips

60-ft span:

Truck loading, $V_{\frac{1}{4}} = (32)(^{45}\!/_{60}) + (8)(^{31}\!/_{60}) = 28.13$ kips

Lane loading, $V_{\frac{1}{4}} = (26)(\frac{3}{4}) + \dfrac{(0.64)(45)^2}{(2)(60)} = 30.3$ kips (*controls*)

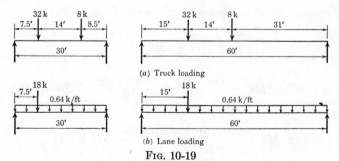

(*a*) Truck loading

(*b*) Lane loading

FIG. 10-19

(*c*) *Maximum bending moment at the quarter point* (Fig. 10-19). Although the criteria in Chap. 9 may be applied, the condition of loading producing maximum bending moment at the quarter point or mid-span can be readily visualized if the influence diagrams are sketched for these sections.

30-ft span:

Truck loading, $M_{\frac{1}{4}} = \dfrac{(8)(8.5) + (32)(22.5)}{4} = 197$ kip-ft (controls)

Lane loading, $M_{\frac{1}{4}} = \dfrac{(18)(22.5)}{4} + \dfrac{(0.64)}{2}(7.5)(22.5)$

$= 155.25$ kip-ft

It is to be noted that the bending moment at a section C, at distances of a and b, respectively, from the left and right supports of a simple beam, which is loaded with a uniform load of w per linear foot over the entire span L, is equal to $M_C = (wL/2)a - (wa^2/2)$ or $M_C = wab/2$.

60-ft span:

Truck loading, $M_{\frac{1}{4}} = \dfrac{(18)(31) + (32)(45)}{4} = 422$ kip-ft (controls)

Lane loading, $M_{\frac{1}{4}} = \dfrac{(18)(45)}{4} + \dfrac{(0.64)}{2}(15)(45)$

$= 418.5$ kip-ft

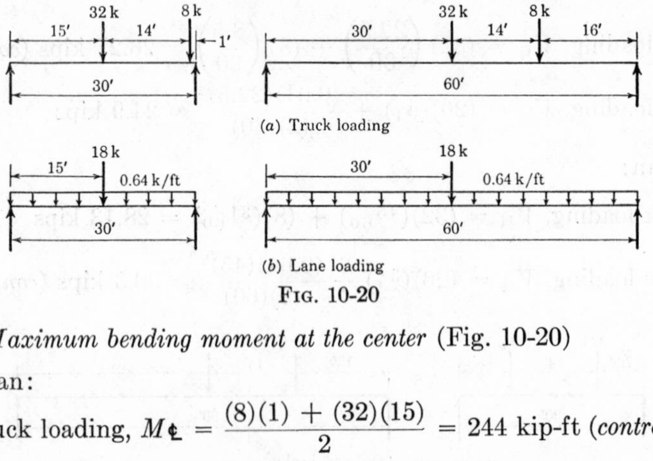

(a) Truck loading

(b) Lane loading

FIG. 10-20

(d) Maximum bending moment at the center (Fig. 10-20)

30-ft span:

Truck loading, $M_{\mathₑ} = \dfrac{(8)(1) + (32)(15)}{2} = 244$ kip-ft (controls)

Lane loading, $M_{\mathₑ} = \dfrac{(18)(30)}{4} + \dfrac{(0.64)(30)^2}{8} = 207$ kip-ft

60-ft span:

Truck loading, $M_{\mathₑ} = \dfrac{(8)(16) + (32)(30)}{2} = 544$ kip-ft

Lane loading, $M_{\mathₑ} = \dfrac{(18)(60)}{4} + \dfrac{(0.64)(60)^2}{8} = 558$ kip-ft (controls)

(e) Absolute maximum bending moment in the span (Fig. 10-21). The absolute maximum bending moment in the span due to the lane loading is the same as the maximum bending moment at the center of the span.

Truck loading:

$$\text{30-ft span, max } M = \frac{(40)(13.6)^2}{30} = 246.6 \text{ kip-ft } (controls)$$

$$\text{60-ft span, max } M = \frac{(40)(28.6)^2}{60} = 545.3 \text{ kip-ft}$$

The results of the above computations are summarized in the following table. Note that generally the truck loading controls in the 30-ft span and, with the exception of maximum bending moment at the quarter

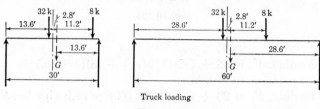

Truck loading

Fig. 10-21

point of the 60-ft span, the lane loading controls in the 60-ft span. For H loadings, the lane loading causes larger end shear for spans over 33 ft and larger bending moment for spans over 56 ft.

Span	30 ft		60 ft	
H20-44 loading	Truck	Lane	Truck	Lane
Max end shear, kips	36.27	35.6	38.13	45.2
Max shear at quarter point, kips	26.27	24.9	28.13	30.3
Max moment at quarter point, kip-ft	197	155.25	422	418.5
Max moment at center, kip-ft	244	207	544	558
Absolute max moment, kip-ft	246.6	207	545.3	558

Example 10-5.   Compute (*a*) the maximum end shear, (*b*) the maximum shear at the quarter point, (*c*) the maximum bending moment at the quarter point, (*d*) the maximum bending moment at the center, and (*e*) the absolute maximum bending moment in the span due to one lane of H20-S16-44 loading on two simple beams. One has a span of 100 ft and the other a span of 160 ft.

SOLUTION

(*a*) *Maximum end shear* (Fig. 10-22)

100-ft span:

$$\text{Truck loading, } V_e = 32 + (32)(^{86.4}\!/_{100}) + (8)(^{72.4}\!/_{100})$$
$$= 65.28 \text{ kips } (controls)$$
$$\text{Lane loading, } V_e = 26 + (\tfrac{1}{2})(0.64)(100) = 58 \text{ kips}$$

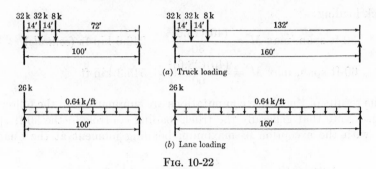

(a) Truck loading

(b) Lane loading

FIG. 10-22

160-ft span:

Truck loading, $V_e = 32 + (32)(^{146}\!/_{160}) + (8)(^{132}\!/_{160})$
$= 67.8$ kips
Lane loading, $V_e = 26 + (\frac{1}{2})(0.64)(160) = 77.2$ kips (*controls*)

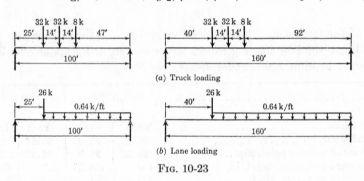

(a) Truck loading

(b) Lane loading

FIG. 10-23

(b) *Maximum shear at the quarter point* (Fig. 10-23)

100-ft span:

Truck loading, $V_{\frac{1}{4}} = (32)(^{75}\!/_{100}) + (32)(^{61}\!/_{100}) + (8)(^{47}\!/_{100})$
$= 47.28$ kips (*controls*)
Lane loading, $V_{\frac{1}{4}} = (26)(\frac{3}{4}) + \dfrac{(0.64)(75)^2}{(2)(100)} = 37.5$ kips

160-ft span:

Truck loading, $V_{\frac{1}{4}} = (32)(^{120}\!/_{160}) + (32)(^{106}\!/_{160}) + (8)(^{92}\!/_{160})$
$= 49.8$ kips (*controls*)
Lane loading $V_{\frac{1}{4}} = (26)(\frac{3}{4}) + \dfrac{(0.64)(120)^2}{(2)(160)} = 48.3$ kips

(c) *Maximum bending moment at the quarter point* (Fig. 10-24). Both truck loading positions, as shown in Fig. 10-24a and b, satisfy the criterion for maximum moment, with $G/4 = 7\frac{2}{4} = 18$ in either case and $G_1 = 8$ to 40 for truck heading to the left and $G_1 = 0$ to 32 for truck heading to the right. To determine which truck loading position gives the larger moment, it is only necessary to compare the ordinates y_1 and

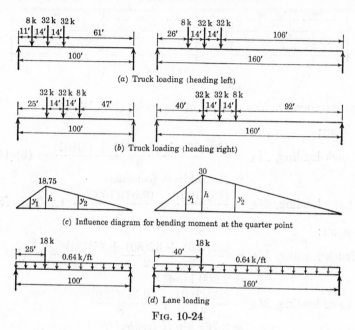

(a) Truck loading (heading left)

(b) Truck loading (heading right)

(c) Influence diagram for bending moment at the quarter point

(d) Lane loading

Fig. 10-24

y_2 as shown in the influence diagrams of Fig. 10-24c. Thus it is seen that the second condition (Fig. 10-24b) is critical.

100-ft span:

$$\text{Truck loading, } M_{\frac{1}{4}} = \frac{(8)(47) + (32)(61) + (32)(75)}{4}$$
$$= 1,182 \text{ kip-ft } (controls)$$
$$\text{Lane loading, } M_{\frac{1}{4}} = \frac{(18)(75)}{4} + \frac{(0.64)}{2}(25)(75)$$
$$= 937.5 \text{ kip-ft}$$

160-ft span:

$$\text{Truck loading, } M_{\frac{1}{4}} = \frac{(8)(92) + (32)(106) + (32)(120)}{4}$$
$$= 1,992 \text{ kip-ft}$$
$$\text{Lane loading, } M_{\frac{1}{4}} = \frac{(18)(120)}{4} + \frac{(0.64)}{2}(40)(120)$$
$$= 2,076 \text{ kip-ft } (controls)$$

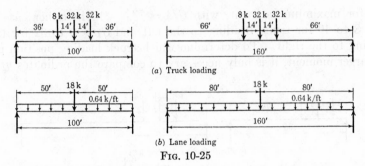

(a) Truck loading

(b) Lane loading

FIG. 10-25

(d) *Maximum bending moment at the center* (Fig. 10-25)

100-ft span:

$$\text{Truck loading, } M_{\mathbb{C}} = \frac{(32)(36) + (32)(50) + (8)(64)}{2} - (8)(14)$$

$$= 1{,}520 \text{ kip-ft } (controls)$$

$$\text{Lane loading, } M_{\mathbb{C}} = \frac{(18)(100)}{4} + \frac{(0.64)(100)^2}{8} = 1{,}250 \text{ kip-ft}$$

160-ft span:

$$\text{Truck loading, } M_{\mathbb{C}} = \frac{(32)(66) + 32(80) + (8)(94)}{2} - (8)(14)$$

$$= 2{,}600 \text{ kip-ft}$$

$$\text{Lane loading, } M_{\mathbb{C}} = \frac{(18)(160)}{4} + \frac{(0.64)(160)^2}{8}$$

$$= 2{,}768 \text{ kip-ft } (controls)$$

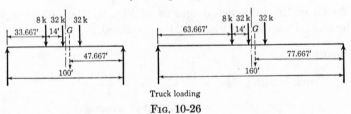

Truck loading

FIG. 10-26

(e) *Absolute maximum bending moment in the span* (Fig. 10-26). The absolute maximum bending moment in the span due to the lane loading is the same as the maximum bending moment at the center. Truck loading:

$$\text{100-ft span, max } M = \frac{(72)(47.667)^2}{100} - (8)(14)$$

$$= 1{,}523.9 \text{ kip-ft } (controls)$$

$$\text{160-ft span, max } M = \frac{(72)(77.667)^2}{160} - (8)(14)$$

$$= 2{,}602.5 \text{ kip-ft}$$

The results of the above computations are summarized in the following table. Note that generally the truck loading controls in the 100-ft span and, with the exception of maximum shear at the quarter point of the 160-ft span, the lane loading controls in the 160-ft span. For H-S loadings, the lane loading causes larger end shear for spans over 120 ft and larger bending moment for spans over 140 ft.

Span..................................	100 ft		160 ft	
H20-S16-44 loading...................	Truck	Lane	Truck	Lane
Max end shear, kips..................	65.28	58	67.8	77.2
Max shear at quarter point, kips.......	47.28	37.5	49.8	48.3
Max moment at quarter point, kip-ft....	1,182	937.5	1,992	2,076
Max moment at center, kip-ft..........	1,520	1,250	2,600	2,768
Absolute max moment, kip-ft..........	1,523.9	1,250	2,602.5	2,768

Example 10-6.   A 120-ft highway girder bridge has floor beams spaced 20 ft on centers as shown in Fig. 10-27.   Compute the maximum shears in panels 0-1, 1-2, and 2-3 and the maximum bending moments at points 1, 2, and 3, due to one lane of H20-S16-44 loading per girder.

SOLUTION.   Inasmuch as there are only three concentrated loads in the H-S truck loading, and a uniform

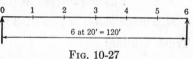

FIG. 10-27

load with one concentrated load in the lane loading, the position of loads for maximum shears or bending moments can be readily determined by inspection of the influence diagrams.   The values of maximum shears or bending moments may be computed by the use of either the influence diagrams or the free-body diagrams of the girder.

Maximum shear in panel 0-1 (Fig. 10-28)

Truck loading:

$$V_{0\text{-}1} = (32)(\tfrac{5}{6}) + (32)(\tfrac{5}{6})(^{86}\!/_{100}) + (8)(\tfrac{5}{6})(^{72}\!/_{100})$$
$$= 54.4 \text{ kips } (controls)$$

Lane loading:

$$V_{0\text{-}1} = (26)(\tfrac{5}{6}) + (0.64)(\tfrac{1}{2})(\tfrac{5}{6})(120) = 53.67 \text{ kips}$$

Maximum shear in panel 1-2 (Fig. 10-29)

Truck loading:

$$V_{1\text{-}2} = (32)(\tfrac{4}{6}) + (32)(\tfrac{4}{6})(^{66}\!/_{80}) + (8)(\tfrac{4}{6})(^{52}\!/_{80})$$
$$= 42.4 \text{ kips } (controls)$$

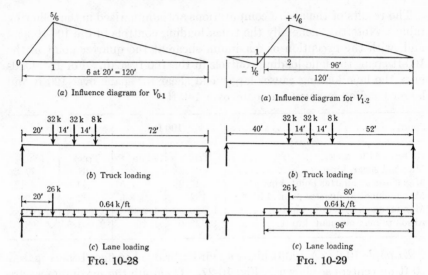

(a) Influence diagram for $V_{0\text{-}1}$

(b) Truck loading

(c) Lane loading

Fig. 10-28

(a) Influence diagram for $V_{1\text{-}2}$

(b) Truck loading

(c) Lane loading

Fig. 10-29

Lane loading:

$$V_{1\text{-}2} = (26)(\tfrac{4}{6}) + (0.64)(\tfrac{1}{2})(\tfrac{4}{6})(96) = 37.81 \text{ kips}$$

Maximum shear in panel 2-3 (Fig. 10-30)

Truck loading:

$$V_{2\text{-}3} = (32)(\tfrac{3}{6}) + (32)(\tfrac{3}{6})(\tfrac{46}{60}) + (8)(\tfrac{3}{6})(\tfrac{32}{60})$$
$$= 30.4 \text{ kips } (controls)$$

Lane loading:

$$V_{2\text{-}3} = (26)(\tfrac{3}{6}) + (0.64)(\tfrac{1}{2})(\tfrac{3}{6})(72) = 24.52 \text{ kips}$$

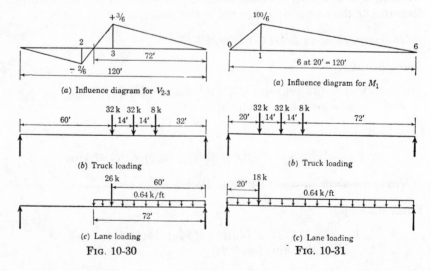

(a) Influence diagram for $V_{2\text{-}3}$

(b) Truck loading

(c) Lane loading

Fig. 10-30

(a) Influence diagram for M_1

(b) Truck loading

(c) Lane loading

Fig. 10-31

• *Maximum bending moment at point* 1 (Fig. 10-31)

Truck loading:

$$M_1 = \frac{(8)(72) + (32)(86) + (32)(100)}{6} = 1,088 \text{ kip-ft } (controls)$$

Lane loading:

$$M_1 = \frac{(18)(100)}{6} + \frac{1}{2}(0.64)(20)(100) = 940 \text{ kip-ft}$$

Maximum bending moment at point 2 (Fig. 10-32)

Truck loading:

$$M_2 = \frac{1}{6}[(8)(52) + (32)(66) + (32)(80)](2)$$
$$= 1,696 \text{ kip-ft } (controls)$$

Lane loading:

$$M_2 = \frac{(18)(80)}{3} + \frac{(0.64)}{2}(40)(80) = 1,504 \text{ kip-ft}$$

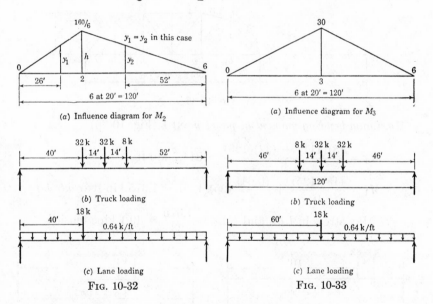

(a) Influence diagram for M_2

(a) Influence diagram for M_3

(b) Truck loading

(b) Truck loading

(c) Lane loading

(c) Lane loading

FIG. 10-32

FIG. 10-33

Maximum bending moment at point 3 (Fig. 10-33)

Truck loading:

$$M_3 = \frac{(32)(46) + (32)(60) + (8)(74)}{2} - (8)(14)$$
$$= 1,880 \text{ kip-ft } (controls)$$

Lane loading:

$$M_3 = \frac{(18)(120)}{4} + \frac{(0.64)(120)^2}{8} = 1,692 \text{ kip-ft}$$

Example 10-7. Compute the maximum and minimum stresses in all members of the six-panel through-truss highway bridge shown in Fig. 10-34 due to the passage of one lane of H20-44 loading per truss.

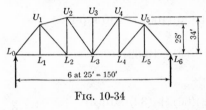

FIG. 10-34

SOLUTION. The lengths of all inclined members and the point of intersection of the prolongation of the inclined upper chord with the lower chord should first be determined as shown in Fig. 10-35.

The maximum stresses in the chord members are easily found after the maximum bending moments at panel points 1, 2, and 3 have been determined. The minimum chord stresses are, of course, zero.

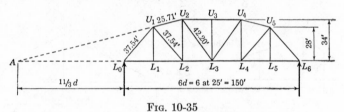

FIG. 10-35

Maximum bending moment at panel point 1 (Fig. 10-36)

$$M_1 = \frac{(8)(111) + (32)(125)}{6} = 814.7 \text{ kip-ft}$$

or $\quad M_1 = \frac{(18)(125)}{6} + \frac{(0.64)}{2}(25)(125) = 1,375 \text{ kip-ft} \text{ (controls)}$

Max stress in L_0L_1 and $L_1L_2 = \dfrac{1,375}{28} = 49.1 \text{ kips tension}$

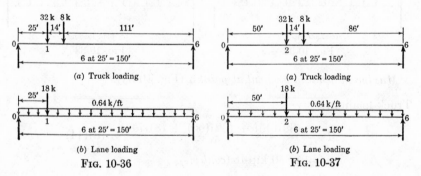

FIG. 10-36   FIG. 10-37

Maximum bending moment at panel point 2 (Fig. 10-37)

$$M_2 = \frac{(8)(86) + (32)(100)}{3} = 1{,}296 \text{ kip-ft}$$

or $$M_2 = \frac{(18)(100)}{3} + \frac{(0.64)}{2}(50)(100) = 2{,}200 \text{ kip-ft } (controls)$$

Max stress in $L_2L_3 = \dfrac{2{,}200}{34} = 64.7$ kips tension

Max stress in $U_1U_2 = \dfrac{2{,}200}{34}\left(\dfrac{25.71}{25}\right) = 66.5$ kips compression

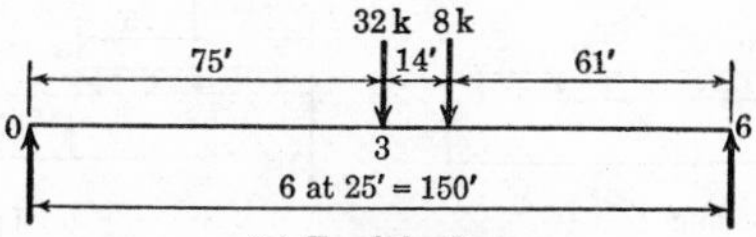

(a) Truck loading

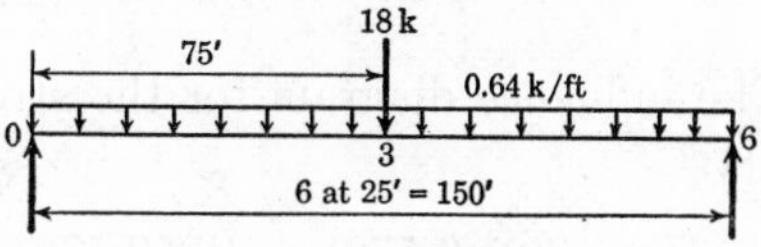

(b) Lane loading

Fig. 10-38

Maximum bending moment at panel point 3 (Fig. 10-38)

$$M_3 = \frac{(8)(61) + (32)(75)}{2} = 1{,}444 \text{ kip-ft}$$

or $$M_3 = \frac{(18)(150)}{4} + \frac{(0.64)(150)^2}{8} = 2{,}475 \text{ kip-ft } (controls)$$

Max stress in $U_2U_3 = \dfrac{2{,}475}{34} = 72.8$ kips compression

The maximum and minimum stresses in the web members can best be determined by use of the influence diagrams.

Maximum stress in L_0U_1 (Fig. 10-39)

Max stress in $L_0U_1 = 1.117[(32)(1) + (8)(^{111}\!\!\!/_{125})]$
    $= 43.7$ kips compression

or   Max stress in $L_0U_1 = (26)(1.117) + (0.64)(\frac{1}{2})(1.117)(150)$
    $= 82.7$ kips compression (*controls*)

Maximum stress in U_1L_1 (Fig. 10-40)

Max stress in $U_1L_1 = (32)(1) + (8)(^{11}\!\!\!/_{25})$
    $= 35.5$ kips tension

or   Max stress in $U_1L_1 = (26)(1) + (0.64)(\frac{1}{2})(1)(50)$
    $= 42.0$ kips tension (*controls*)

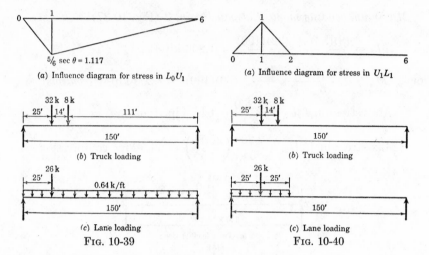

(a) Influence diagram for stress in L_0U_1

(a) Influence diagram for stress in U_1L_1

(b) Truck loading

(b) Truck loading

(c) Lane loading

(c) Lane loading

FIG. 10-39

FIG. 10-40

Member U_1L_2. The influence diagram for the stress in member U_1L_2 is shown in Fig. 10-41.

$$\text{Max stress in } U_1L_2 = (32)(0.578) + (8)(0.578)(^{86}\!/_{100})$$
$$= 22.5 \text{ kips tension}$$

or $\text{Max stress in } U_1L_2 = (26)(0.578) + (0.64)(\tfrac{1}{2})(0.578)(115.07)$
$$= 33.5 \text{ kips tension } (controls)$$

$$\text{Min stress in } U_1L_2 = (32)(0.381) + (8)(0.381)(^{11}\!/_{25})$$
$$= 13.5 \text{ kips compression}$$

or $\text{Min stress in } U_1L_2 = (26)(0.381) + (0.64)(\tfrac{1}{2})(0.381)(34.93)$
$$= 14.2 \text{ kips compression } (controls)$$

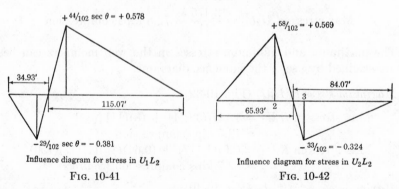

Influence diagram for stress in U_1L_2

FIG. 10-41

Influence diagram for stress in U_2L_2

FIG. 10-42

Member U_2L_2. The influence diagram for the stress in member U_2L_2 is shown in Fig. 10-42.

$$\text{Max stress in } U_2L_2 = (32)(0.569) + (8)(0.569)(^{36}\!/_{50})$$
$$= 21.5 \text{ kips tension}$$

or   Max stress in $U_2L_2 = (26)(0.569) + (0.64)(\frac{1}{2})(0.569)(65.93)$
         $= 26.8$ kips tension (*controls*)
      Min stress in $U_2L_2 = (32)(0.324) + (8)(0.324)(^{61}\!/_{75})$
         $= 12.5$ kips compression
or   Min stress in $U_2L_2 = (26)(0.324) + (0.64)(\frac{1}{2})(0.324)(84.07)$
         $= 17.1$ kips compression (*controls*)

Usually the kind of live-load stress having the same sign as the dead-load stress is called the maximum live-load stress.   In this case (member U_2L_2), the dead-load stress is tensile because the area of the influence diagram above the base line is larger than that below the base line.

Member U_2L_3.   The influence diagram for the stress in member U_2L_3 is shown in Fig. 10-43.

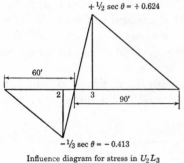

Influence diagram for stress in U_2L_3

FIG. 10-43

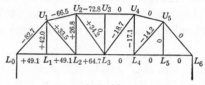

Max. and min. *LL* stresses

FIG. 10-44

      Max stress in $U_2L_3 = (32)(0.624) + (8)(0.624)(^{61}\!/_{75})$
         $= 24.0$ kips tension
or   Max stress in $U_2L_3 = (26)(0.624) + (0.624)(\frac{1}{2})(0.64)(90)$
         $= 34.2$ kips tension (*controls*)
      Min stress in $U_2L_3 = (32)(0.413) + (8)(0.413)(^{36}\!/_{50})$
         $= 15.6$ kips compression
or   Min stress in $U_2L_3 = (26)(0.413) + (0.64)(\frac{1}{2})(0.413)(60)$
         $= 18.7$ kips compression (*controls*)

The maximum and minimum stresses in all members of the truss due to the passage of one lane of H20-44 loading per truss are entered on the left and right sides, respectively, of the truss diagram in Fig. 10-44.

10-4. Live Load on Railway Bridges.   The live load to be used in the design of railway bridges is given in the "Specifications for Steel Railway Bridges," 1952 edition, of the American Railway Engineering Association.   It is stated in Art. 203 of this specification that the recommended live load for each track is the Cooper E72 load as shown in Fig. 10-45, but the engineer may specify the live load to be used, such load to be proportional to the recommended load, with the same axle spacing.

Naturally the live load to be used in the design of a railway bridge should be determined from the weights of the heaviest locomotives and train loads which may be expected to pass over the bridge during its lifetime. It would, however, be rather tedious and probably unwarranted to compare the effects of the numerous types of locomotive loadings, each with different axle loads at different spacings. In 1894, Theodore Cooper suggested the use of Cooper E40 load, a standard which was supposed to be the equivalent of the various types of locomotive loadings. The letter E means engine; 40 is the weight of the driver axle in kips; and the train load is 4 kips per lin ft. The present Cooper E72 load is $^{72}\!/_{40}$ times the original Cooper E40 load, with the same axle spacings.

In 1923, D. B. Steinman proposed the M60 loading, which is approximately equivalent to Cooper E75 for short spans and to Cooper E60 for long spans. A summary of Steinman's extensive studies may be found in Locomotive Loadings for Railway Bridges, vol. 86 of the *Transactions of the American Society of Civil Engineers.*

The discussion in this text will not deal with the choice of loading in a particular situation. The E loading will be used to illustrate the typical

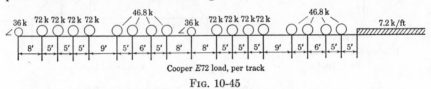

Cooper *E*72 load, per track

Fig. 10-45

method of computing maximum shears, bending moments, and stresses due to locomotive and train loading.

Before numerical examples for finding the maximum shears and bending moments in a girder, or the maximum and minimum stresses in members of a truss due to Cooper's loadings are shown, it will be advisable to look at the content of Fig. 10-46. The wheels of the first locomotive are numbered from 1 to 9, and those of the second locomotive from 10 to 18. The axle loads and their spacings are respectively shown above and below the wheel designations. The train load follows the wheel concentrations and is a uniform load of indefinite length. As typical for the other wheel concentrations, the five values listed in the column under wheel 13 are noted to be 74, 763.2, 727.2, 27,604.8, and 24,940.8. The distance from wheel 1 to wheel 13 is 74 ft; the summation of axle loads 1 through 13 is 763.2 kips; the summation of axle loads 2 through 13 is 727.2 kips; the summation of moments of axle loads 1 through 13 about wheel 13 is 27,604.8 kip-ft; and the summation of moments of axle loads 2 through 13 about wheel 13 is 24,940.8 kip-ft. The reader is advised to compute and construct Fig. 10-46 independently on a separate sheet of paper. This table will be of use in solving the problems which follow.

Spacing	36 ⟨1⟩	8'	72 ⟨2⟩	5'	72 ⟨3⟩	5'	72 ⟨4⟩	5'	72 ⟨5⟩	9'	46.8 ⟨6⟩	5'	46.8 ⟨7⟩	6'	46.8 ⟨8⟩	5'	46.8 ⟨9⟩	5'	36 ⟨10⟩	8'	46.8 ⟨11⟩	8'	72 ⟨12⟩	5'	72 ⟨13⟩	5'	72 ⟨14⟩	5'	46.8 ⟨15⟩	9'	46.8 ⟨16⟩	5'	46.8 ⟨17⟩	6'	46.8 ⟨18⟩	5'	109	7.2 k/ft
Σ distance	0		8		13		18		23		32		37		43		48		56		64		69		74		79		88		93		99		104		109	
ΣP (incl. P₁)	36		108		180		252		324		370.8		417.6		464.4		511.2		547.2		619.2		691.2		763.2		835.2		882		928.8		975.6		1,022.4		1,022.4	
ΣP (not incl. P₁)	—		72		144		216		288		334.8		381.6		428.4		475.2		511.2		583.2		655.2		727.2		799.2		846		892.8		939.6		986.4		986.4	
ΣM (incl. P₁)	0		288		828		1,728		2,988		5,904		7,758		10,263.6		12,585.6		16,675.2		21,052.8		24,148.8		27,604.8		31,420.8		38,937.6		43,347.6		48,920.4		53,798.4		58,910.4	
ΣM (not incl. P₁)	—		0		360		1,080		2,160		4,752		6,426		8,715.6		10,857.6		14,659.2		18,748.8		21,664.8		24,940.8		28,576.8		35,769.6		39,999.6		45,356.4		50,054.4		54,986.4	

Cooper E72 load, per track

Fig. 10–46

Another statement which may become quite useful later in moment computations will be made and proved here. The moment of forces P_1 to P_n, inclusive, about point B (Fig. 10-47) is equal to the moment of forces P_1 to P_n, inclusive, about point A plus the product of the sum of

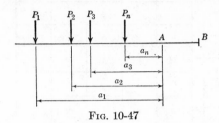

FIG. 10-47

P_1 to P_n, inclusive, and the distance AB. Expressed in a formula, the above statement becomes

$$M_B = M_A + [\Sigma(P_1 \text{ to } P_n, \text{ inclusive})](AB) \qquad (10\text{-}5)$$

Proof

$$M_A = P_1a_1 + P_2a_2 + P_3a_3 + \cdots + P_na_n$$
$$M_B = P_1(a_1 + AB) + P_2(a_2 + AB) + P_3(a_3 + AB) + \cdots$$
$$+ P_n(a_n + AB)$$
$$= P_1a_1 + P_2a_2 + P_3a_3 + \cdots + P_na_n$$
$$+ (P_1 + P_2 + P_3 + \cdots + P_n)(AB)$$
$$= M_A + [\Sigma(P_1 \text{ to } P_n, \text{ inclusive})] (AB)$$

Thus, in Fig. 10-46,

$$\Sigma[M \text{ of } 1 \text{ to } (12 \text{ or } 13) \text{ about } 13] = \Sigma(M \text{ of } 1 \text{ to } 12 \text{ about } 12)$$
$$+ [\Sigma P(1 \text{ to } 12)](5)$$
$$= 24{,}148.8 + (691.2)(5)$$
$$= 27{,}604.8 \text{ kip-ft}$$

or

$$\Sigma[M \text{ of } 2 \text{ to } (12 \text{ or } 13) \text{ about } 13] = \Sigma(M \text{ of } 2 \text{ to } 12 \text{ about } 12)$$
$$+ [\Sigma P(2 \text{ to } 12)](5)$$
$$= 21{,}664.8 + (655.2)(5)$$
$$= 24{,}940.8 \text{ kip-ft}$$

This procedure suggests a convenient way of computing the values shown in the last two lines of Fig. 10-46. Each subsequent value may be computed from the preceding one along the horizontal line.

Example 10-8. Two 72-ft plate girders support a single-track railway bridge. Compute (*a*) the maximum end shear, (*b*) the maximum shear at the quarter point, (*c*) the maximum bending moment at the quarter point, and (*d*) the maximum bending moment at the center due to the

Cooper E72 load. In each case, find the equivalent uniform load which will cause the same maximum effect.

SOLUTION

(a) *Maximum end shear* (Fig. 10-48). Compare P_1 at A with P_2 at A. When P_1 is at A, G on span (not including P_1) = 655.2 kips and Gb/L = (655.2)(8)/72 = 72.8 kips > P_1. When P_2 is at A, G on span (including P_2) = 799.2 kips, and

$$\frac{Gb}{L} = \frac{(799.2)(8)}{72} = 88.8 \text{ kips} > P_1.$$

Thus, by moving P_2 forward to A, the gain in the end reaction is between 72.8 and 88.8 kips, while the loss is equal to P_1 = 36 kips. The gain is greater than the loss; therefore P_2 at A will cause a larger end shear or reaction than P_1 at A.

Compare P_2 at A with P_3 at A. When P_2 is at A, G on span (not including P_2) = 727.2 kips and Gb/L = (727.2)(5)/72 = 50.5 kips < P_2. When P_3 is at A, G on span (including P_3) = 727.2 kips, and

$$\frac{Gb}{L} = 50.5 \text{ kips} < P_2.$$

Thus P_2 at A will cause the maximum end shear.

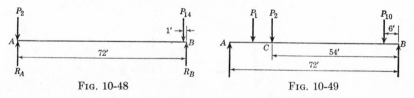

FIG. 10-48   FIG. 10-49

Since the single track is supported by two girders, only one-half of the load per track goes to each girder. Referring to Fig. 10-48,

$$\text{Max } R_A = \left(\frac{1}{2}\right)\frac{28{,}576.8 + (799.2)(1)}{72} = 204 \text{ kips}$$

Let w = equivalent uniform load per track

$$\text{Max } R_A = \frac{1}{2}(\tfrac{1}{2}w)(72) = 204$$
$$w = 11.33 \text{ kips per ft}$$

(b) *Maximum shear at the quarter point* (Fig. 10-49). Compare P_1 at C with P_2 at C. When P_1 is at C, G on span = 511.2 kips, and Gb/L = (511.2)(8)/72 = 56.8 kips > P_1. When P_2 is at C,

$$G \text{ on span} = 547.2 \text{ kips}$$

and Gb/L = (547.2)(8)/72 = 60.8 kips > P_1.

Compare P_2 at C with P_3 at C. When P_2 is at C,

$$G \text{ on span} = 547.2 \text{ kips}$$

and $Gb/L = (547.2)(5)/72 = 38.0$ kips $< P_2$. When P_3 is at C,

$$G \text{ on span} = 619.2 \text{ kips}$$

and $Gb/L = (619.2)(5)/72 = 43.0$ kips $< P_2$. Thus P_2 at C will cause the maximum shear at C. Referring to Fig. 10-49,

$$\text{Max } V_C = \left(\frac{1}{2}\right)\left[\frac{16{,}675.2 + (547.2)(6)}{72} - 36\right] = 120.6 \text{ kips}$$

Let w = equivalent uniform load per track

$$\text{Max } V_C = \frac{(\tfrac{1}{2}w)(54)^2}{(2)(72)} = 120.6$$

$$w = 11.91 \text{ kips per ft}$$

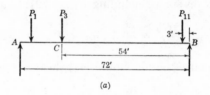

(a)

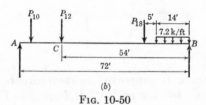

(b)

Fig. 10-50

(c) *Maximum bending moment at the quarter point* (Fig. 10-50)

Wheel at C	G_1	$\dfrac{G}{4}$	G_1	Yes or no
P_2	36	$547.2/4 = 136.8$	108	x
P_3	108	$619.2/4 = 154.8$	180	✓
P_4	180	$\begin{pmatrix} 691.2/4 = 172.8 \\ 655.2/4 = 163.8 \end{pmatrix}$	216	x
P_{11}	82.8	$622.8/4 = 155.7$	154.8	x
P_{12}	108	$612/4 = 153$	180	✓
P_{13}	180	$648/4 = 162$ $612/4 = 153$	216	x

In this case both P_3 at C and P_{12} at C satisfy the criterion for maximum bending moment at C so calculations must be made for both conditions of loading.

With P_3 at C (Fig. 10-50a),

$$M_C = \frac{1}{2}\left[\frac{21{,}052.8 + (619.2)(3)}{4} - 828\right] = 2{,}449.8 \text{ kip-ft}$$

With P_{12} at C (Fig. 10-50b),

$$M_C = \frac{1}{2}\left[\frac{12{,}585.6 + (511.2)(19) + (\frac{1}{2})(7.2)(14)^2}{4} - 828\right]$$
$$= 2{,}461.5 \text{ kip-ft } (controls)$$

Note that the moment of wheels 10 to 18 about 18 is identical with the moment of wheels 1 to 9 about 9, which is found to be 12,585.6 kip-ft in Fig. 10-46. Similarly the moment of wheels 10 to 12 about 12 is the same as the moment of wheels 1 to 3 about 3.

Thus P_{12} at C gives a larger bending moment at C than does P_3.

Let w = equivalent uniform load per track

$$\frac{1}{2}(\frac{1}{2}w)(18)(54) = 2{,}461.5$$
$$w = 10.13 \text{ kips per ft}$$

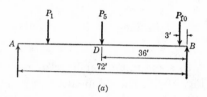

(a)

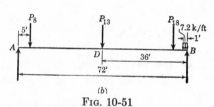

(b)

Fig. 10-51

(d) *Maximum bending moment at the center* (Fig. 10-51)

Wheel at D	G_1	$\dfrac{G}{2}$	G_1	Yes or no
P_4	180	511.2/2 = 255.6	252	x
P_5	252	547.2/2 = 273.6	324	✓
P_6	324	619.2/2 = 309.6	370.8	x
P_{12}	248.4	651.6/2 = 352.8	320.4	x
P_{13}	273.6	612/2 = 306	345.6	✓
P_{14}	345.6	$\left(\begin{array}{c} 648/2 = 324 \\ 601.2/2 = 300.6 \end{array}\right)$	370.8	x

In this case, both P_5 and P_{13} satisfy the criterion for maximum bending moment at D; consequently calculations must be made for both conditions of loading.

With P_5 at D (Fig. 10-51a),

$$M_D = \frac{1}{2}\left[\frac{16,675.2 + (547.2)(3)}{2} - 2,988\right] = 3,085.2 \text{ kip-ft}$$

With P_{13} at D (Fig. 10-51b),

Moment of P_1 to P_{18} about B = 58,910.4 + (1,022.4)(1)
= 59,932.8 kip-ft
Moment of P_1 to P_7 about B = 7,758 + (417.6)(73)
= 38,242.8 kip-ft
Moment of P_8 to P_{18} about B = 59,932.8 − 38,242.8
= 21,690 kip-ft
Moment of P_1 to P_{13} about D = 27,604.8 kip-ft
Moment of P_1 to P_7 about D = 7,758 + (417.6)(37)
= 23,209.2 kip-ft
Moment of P_8 to P_{13} about D = 27,604.8 − 23,209.2
= 4,395.6 kip-ft

In some more elaborate moment tables[1] other than the one shown in Fig. 10-46, the moment of any group of axle loads about the first and last axle load in this group may be read off directly.

$$M_D = \frac{1}{2}\left[\frac{21,690 + (\frac{1}{2})(7.2)(1)^2}{2} - 4,395.6\right] = 3,225.6 \text{ kip-ft}$$

It is seen that P_{13} causes a greater bending moment at D than does P_5. Let w = equivalent uniform load per track

$$\tfrac{1}{8}(\tfrac{1}{2}w)(72)^2 = 3,225.6$$
$$w = 9.96 \text{ kips per ft}$$

The significance of the equivalent uniform load will now be examined. In the Cooper E72 load, the uniform load which follows the two locomotives is 7.2 kips per lin ft; the average load per linear foot under each locomotive is 511.2/48 = 10.65 kips per ft, while the average load under the driver axles is $7\frac{2}{5}$ = 14.4 kips per lin ft. Thus the equivalent uniform load must, in the first place, be larger than 7.20 kips per ft; and it increases as the loaded length decreases. Generally, it is larger for shear than for bending moment.

Attention of interested readers should be called to the fact that tables or curves giving equivalent uniform loads for Cooper's loading are available.[2] Once the equivalent uniform load for a special situation is known,

---

[1] Hool and Kinne, "Stresses in Framed Structures," 2d ed., p. 106, McGraw-Hill Book Company, Inc., New York, 1942.

[2] *Ibid.*, p. 134; *Trans. ASCE*, vol. 86, p. 610, 1923.

the desired maximum shear, bending moment, or stress in a member may be readily computed.

Example 10-9.   The floor beams of a 120-ft single-track railway girder bridge are spaced 20 ft center to center as shown in Fig. 10-52.   Compute the maximum shears in panels 0-1, 1-2, and 2-3 and the maximum bending moments at points 1, 2, and 3 due to the Cooper E72 load.   In each case, find the equivalent uniform load which will give the same maximum effect.

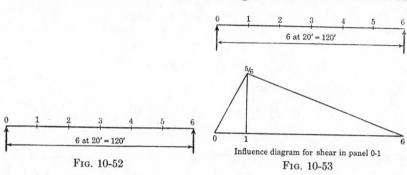

FIG. 10-52

Influence diagram for shear in panel 0-1

FIG. 10-53

SOLUTION

(a) *Maximum shear in panel* 0-1 (Fig. 10-53)

Wheel at 1	G_2	$\dfrac{G}{6}$	G_2	Yes or no
P_2	36	$1,022.4/6 = 170.4$	108	x
P_3	108	$1,051.2/6 = 175.2$	180	✓
P_4	180	$1,087.2/6 = 181.2$	252	✓
P_5	216	$1,087.2/6 = 181.2$	288	x

With P_3 at 1,

$$V_{0\text{-}1} = \frac{1}{2}\left[\frac{58,910.4 + (1,022.4)(4) + 3.6(4)^2}{120} - \frac{828}{20}\right]$$
$$= 242.04 \text{ kips } (controls)$$

With P_4 at 1,

$$V_{0\text{-}1} = \frac{1}{2}\left[\frac{58,910.4 + (1,022.4)(9) + 3.6(9)^2}{120} - \frac{1,728}{20}\right]$$
$$= 241.82 \text{ kips}$$

Let w = equivalent uniform load per track

$$(\tfrac{1}{2}w)(\tfrac{1}{2})(\tfrac{5}{6})(120) = 242.04$$
$$w = 9.68 \text{ kips per ft}$$

(b) *Maximum shear in panel* 1-2 (Fig. 10-54)

Wheel at 2	G_2	$\dfrac{G}{6}$	G_2	Yes or no
P_2	36	$\left(\begin{array}{l}835.2/6 = 139.2\\ 882/6 = 147\end{array}\right)$	108	x
P_3	108	$\left(\begin{array}{l}822/6 = 147\\ 928.8/6 = 154.8\end{array}\right)$	180	✓
P_4	180	$928.8/6 = 154.8$	252	x

$$V_{1\text{-}2} = \frac{1}{2}\left[\frac{43{,}347.6}{120} - \frac{828}{20}\right] = 159.92 \text{ kips}$$

Let w = equivalent uniform load per track

$$(\tfrac{1}{2}w)(\tfrac{1}{2})(\tfrac{4}{6})(96) = 109.92$$
$$w = 9.99 \text{ kips per ft}$$

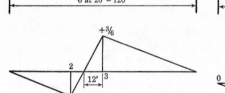

Influence diagram for shear in panel 2-3

FIG. 10-54

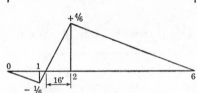

Influence diagram for shear in panel 1-2

FIG. 10-55

(c) *Maximum shear in panel* 2-3 (Fig. 10-55)

Wheel at 3	G_2	$\dfrac{G}{6}$	G_2	Yes or no
P_1	0	$547.2/6 = 91.2$	36	x
P_2	36	$619.2/6 = 103.2$	108	✓
P_3	108	$691.2/6 = 115.2$	180	✓
P_4	180	$763.2/6 = 127.2$	252	x

With P_2 at 3,

$$V_{2\text{-}3} = \frac{1}{2}\left[\frac{21{,}052.8 + (619.2)(4)}{120} - \frac{288}{20}\right] = 90.84 \text{ kips}$$

With P_3 at 3,

$$V_{2\text{-}3} = \frac{1}{2}\left[\frac{24{,}148.8 + (691.2)(4)}{120} - \frac{828}{20}\right]$$
$$= 91.44 \text{ kips } (controls)$$

Let w = equivalent uniform load per track

$$(\tfrac{1}{2}w)(\tfrac{1}{2})(\tfrac{3}{6})(72) = 91.44$$
$$w = 10.16 \text{ kips per ft}$$

(d) *Maximum bending moment at point* 1

Wheel at 1	G_1	$\dfrac{G}{6}$	G_1	Yes or no
P_2	36	$1{,}022.4/6 = 170.4$	108	x
P_3	108	$1{,}051.2/6 = 175.2$	180	✓
P_4	180	$1{,}087.2/6 = 181.2$	252	✓
P_5	216	$1{,}087.2/6 = 181.2$	288	x

With P_3 at 1,

$$M_1 = \frac{1}{2}\left[\frac{58{,}910.4 + (1{,}022.4)(4) + 3.6(4)^2}{6} - 828\right]$$
$$= 4{,}840.8 \text{ kip-ft } (controls)$$

With P_4 at 1,

$$M_1 = \frac{1}{2}\left[\frac{58{,}910.4 + (1{,}022.4)(9) + 3.6(9)^2}{6} - 1{,}728\right]$$
$$= 4{,}836.3 \text{ kip-ft}$$

Wheel at 5	G_1	$\tfrac{5}{6}G$	G_1	Yes or no
P_{13}	691.2	$(\tfrac{5}{6})(928.8) = 774.0$	763.2	x
P_{14}	763.2	$\left[\begin{array}{l}(\tfrac{5}{6})(928.8) = 774.0\\(\tfrac{5}{6})(975.6) = 813.0\end{array}\right]$	835.2	✓
P_{15}	835.2	$(\tfrac{5}{6})(1{,}022.4) = 852.0$	882	✓
P_{16}	882	$(\tfrac{5}{6})(1{,}051.2) = 876.0$	928.8	x

With P_{14} at 5,

$$M_5 = \frac{1}{2}\left[\frac{(48{,}920.4)(5)}{6} - 31{,}420.8\right] = 4{,}673.1 \text{ kip-ft}$$

With P_{15} at 5,

$$M_5 = \frac{1}{2}\left[\frac{53{,}798.4 + (1{,}022.4)(4)}{6}(5) - 38{,}937.6\right] = 4{,}651.2 \text{ kip-ft}$$

Let w = equivalent uniform load per track

$$\tfrac{1}{2}(\tfrac{1}{2}w)(20)(100) = 4{,}840.8$$
$$w = 9.68 \text{ kips per ft}$$

(e) Maximum bending moment at point 2

Wheel at 2	G_1	$\dfrac{G}{3}$	G_1	Yes or no
P_5	252	$975.6/3 = 325.2$	324	x
P_6	324	$1,044/3 = 348$	370.8	✓
P_7	370.8	$1,080/3 = 360$	417.6	x

With P_6 at 2,

$$M_2 = \frac{1}{2}\left[\frac{58,910.4 + (1,022.4)(3) + 3.6(3)^2}{3} - 5,904\right]$$
$$= 7,383 \text{ kip-ft}$$

Wheel at 4	G_1	$\frac{2}{3}G$	G_1	Yes or no
P_{11}	547.2	$\begin{array}{l}(\frac{2}{3})(975.6) = 650.4\\ (\frac{2}{3})(1,022.4) = 681.6\end{array}$	619.2	x
P_{12}	619.2	$(\frac{2}{3})(1,022.4) = 681.6$	619.2	✓
P_{13}	691.2	$(\frac{2}{3})(1,058.4) = 705.6$	763.2	✓
P_{14}	763.2	$(\frac{2}{3})(1,094.4) = 729.6$	835.2	x

With P_{12} at 4,

$$M_4 = \frac{1}{2}\left[\frac{(58,910.4)(2)}{3} - 24,148.8\right] = 7,562.4 \text{ kip-ft}$$

With P_{13} at 4,

$$M_4 = \frac{1}{2}\left[\frac{58,910.4 + (1,022.4)(5) + 3.6(5)^2}{3}(2) - 27,604.8\right]$$
$$= 7,568.4 \text{ kip-ft } (controls)$$

Let w = equivalent uniform load per track

$$\tfrac{1}{2}(\tfrac{1}{2}w)(40)(80) = 7,568.4$$
$$w = 9.46 \text{ kips per ft}$$

(f) Maximum bending moment at point 3

Wheel at 3	G_1	$\dfrac{G}{2}$	G_1	Yes or no	M_3, kip-ft
P_8	417.6	$975.6/2 = 487.8$	464.4	x	
P_9	464.4	$1,022.4/2 = 511.2$	511.2	✓	8,179.2
P_{10}	511.2	$1,072.8/2 = 536.4$	547.2	✓	8,223.2
P_{11}	511.2	$1,094.4/2 = 547.2$	583.2	✓	8,273.7
P_{12}	511.2	$1,058.4/2 = 529.2$	583.2	✓	8,224.2
P_{13}	511.2	$1,022.4/2 = 511.2$	583.2	✓	8,129.7
P_{14}	511.2	$986.4/2 = 483.2$	583.2	x	

Let w = equivalent uniform load per track

$$\tfrac{1}{8}(\tfrac{1}{2}w)(120)^2 = 8{,}273.7$$
$$w = 9.19 \text{ kips per ft}$$

Example 10-10. Compute the maximum and minimum stresses in members U_1L_1 and U_1L_2 of the six-panel through-truss single track

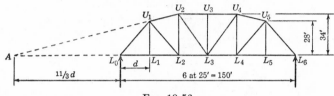

Fig. 10-56

railway bridge shown in Fig. 10-56 due to the Cooper E72 load.   In each case, find the equivalent uniform load which will cause the same maximum or minimum effect.

SOLUTION.   (a) *Maximum and minimum stresses in member* U_1L_1 (Fig. 10-57).   The maximum floor-beam reaction at L_1 will be the maximum stress in hanger U_1L_1. As shown in Chap. 8, a comparison of the influence diagram for stress in member U_1L_1 with that for bending moment at the center of a simple beam with span equal to two times the panel length (Fig. 10-57) indicates that the maximum stress in U_1L_1 is equal to $2/d$ times the maximum bending moment at the middle of the simple beam.

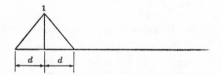

(a) Influence diagram for stress in U_1L_1

(b) Influence diagram for bending moment at center of a simple beam with span = 2d

Fig. 10-57

In the present case, the maximum bending moment at the middle of a 50-ft beam will be computed first.

Wheel at center	G_1	$\dfrac{G}{2}$	G	Yes or no
P_3	108	$417.6/2 = 208.8$	180	x
P_4	180	$\left(\begin{array}{l}417.6/2 = 208.8\\464.4/2 = 232.2\end{array}\right)$	252	✓
P_5	252	$464.4/2 = 232.2$	324	x
		$511.2/2 = 255.6$		

$$M \text{ at middle of 50-ft beam} = \frac{1}{2}\left[\frac{10{,}263.6}{2} - 1{,}728\right]$$
$$= 1{,}701.8 \text{ kip-ft}$$
$$\text{Max stress in } U_1L_1 = (1{,}701.9)(\tfrac{2}{25})$$
$$= 136.2 \text{ kips tension}$$

Let w = equivalent uniform load per track

$$(\tfrac{1}{2}w)(25) = 136.2$$
$$w = 10.90 \text{ kips per ft}$$

Min stress in $U_1L_1 = 0$

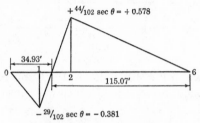

$+\,{}^{44}/_{102} \sec \theta = +\,0.578$

34.93′

0   2   115.07′   6

$-\,{}^{29}/_{102} \sec \theta = -\,0.381$

Influence diagram for stress in U_1L_2

Fig. 10-58

(b) *Maximum and minimum stresses in member* U_1L_2 (Fig. 10-58). With traffic coming on the span from the right, for maximum tensile stress in member U_1L_2, a load should be placed at L_2.

Wheel at L_2	G_2	$\dfrac{15.07}{115.07}G = 0.131G$	G_2	Yes or no
P_2	36	$(0.131)(1{,}022.4) = 133.9$	108	x
P_3	108	$(0.131)(1{,}051.2) = 137.7$	180	✓
P_4	180	$(0.131)(1{,}087.2) = 142.4$	252	x

With P_3 at L_2,

$$R_0 = \frac{1}{2}\left[\frac{58{,}910.4 + (1{,}022.4)(4) + 3.6(4)^2}{150}\right]$$
$$= 210.19 \text{ kips}$$

Panel load at $L_1 = \tfrac{1}{2}(8{\,}2{}^{8}/_{25}) = 16.56 \text{ kips}$

$$\text{Max stress in } U_1L_2 = \frac{(210.19)(11\tfrac{1}{3}) - (16.56)(14\tfrac{1}{3})}{(17\tfrac{1}{3})} \times \frac{37.54}{28}$$
$$= 164.1 \text{ kips tension}$$

Let w = equivalent uniform load per track

$$(\tfrac{1}{2}w)(\tfrac{1}{2})(0.578)(115.07) = 164.1$$
$$w = 9.87 \text{ kips per ft}$$

With traffic coming on the span from the left, for maximum compressive stress in member U_1L_2, a load should be placed at L_1.

Wheel at L_1	G_2	$\dfrac{9.93}{34.93} G = 0.284G$	G_2	Yes or no
P_1	0	$(0.284)(324) = 92.0$	36	x
P_2	36	$(0.284)(370.8) = 105.3$	108	✓
P_3	108	$(0.284)(417.6) = 118.6$	180	✓
P_4	180	$\begin{bmatrix}(0.284)(417.6) = 118.6 \\ (0.284)(464.4) = 131.9\end{bmatrix}$	252	x

With P_2 at L_1,

$$R_6 = \frac{1}{2}\left[\frac{5{,}904 + (370.8)(1)}{150}\right] = 20.92 \text{ kips}$$

Panel load at $L_2 = \frac{1}{2}(288\tfrac{2}{5}) = 5.76$ kips

Min stress in $U_1L_2 = \dfrac{(20.92)(11\tfrac{1}{3} + 6) - 5.76(11\tfrac{1}{3} + 2)}{(11\tfrac{1}{3} + 2)} \times \dfrac{37.54}{28}$

$\qquad\qquad\qquad = 40.1$ kips compression

With P_3 at L_1,

$$R_6 = \frac{1}{2}\left[\frac{7{,}758 + (417.6)(1)}{150}\right] = 27.25 \text{ kips}$$

Panel load at $L_2 = \frac{1}{2}(828\tfrac{2}{5}) = 16.56$ kips

Min stress in $U_1L_2 = \dfrac{(27.25)(11\tfrac{1}{3} + 6) - 16.56(11\tfrac{1}{3} + 2)}{(11\tfrac{1}{3} + 2)} \times \dfrac{27.54}{28}$

$\qquad\qquad\qquad = 40.1$ kips compression

Let w = equivalent uniform load per track

$$(\tfrac{1}{2}w)(\tfrac{1}{2})(0.381)(34.93) = 40.1$$
$$w = 12.05 \text{ kips per ft}$$

10-5. Impact. When highway or railway traffic passes over a bridge, the structure is subjected to not only the static but also the dynamic effect of the moving loads. The static effect has been treated in Arts. 10-3 and 10-4. The dynamic effect, usually called impact, is ordinarily considered to be proportional to the static effect. Thus the shears, bending moments, or stresses due to impact are found by multiplying those due to live load by a fraction, known as the impact factor.

The impact factor is usually expressed in terms of the loaded length of the bridge; the shorter this length, the larger the value of the factor. It should not be hard to surmise that the impact factor for railway bridges is much larger than that for highway bridges.

In the 1953 "Standard Specifications for Highway Bridges" of the American Association of State Highway Officials, the impact formula is given in Art. 3.2.12 (c) as

$$I = \frac{50}{L + 125} \qquad (10\text{-}6)$$

in which I = impact fraction (maximum 30 per cent)

L = length, ft, of the portion of the span which is loaded to produce the maximum stress in the member

For computing truck-load moments, L is the length of span; except for cantilever arms, L is the distance from the moment center to the far end of truck. For shear due to truck loads, L is the length of the loaded portion of the span from the point under consideration to the reaction; except for cantilever arms, I = 30 per cent is used.

In the 1952 "Specifications for Steel Railway Bridges" of the American Railway Engineering Association, the impact formula is given in Art. 206 as:

(a) The rolling effect: Vertical forces due to the rolling of the train from side to side, acting downward on one rail and upward on the other, the forces on each rail being equal to 10 per cent of the axle loads.

(b) The direct vertical effect: Downward forces, distributed equally to the two rails and acting normal to the top-of-rail plane, due, in the case of steam locomotives, to hammer blow, track irregularities, speed effect, and car impact, and equaling the following percentage of the axle loads:

(1) For beam spans, stringers, girders, floor beams, posts of deck truss spans carrying load from floor beam only, and floor beam hangers,

For L less than 100 ft: $\qquad I = 60 - \dfrac{L^2}{500} \qquad (10\text{-}7)$

For L 100 ft or more: $\qquad I = \dfrac{1,800}{L - 40} + 10 \qquad (10\text{-}8)$

(2) For truss spans: $\qquad I = \dfrac{4,000}{L + 25} + 15 \qquad (10\text{-}9)$

or due, in the case of rolling equipment without hammer blow (diesels, electric locomotives, tenders alone, etc.), to track irregularities, speed effect, and car impact, and equaling the following percentage of axle loads:

For L less than 80 ft: $\qquad I = 40 - \dfrac{3L^2}{1,600} \qquad (10\text{-}10)$

For L 80 ft or more: $\qquad I = \dfrac{600}{L - 30} + 16 \qquad (10\text{-}11)$

where L = length, ft, center to center of supports for stringers, transverse floor beams without stringers, longitudinal girders and trusses (main members) or L = length, ft, of the longer adjacent supported stringer, longitudinal beam, girder or truss for impact in floor beams, floor beam hangers, subdiagonals of trusses, transverse girders, supports for longitudinal and transverse girders and viaduct columns.

Example 10-11. Compute the maximum and minimum impact stresses in all members of the six-panel through-truss highway bridge shown in Fig. 10-59 due to the passage of one lane of H20-44 loading per truss.

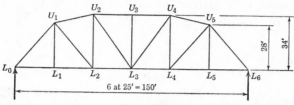

FIG. 10-59

SOLUTION. The maximum and minimum live-load stresses have been previously computed in Example 10-7. The impact formula

$$I = \frac{50}{L + 125}$$

is used to calculate the impact stresses as shown in Table 10-1.

TABLE 10-1

Member	Max stress				Min stress			
	LL stress	Loaded length L	Impact factor I	Impact stress	LL stress	Loaded length L	Impact factor I	Impact stress
L_0L_1	+49.1	150	0.182	+ 8.9	0	.....	......	0
L_1L_2	+49.1	150	0.182	+ 8.9	0	.....	......	0
L_2L_3	+64.7	150	0.182	+11.8	0	.....	......	0
U_1U_2	−66.5	150	0.182	−12.1	0	.....	......	0
U_2U_3	−72.8	150	0.182	−13.2	0	.....	......	0
L_0U_1	−82.7	150	0.182	−15.0	0	.....	......	0
U_1L_1	+42.0	50	0.286	+12.0	0	.....	......	0
U_1L_2	+33.5	115.07	0.208	+ 7.0	−14.2	34.93	0.300*	−4.3
U_2L_2	+26.8	65.93	0.262	+ 7.0	−17.1	84.07	0.239	−4.1
U_2L_3	+34.2	90	0.232	+ 7.9	−18.7	60	0.270	−5.0
U_3L_3	0	......	.....	0	0	.....	......	0

* Max 30 per cent.

Example 10-12. Determine the impact factor for the live-load shears and bending moments in the railway girder described in Example 10-9.

SOLUTION. In accordance with the 1952 AREA specifications, regardless of the loaded length of live load, the same impact factor will be used for shears and moments at all points of the girder.

The distance between center lines of rails will be assumed to be 5 ft and that of girders 8 ft.

The impact due to rolling effect is shown in Fig. 10-60a. It is seen that this may increase the live-load stress in one girder by 12.5 per cent and, at the same time, decrease the live-load stress in the other girder by a like amount. If steam locomotives are assumed to run on this bridge, the direct vertical effect (Fig. 10-60b) can be considered by using Eq. (10-8). Thus

$$I = \left(\frac{1,800}{L-40} + 10\right) \text{per cent} = \left(\frac{1,800}{120-40} + 10\right) \text{per cent}$$
$$= (22.5 + 10) \text{per cent} = 32.5 \text{per cent}$$

The total impact factor including both the rolling and direct vertical effects is (12.5 + 32.5) per cent = 45 per cent.

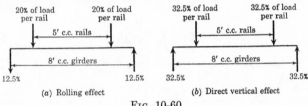

(a) Rolling effect   (b) Direct vertical effect

Fig. 10-60

10-6. Use of Counters in Trusses. The maximum or minimum combined stress in any member of a bridge truss is the sum of stresses due to dead load, live load, and impact. If the sign of the maximum combined stress is opposite to that of the minimum combined stress, the member is subjected to *stress reversal*. Members subjected to stress reversal are usually designed, depending on the governing specifications, to accommodate either stress, increased by 50 per cent of the smaller. Also, regardless of the amount of compressive stress a member has to take, the limiting value of the slenderness ratio must not be exceeded (the maximum value of the slenderness ratio is usually limited to 120 for main compression members). In case of main members subjected to a small reversal in compression, this requirement is quite severe.

The diagonals in bridge trusses, especially those in the panels near the middle of the span, are usually subjected to stress reversals. Designers sometimes provide two diagonals in such a panel so that either takes tension only. When two diagonals are provided in the same panel, the diagonal in which the dead-load stress is tensile is called the main diagonal, and the other diagonal, in which the dead-load stress is compressive, is called the counter. Thus in Fig. 10-61, members U_3L_4 and L_4U_5 are main diagonals, and members L_3U_4 and U_4L_5, if used, are counters.

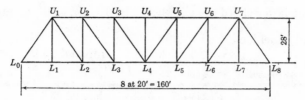

Dead load: 1200 lb/ft of truss; ⅓ at top, ⅔ at bottom
Live load: 3000 lb/ft of truss
Impact factor: 45% for all members

FIG. 10-61

Counters L_3U_4 and U_4L_5 will be needed if the minimum combined stress in members U_3L_4 and L_4U_5, found on the assumption that no counters are used, is compressive, and if compression is not allowed in any diagonal.

It will be appropriate to compare the maximum and minimum stresses in all members of a bridge truss when counters are not used with those of the same truss when counters are used. In the following example, for simplicity of illustration, a Pratt truss subjected to a moving uniform live load will be considered. Complications[1] will necessarily arise in other more elaborate forms of trusses, especially when the live load involves a system of moving concentrated loads; however, a simple example is used to afford an initial understanding of the effect of the use of counters in bridge trusses.

Example 10-13. Compute the maximum and minimum combined stresses in all members of the bridge truss shown in Fig. 10-61, (*a*) when counters are not used, and (*b*) when counters are used.

SOLUTION. Maximum and minimum combined stresses in all members will first be computed when no counters are used.

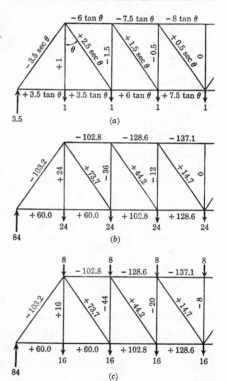

FIG. 10-62

The dead-load stresses can be found conveniently by the method outlined in Fig. 10-62 where stresses in all members due to unit panel

[1] Consult section on bridge trusses, Hool and Kinne, "Stresses in Framed Structures," 2d ed., McGraw-Hill Book Company, Inc., New York, 1942.

loads at the lower chord are shown in Fig. 10-62a. In this problem,

Dead-load panel load $= (1.2)(20)$
$= 24$ kips (8 kips at top and 16 kips at bottom)
$24 \tan \theta = (24)(2\%_{28}) = 17.14$ kips

$$24 \sec \theta = (24) \left(\frac{\sqrt{20^2 + 28^2}}{28} \right) = 29.49 \text{ kips}$$

The dead-load stresses due to panel loads of 24 kips each at the lower chord as shown in Fig. 10-62b are found by multiplying the stresses in Fig. 10-62a by 24. If one-third of the dead load is assumed to go to the top chord, then a compression of 8 kips is added to the stress in each vertical member of Fig. 10-62b to give the corrected dead-load stresses as shown in Fig. 10-62c.

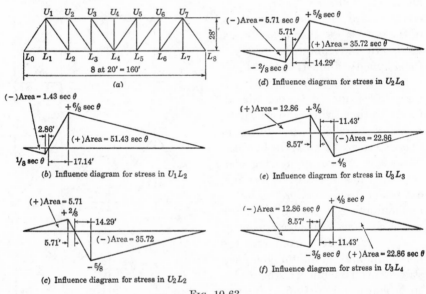

Fig. 10-63

The maximum live-load stresses in all chord members as well as those in the end posts can be found by multiplying the dead-load stresses in these members by the ratio of live load to dead load, or 2.5 in this case. The minimum live-load stresses in these members are, of course, zero. The maximum live-load stress in the hip vertical (U_1L_1) is equal to the live-load panel load, or $(3)(20) = 60$ kips tension; the minimum is zero.

The maximum and minimum live-load stress in the web members U_1L_2, U_2L_2, U_2L_3, U_3L_3, and U_3L_4 can best be found by use of the influence diagrams. Thus, if each area as indicated in Fig. 10-63 is multiplied by the intensity of live load, or 3 kips per ft, the appropriate maximum or minimum live-load stress will be obtained.

The reader is now advised to check the table of stresses (Table 10-2). From Table 10-2 it is noted that, if stress reversal in all diagonal members is to be avoided, counters L_3U_4 and U_4L_5 must be provided in panels L_3L_4 and L_4L_5. When counters are used, the lines preceded by a single asterisk in Table 10-2 are to be replaced by those preceded by double asterisks.

TABLE 10-2

Member	Dead-load stress	Live-load stress		Impact stress		Combined stress	
		Max	Min	Max	Min	Max	Min
U_1U_2	−102.8	−257.1	0	−115.7	0	−475.6	−102.8
U_2U_3	−128.6	−321.4	0	−144.6	0	−594.6	−128.6
U_3U_4	−137.1	−342.8	0	−154.3	0	−634.2	−137.1
L_0L_1	+ 60.0	+150.0	0	+ 67.5	0	+277.5	+ 60.0
L_1L_2	+ 60.0	+150.0	0	+ 67.5	0	+277.5	+ 60.0
L_2L_3	+102.8	+257.1	0	+115.7	0	+475.6	+102.8
L_3L_4	+128.6	+321.4	0	+144.6	0	+594.6	+128.6
L_0U_1	−103.2	−258.1	0	−116.1	0	−477.4	−103.2
U_1L_1	+ 16.0	+ 60.0	0	+ 27.0	0	+103.0	+ 16.0
U_1L_2	+ 73.7	+189.6	− 5.3	+ 85.3	− 2.4	+348.6	+ 66.0
U_2L_2	− 44.0	−107.2	+17.1	− 48.2	+ 7.7	−199.4	− 19.2
U_2L_3	+ 44.2	+131.7	−21.1	+ 59.3	− 9.5	+235.2	+ 13.6
*U_3L_3	− 20.0	− 68.6	+38.6	− 30.9	+17.4	−119.5	+ 36.0
*U_3L_4	+ 14.7	+ 84.3	−47.4	+ 37.9	−21.3	+136.9	− 54.0
*U_4L_4	− 8.0	0	0	0	0	− 8.0	− 8.0
**U_3L_3	− 20.0 − 8.0	− 68.6	0	− 30.9	0	−119.5	− 8.0
**U_3L_4	+ 14.7	+ 84.3	......	+ 37.9	......	+136.9	0
**U_4L_4	+ 4.0 − 8.0	− 38.6	0	− 17.4	0	− 52.0	− 8.0
**L_3U_4	− 14.7	+ 47.4	......	+ 21.3	......	+ 54.0	0

* Counters are not in action.
** Counters are in action.

Even with counters L_3U_4 and U_4L_5 provided, the maximum combined stress in the main diagonals U_3L_4 or L_4U_5 remains the same (+136.9 kips) as when counters are not used, and the minimum combined stress is zero, since the diagonals cannot carry compression. For instance, zero combined stress occurs in member U_3L_4 when the live load covers a short distance at the left end of the span so that the compression caused by the live load plus impact just balances the tension due to dead load. When the chord members U_3U_4 and L_3L_4 are parallel, as in this problem, the maximum combined stress (tension) in the counters L_3U_4 or U_4L_5 is equal to the minimum combined stress (compression) in the main diagonals U_3L_4 or L_4U_5. The minimum combined stress in the counters is, of course, zero.

The effect of the counters on the stresses in other members will be now investigated. The loading diagram when counter L_3U_4 is in action is shown in Fig. 10-64, but it should be noted that the live load does not actually come to the lower chord as uniform load. The moment center for U_3U_4 is now at L_3 and that for L_3L_4, at U_4. However, the combined stresses in the chord members due to this partial live load will not be critical when compared with those due to full live load, when the main diagonals are in action. Thus the use of counters will not affect the chord stresses at all. The maximum combined stress in U_3L_3 is -119.5 kips and occurs when main diagonal U_3L_4 is in action; and, as can be observed from Fig. 10-64, when the counter is in action, the minimum combined stress is -8 kips. When the main diagonals are in action, the stress in U_4L_4 is -8 kips, and when counter L_3U_4 is in action (Fig. 10-64), the

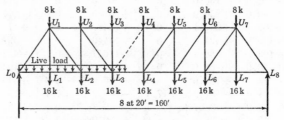

Loading diagram—counter L_3U_4 in action

FIG. 10-64

dead-load stress is $+4$ kips, the live-load stress is -38.6 kips with impact of -17.4 kips; so the combined stress is -52.0 kips.

10-7. Analysis of Bridge Portals. A sketch showing the typical construction of a single-track through-truss railway bridge is shown in Fig. 10-65. The upper and lower lateral trusses serve not only to tie the two main trusses together as a matter of practical necessity for lateral stability, but they are also subjected to lateral forces. These lateral forces include wind pressure on the vertical surface of the structure and the live load (the train, for instance), and the sway or vibratory forces in the lateral direction during the passage of the live load. The amount of these wind or lateral forces to be considered in the design is usually given in the governing specifications, such as the AREA specifications for railway bridges or the AASHO specifications for highway bridges. Generally, when the stresses due to lateral loads are combined with those due to vertical loads (dead, live, and impact) the allowable design stresses are increased by a percentage, which is also dictated by the appropriate specifications.

The lateral forces are usually treated as moving concentrated loads on the upper or lower lateral trusses. The cross diagonals may be assumed to take tension only, or if appropriately designed each diagonal may be assumed to take half the shear in the panel with one diagonal in compression and the other in tension. The lateral forces will also cause stresses in some members of the main truss, especially in the lower chord. The complete analysis of bridge trusses under the action of lateral forces will not be treated here; interested readers are referred to Hool and Kinne's "Stresses in Framed Structures," 2d ed., p. 334, McGraw-Hill Book Company, Inc., New York, 1942.

Wind forces acting on the vertical surface of the structure are usually assumed to be equally divided between the upper and lower lateral

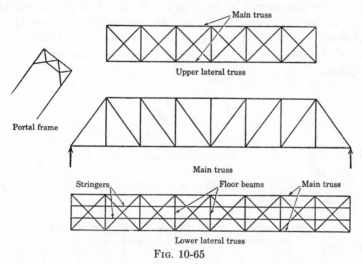

Fig. 10-65

systems. The loads on the lower lateral truss go directly into the end supports, while those on the upper lateral truss are carried by the portal frames to the supports at either end of the bridge. The stress analysis of the portal frames (such as the one shown in Fig. 10-65) will be considered in the subsequent discussion.

The analysis of a bridge portal is very similar to that of a building bent. The end posts of the bridge truss are analogous to the columns of the bent. Take, for example, a typical portal frame shown in Fig. 10-66. The load P is one-half of the total wind or other lateral forces assumed to act on the top chord of the through truss. The lower ends of the end posts are assumed to be either fixed or partially fixed because of the large compressive stresses in the end posts due to vertical loads. In the case of the combination of large wind pressure with small vertical loads (such as a wind of 50 psf on an unloaded highway bridge), the end

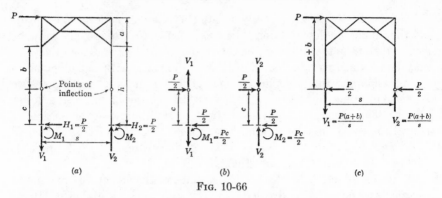

Fig. 10-66

posts may be considered to be partially fixed. The usual assumptions used in the analysis are, then, that the horizontal reactions are equal and that the points of inflection are at one-third to one-half of the clear length of the end post from the end joint of the main truss. Thus, in Fig. 10-66a,

$$H_1 = H_2 = \frac{P}{2}$$

In Fig. 10-66b:

$$M_1 = M_2 = \frac{Pc}{2}$$

In Fig. 10-66c:

$$V_1 = V_2 = \frac{P(a+b)}{s}$$

In Fig. 10-66c, the end posts are three-force members, but all others are two-force members.

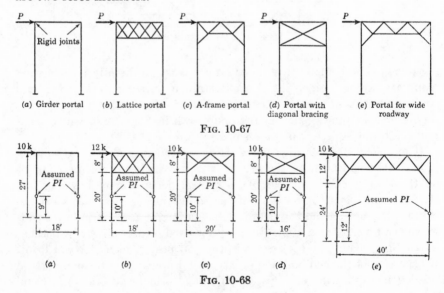

(a) Girder portal (b) Lattice portal (c) A-frame portal (d) Portal with diagonal bracing (e) Portal for wide roadway

Fig. 10-67

Fig. 10-68

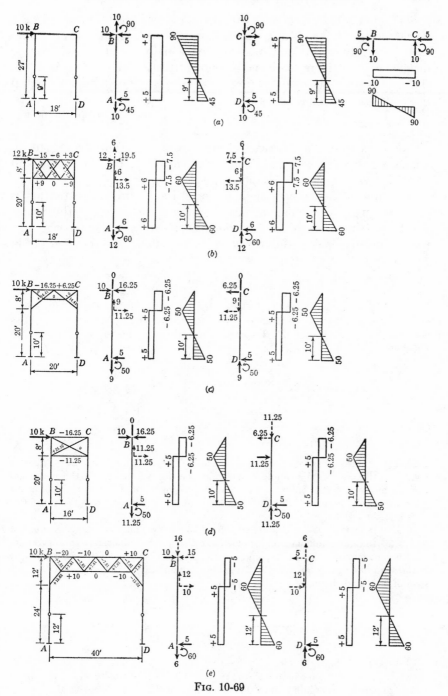

FIG. 10-69

Some common types of portal frames are shown in Fig. 10-67.   In the lattice portal of Fig. 10-67b wherein two short diagonals are present in each panel, the vertical shear may be equally divided between the diagonals, each taking an equal amount of tension or compression.   In the portal with diagonal bracing (Fig. 10-67d), it will be economical in the design to assume that the diagonals take tension only.   It is to be noted that the transverse load P may act in either direction, to the left or right.

Example 10-14.   Analyze each of the portal frames shown in Fig. 10-68. Direct stresses, shears, and bending moments in the end posts and direct stresses in all other members are required.

SOLUTION.   The analysis of these portal frames involves little that is new; so complete details will not be given.   Nevertheless, the reader will benefit by a review of the principles of statics by analyzing these frames independently and checking the results with those given in Fig. 10-69.

PROBLEMS

10-1. Estimate the dead load on a single-track 160-ft through-truss railway bridge with floor beams spaced 20 ft apart.   The truss is an eight-panel Parker truss as shown.   Assume Cooper's E60 loading.   Determine the dead-load stresses in all members.

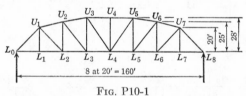

FIG. P10-1

10-2. Compute (a) the maximum end shear, (b) the maximum shear at the quarter point, (c) the maximum bending moment at the quarter point, (d) the maximum bending moment at the center, and (e) the absolute maximum bending moment in the span due to one lane of H20-44 loading on two simple beams.   One has a span of 40 ft and the other a span of 80 ft.

10-3. Compute (a) the maximum end shear, (b) the maximum shear at the quarter point, (c) the maximum bending moment at the quarter point, (d) the maximum bending moment at the center, and (e) the absolute maximum bending moment in the span due to one lane of H20-S16-44 loading on two simple beams.   One has a span of 80 ft and the other a span of 180 ft.

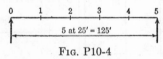

FIG. P10-4

10-4. A 125-ft highway girder bridge has floor beams spaced 25 ft on centers as shown.   Compute the maximum shears in panels 0-1, 1-2, and 2-3 and the maximum bending moments at points 1 and 2 due to one lane of H20-S16-44 loading.

10-5. Compute the maximum and minimum stresses in all members of the eight-panel through-truss highway bridge due to the passage of one lane of H20-44 loading per truss.   For dimensions of the truss, see sketch for Prob. 10-1.

10-6. Two 60-ft plate girders support a single-track railway bridge.   Compute (a) the maximum end shear, (b) the maximum shear at the quarter point, (c) the maximum bending moment at the quarter point, and (d) the maximum bending moment at the center due to the Cooper E72 load.   In each case, find the equivalent uniform load which will cause the same maximum effect.

10-7. The floor beams of a 160-ft single-track railway girder bridge are spaced 20 ft on centers as shown.   Compute the maximum shears in panels 0-1, 1-2, 2-3, and 3-4

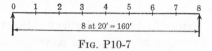

Fig. P10-7

and the maximum bending moments at points 1, 2, 3, and 4 due to the Cooper E72 load.   In each case, find the equivalent uniform load which will give the same maximum effect.

10-8. Compute the maximum and minimum stresses in members U_1L_1, U_2L_2, and U_2L_3 of the eight-panel through-truss single-track railway bridge as shown in Prob. 10-1 due to the Cooper E72 load.   In each case, find the equivalent uniform load which will cause the same maximum or minimum effect.

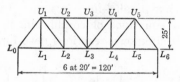

Fig. P10-9

10-9. The bridge truss shown is subjected to a dead load of 1,200 lb per foot of truss (one-third at top; two-thirds at bottom) and a live load of 3,000 lb per foot of truss with 45 per cent impact.   Compute the maximum and minimum combined stresses in all members (a) when counters are not used, and (b) when counters are used.

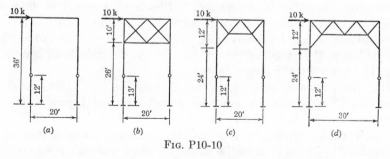

Fig. P10-10

10-10. Analyze each of the portal frames as shown.   Required are the direct stresses, shears, and bending moments in the end posts and the direct stresses in all other members.

CHAPTER 11

ANALYSIS OF STATICALLY INDETERMINATE BEAMS

11-1. Statically Determinate vs. Statically Indeterminate Beams.
A beam is a structural member subjected to transverse forces only. A
beam is completely analyzed when the variation in shear and bending
moment along its length has been found. The shear and bending-
moment diagrams may be drawn when the magnitudes of all the forces
(or couples) acting on the beam have been determined. The known forces
(or couples) are usually called the loads, and the unknown forces the
reactions. The reactions may be determined by use of the laws of
statics; viz., $\Sigma F = 0$ and $\Sigma M = 0$, which are the two conditions for
equilibrium of a coplanar-parallel-force system. Inasmuch as there are
only two independent equations of equilibrium in a coplanar-parallel-
force system, only two unknown reactions (two forces or one force and

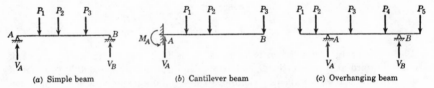

(a) Simple beam   (b) Cantilever beam   (c) Overhanging beam

FIG. 11-1. Statically Determinate Beams.

one couple) can thus be determined. Beams with only two unknown
reactions are therefore said to be statically determinate because these
two unknowns can be determined by the laws of statics alone. The
simple beam, the cantilever beam, and the overhanging beam shown in
Fig. 11-1 are statically determinate.

If a beam rests on three supports, as shown in Fig. 11-2a, the three
reactions V_A, V_B, and V_C must be first determined before shear and
bending-moment diagrams can be constructed. Since statics offers only
two independent equations, the three reactions on this beam cannot be
determined by the laws of statics alone; therefore a beam resting on three
supports is said to be statically indeterminate to the first degree because
it lacks one condition other than the two of statics to give the three
necessary equations for solving the three unknown reactions. This third

condition, which cannot be an equation of statics, is furnished by the geometry of the elastic curve of the beam.   This elastic curve is sketched in dotted form in Fig. 11-2a.   A simple beam supported at A and C and subjected to the known forces P_1 and P_2, and the unknown force V_B, is shown in Fig. 11-2b.   If the physical properties of the beam and the forces in Fig. 11-2b are identical with those in Fig. 11-2a, the elastic curves must also be identical.   The value of V_B in Fig. 11-2b may be determined from the condition that, when the simple beam AC is subjected to the loads P_1 and P_2 and an unknown force V_B, the vertical deflection of point B is equal

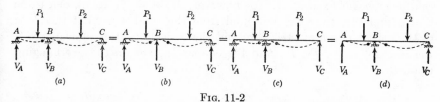

FIG. 11-2

to zero.   In other words, the downward deflection Δ_B (Fig. 11-3b) at B due to P_1 and P_2 must be equal to the upward deflection $V_B\delta_B$ (Fig. 11-3c) at B due to V_B, wherein δ_B is the upward deflection at B due to a unit upward load acting at point B of simple beam AC.   Thus $V_B\delta_B = \Delta_B$ and V_B is equal to Δ_B/δ_B.   Once V_B in Fig. 11-3a or Fig. 11-2b has thus been determined from a condition of geometry, V_A and V_C may be found from the two laws of statics.   A beam, then, may be statically indeterminate, but statically and geometrically, no beam is indeterminate.

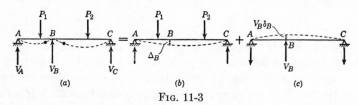

FIG. 11-3

By inspection of Fig. 11-2c, it is seen that V_C may first be determined by the geometrical condition that the vertical deflection of point C of the overhanging beam AB, subjected to the loads P_1 and P_2 and an unknown force V_C, is equal to zero.   Likewise, as shown in Fig. 11-2d, V_A may first be determined independently by applying this same condition of geometry at A.   The unknown reaction, which is selected to be first found by the condition of geometry, is said to be redundant.   If the beam has only one redundant, it is statically indeterminate to the first degree.   This is true even though any one of the three unknown reactions may be chosen as the redundant and must consequently be determined by a condition of geometry.

Additional statically indeterminate beams are shown in Fig. 11-4. The beams of Fig. 11-4a and b are statically indeterminate to the first degree; those of Fig. 11-4c and d, to the second degree; and those of Fig. 11-4e and f, to the third degree. The *degree* of indeterminacy is equal to the number of redundants, the numerical values of which can be found only from the conditions of the geometry of the elastic curve. For example, V_B and V_C in Fig. 11-4d, when chosen to be the redundants, can be found from the two conditions that the deflections at B and C must both be zero when the five forces V_B, V_C, P_1, P_2, and P_3 act on the cantilever beam fixed at A. If, however, M_A and V_A are chosen to be the redundants, they can be found from the conditions that the slope of the elastic curve at A is zero and the deflection at A is also zero if the forces V_A, P_1, P_2, P_3, and the couple M_A act on the overhanging beam BC.

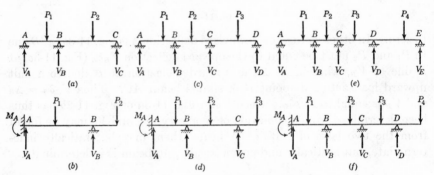

FIG. 11-4. Statically Indeterminate Beams.

Before going into the analysis of statically indeterminate beams by the method of consistent deformation as described above, it will be necessary to discuss methods of finding the vertical deflections or the slopes at various points on the elastic curve of a statically determinate beam. For instance, in Fig. 11-3b and c, Δ_B and δ_B are merely deflections in a statically determinate simple beam.

11-2. Deflections and Slopes in Statically Determinate Beams. There are generally three methods by which the geometry of the elastic curve (deflections and slopes) of statically determinate beams may be computed; viz., the double-integration method, the moment-area method, and the unit-load method. The double-integration method is usually treated in the texts on strength of materials. Since this method is not as convenient in its application as the other two, it will not be treated here. The moment-area method, with the conjugate-beam method as its special case, is perhaps the most powerful method of all. The unit-load method is classic in nature and is especially useful as a spot check

on the results found by the moment-area method.   Actually the method
to be used is a matter of individual preference.

11-3. The Moment-area Method.   A simple beam AB subjected to
two concentrated loads is shown in Fig. 11-5a.   The bending-moment
diagram for this loading is shown in Fig. 11-5b.   Points 1 and 2 at a
distance dx apart on the neutral axis of the unloaded beam will deflect
vertically downward to positions 1′ and 2′ when the beam is loaded.   An
enlarged view of the deformed beam between the sections 1′ and 2′ is

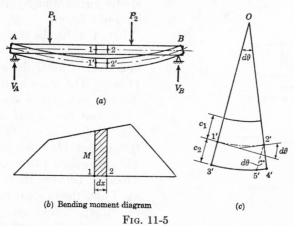

(a)

(b) Bending moment diagram   (c)

FIG. 11-5

shown in Fig. 11-5c.   It will be shown that the angle $d\theta$ between the
tangents to the elastic curve at points 1′ and 2′ is equal to the area of the
moment diagram between sections 1 and 2 divided by EI; or

$$d\theta = \frac{M \, dx}{EI} \tag{11-1}$$

Because points 1 and 2 are at an infinitesimal distance dx apart, the
bending moment may be assumed to be constant on this short segment
and the elastic curve 1′-2′ becomes a circular arc.   In Fig. 11-5c, 2′-5′
is drawn parallel to 1′-3′.   Angle 5′-2′-4′ is equal to the central angle at
O, which in turn is equal to the angle between the tangents at 1′ and 2′.
Thus

$$d\theta = \text{angle } 5'\text{-}2'\text{-}4' = \frac{\text{arc } 4'\text{-}5'}{2'\text{-}5'} = \frac{\text{arc } 4'\text{-}5'}{c_2} \tag{11-2}$$

But arc 4′-5′ is the total elongation of the lower extreme fiber, the original length of which was 3′-5′ or 1′-2′.   The unit tensile stress at the
lower extreme fiber, from the flexure formula, is Mc_2/I.   By Hooke's law,

$$\text{arc } 4'\text{-}5' = \frac{Mc_2/I}{E} (3'\text{-}5') = \frac{Mc_2}{EI} dx \tag{11-3}$$

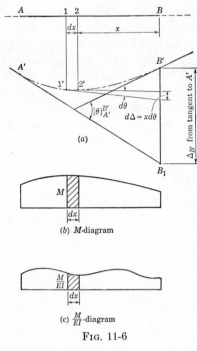

(a)

(b). M-diagram

(c) $\frac{M}{EI}$-diagram

FIG. 11-6

Substituting (11-3) in (11-2)

$$d\theta = \frac{\text{arc } 4'\text{-}5'}{c_2} = \frac{Mc_2}{EI}\frac{dx}{c_2} = \frac{M\,dx}{EI}$$

which is Eq. (11-1). It is to be noted that Eq. (11-1) is general in nature; points 1' and 2' at a distance dx apart may be on the elastic curve of any portion of any kind of a beam.

Let $A'B'$ be the elastic curve of an originally straight portion AB in a beam of indefinite length (Fig. 11-6a). The moment diagram for the segment AB is shown in Fig. 11-6b and the M/EI diagram in Fig. 11-6c. Unless I is constant throughout AB, the M/EI diagram will not be similar to the M diagram. The angle between the tangent $A'B_1$ at A and the tangent at B' will be called $[\theta]_{A'}{}^{B'}$, and the *vertical* distance $B'B_1$ of point B' from the tangent at A' will be called [$\Delta_{B'}$ from tangent at A']. It can be proved that

$$[\theta]_{A'}{}^{B'} = \frac{M}{EI} \text{ diagram between } A \text{ and } B \qquad (11\text{-}4)$$

and

[$\Delta_{B'}$ from tangent at A']

$$= \text{moment of } \frac{M}{EI} \text{ diagram between } A \text{ and } B \text{ about } B \qquad (11\text{-}5)$$

Equations (11-4) and (11-5) are the two moment-area theorems.

Moment-area Theorem 1. The angle in radians or change in slope between the tangents at any two points on a continuous elastic curve is equal to the area of the M/EI diagram between these two points.

Moment-area Theorem 2. The deflection of a second point on a continuous elastic curve, measured in a direction perpendicular to the original straight axis of the member, from the tangent at a first point on the elastic curve, is equal to the moment of the M/EI diagram between these two points about the second point.

Equations (11-4) and (11-5) will now be proved. Take points 1' and 2' at dx apart on the elastic curve $A'B'$ shown in Fig. 11-6a. Draw tangents at 1' and 2' and prolong them to intercept a distance $d\Delta$ on

$B'B_1$. Let $d\theta$ equal the angle between the tangents at 1' and 2' and x the distance from 1' or 2' to B. In Fig. 11-6a, it should be noted that the vertical dimensions are very small when compared with the horizontal distance and that the length of any curve or inclined distance may be considered to be equal to its horizontal projection. By applying Eq. (11-1) to Fig. 11-6abc, it is seen that

$$d\theta = \frac{M}{EI}\,dx$$

Integrating both sides between the limits A and B,

$$\int_A^B d\theta = \int_A^B \frac{M}{EI}\,dx$$

or $\qquad [\theta]_{A'}{}^{B'} = \dfrac{M}{EI}$ diagram between A and B $\qquad$ (11-4)

Also, from Fig. 11-6a,

$$d\Delta = x\,d\theta$$

Integrating both sides between the limits A and B

$$\int_A^B d\Delta = \int_A^B x\,d\theta = \int_A^B x\,\frac{M\,dx}{EI}$$

or

$[\Delta_{B'}$ from tangent at $A']$

$\qquad = $ moment of $\dfrac{M}{EI}$ diagram between A and B about B $\qquad$ (11-5)

The application of the two moment-area theorems to finding slopes and deflections of the elastic curves of statically determinate beams will be illustrated by the following examples.

Example 11-1. By the moment-area method find the slopes of the elastic curve at A and B and the vertical deflection at the center of the beam shown in Fig. 11-7a.

SOLUTION. Since I is constant, the M/EI diagram is similar to the M diagram and is as shown in Fig. 11-7b. Applying the first moment-area theorem,

$$\theta_A = \theta_B = [\theta]_{C'}{}^B = \text{area } A_1 = \frac{2}{3}\frac{wL^2}{8EI}\frac{L}{2} = \frac{wL^3}{24EI}$$

Thus $\qquad\qquad \theta_A = \dfrac{wL^3}{24EI}$ clockwise

$$\theta_B = \frac{wL^3}{24EI} \text{ counterclockwise}$$

Applying the second moment-area theorem,

$$\Delta_C = B_1B = \Delta_B \text{ from tangent at } C' = \text{moment of area } A_1 \text{ about } B$$

$$= \frac{wL^3}{24EI} \frac{5}{16} L = \frac{5wL^4}{384EI} \text{ downward}$$

Example 11-2.   By the moment-area method find the slope and deflection at B of the cantilever beam shown in Fig. 11-8a.

SOLUTION.   The tangent at A is observed to be horizontal.   Thus

$$\theta_B = [\theta]_A{}^{B'} = \text{area } A_1 = \frac{1}{3} \frac{wL^2}{2EI} L = \frac{wL^3}{6EI} \text{ clockwise}$$

$$\Delta_B = BB' = \Delta_{B'} \text{ from tangent at } A = \text{moment of area } A_1 \text{ about } B$$

$$= \frac{wL^3}{6EI} \frac{3}{4} L = \frac{wL^4}{8EI} \text{ downward}$$

Example 11-3.   By the moment-area method find the slope and deflection under the load in the beam shown in Fig. 11-9a.

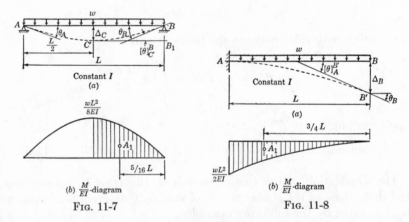

Constant I
(a)

$\frac{wL^2}{8EI}$

$5/16 L$

(b) $\frac{M}{EI}$-diagram

FIG. 11-7

Constant I

L

(a)

$3/4 L$

$\frac{wL^2}{2EI}$

(b) $\frac{M}{EI}$-diagram

FIG. 11-8

SOLUTION.   The bending-moment diagram for the beam in Fig. 11-9a is shown in Fig. 11-9b.   In applying the moment-area theorems, the M/EI diagram, not the M diagram, must be used; however, if I is constant throughout the member, the M/EI diagram is similar to the M diagram.   In such a case, it is convenient to work with the properties of the M diagram and divide the results by EI to determine the correct slope or deflection.   If I is not constant, it is advisable to choose the smallest value of I as a standard unit (I_c) and then express the other I's in terms of this unit (nI_c).   Thus the ordinates of the M diagram are divided by the values of n to give the modified M diagram.   In slope and deflection calculations, the properties of the modified M diagram can be used, except that the results should be divided by EI_c.

In this problem, $I_c = 200$ in.^{4}; $n = 1$ for portions AC and EB, and $n = 1.5$ for portion CE. The modified M diagram is shown in Fig. 11-9c. The location of the horizontal tangent is unknown in this beam, but the deflections at A and B are known to be zero.

$$BB_1 = \Delta_B \text{ from tangent at } A$$

$$= \frac{1}{EI_c} \text{ (moment of areas 1 to 6, inclusive, about } B)$$

$$= \frac{1}{EI_c} [(180)(20) + (60)(17) + (90)(16) + (270)(12) + (108)(9)$$
$$+ (108)(4)]$$

$$= 10{,}704 \frac{\text{kip-cu ft}}{EI_c}$$

$$\theta_A = \frac{BB_1}{AB} = \frac{10{,}704}{24} = 446 \frac{\text{kip-sq ft}}{EI_c} \text{ clockwise}$$

In tracing the elastic curve from A to D', the tangent rotates progressively in the counterclockwise direction an amount equal to the sum

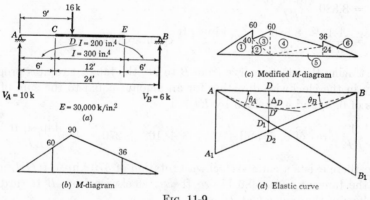

(c) Modified M-diagram

(b) M-diagram

(d) Elastic curve

Fig. 11-9

of the areas 1, 2, and 3, divided by EI_c, or

$$\frac{1}{EI_c} \text{ (area of 1, 2, and 3)} = 180 + 60 + 90 = 330 \frac{\text{kip-sq ft}}{EI_c}$$

Since there exists a clockwise rotation of 446 kip-sq ft/EI_c at A, and the tangent rotates 330 kip-sq ft/EI_c counterclockwise from A toward D', the slope of the tangent at D' is

$$\theta_D = 446 - 330 = 116 \frac{\text{kip-sq ft}}{EI_c} \text{ clockwise}$$

$$= \frac{(116)(144)}{(30{,}000)(200)} = 2.784 \times 10^{-3} \text{ radians clockwise}$$

$$\Delta_D = DD' = DD_1 - D_1D' = (AD)(\theta_A) - [\Delta_{D'} \text{ from tangent at } A]$$

$$= (AD)(\theta_A) - \frac{1}{EI_c} \text{(moment of 1, 2, and 3 about } D)$$

$$= (9)(446) - [(180)(5) + (60)(2) + (90)(1)]$$

$$= 2,904 \frac{\text{kip-cu ft}}{EI_c} \text{ downward}$$

$$= \frac{(2,904)(1,728)}{(30,000)(200)} = 0.836 \text{ in. downward}$$

For purpose of demonstration, θ_D and Δ_D will be found by tracing the elastic curve from end B toward D.

$$AA_1 = \Delta_A \text{ from tangent at } B$$

$$= \frac{1}{EI_c} \text{(moment of areas 1 to 6, inclusive, about } A)$$

$$= \frac{1}{EI_c} [(180)(4) + (60)(7) + (90)(8) + (270)(12) + (108)(15)$$
$$+ (108)(20)]$$

$$= 8,880 \frac{\text{kip-cu ft}}{EI_c}$$

$$\theta_B = \frac{AA_1}{AB} = \frac{8,880}{24} = 370 \frac{\text{kip-sq ft}}{EI_c} \text{ counterclockwise}$$

In tracing the elastic curve from B to D', the tangent rotates progressively in the clockwise direction for an amount equal to the sum of the areas of 6, 5, and 4, divided by EI_c, or

$$\frac{1}{EI_c} \text{(area of 6, 5, and 4)} = 108 + 108 + 270 = 486 \frac{\text{kip-sq ft}}{EI_c}$$

Since there exists a counterclockwise rotation of 370 kip-sq ft/EI_c at B, and the tangent turns 486 kip-sq ft/EI_c clockwise from B toward D', the slope of the tangent at D' is

$$\theta_D = 486 - 370 = 116 \frac{\text{kip-sq ft}}{EI_c} \text{ clockwise} \qquad (check)$$

$$\Delta_D = DD' = DD_2 - D_2D' = (BD)(\theta_B) - [\Delta_{D'} \text{ from tangent at } B]$$

$$= (BD)(\theta_B) - \frac{1}{EI_c} \text{(moment of 6, 5, and 4 about } D)$$

$$= (15)(370) - [(108)(11) + (108)(6) + (270)(3)]$$

$$= 2,904 \frac{\text{kip-cu ft}}{EI_c} \text{ downward} \qquad (check)$$

Example 11-4.   By the moment-area method find the slope and deflection at the free end of the overhanging beam shown in Fig. 11-10a.

SOLUTION.   The shear and bending-moment diagrams are shown in Fig. 11-10b and c. Since I is constant for the entire length of the beam, the M diagram need not be modified. For convenience the M diagram in Fig. 11-10c will be decomposed into the elements shown in Fig. 11-10d. The free-body diagram for BC (Fig. 11-10d) indicates that this segment is subjected to a uniform load and an end moment, and for each of these the M diagram is drawn separately as parts 1 and 2.

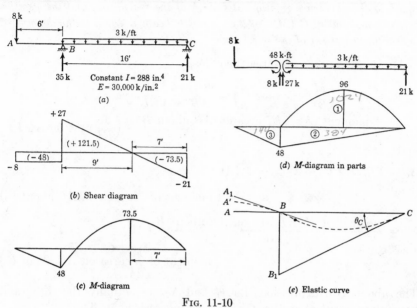

(a)

(b) Shear diagram

(c) M-diagram

(d) M-diagram in parts

(e) Elastic curve

Fig. 11-10

$BB_1 = \Delta_B$ from tangent at C

$\qquad = \dfrac{1}{EI}$ [(moment of 1) $-$ (moment of 2)]

$\qquad = \dfrac{1}{EI} [\tfrac{2}{3}(96)(16)(8) - \tfrac{1}{2}(48)(16)(\tfrac{16}{3})] = 6{,}144 \dfrac{\text{kip-cu ft}}{EI}$

$\theta_c = \dfrac{BB_1}{BC} = \dfrac{6{,}144}{16} = 384 \dfrac{\text{kip-sq ft}}{EI}$ counterclockwise

$\theta_B = \dfrac{384}{EI} - \dfrac{1}{EI} (\text{area 1}) + \dfrac{1}{EI} (\text{area 2}) = \dfrac{1}{EI} (384 - 1{,}024 + 384)$

$\qquad = -256 \dfrac{\text{kip-sq ft}}{EI}$ counterclockwise $= 256 \dfrac{\text{kip-sq ft}}{EI}$ clockwise

$\theta_A = \dfrac{256}{EI} - \dfrac{\text{area 3}}{EI} = \dfrac{1}{EI} (256 - 144) = 112 \dfrac{\text{kip-sq ft}}{EI}$ clockwise

$\qquad = \dfrac{(112)(144)}{(30{,}000)(288)} = 1.867 \times 10^{-3}$ radians clockwise

$\Delta_A = AA' = AA_1 - A_1A' = \dfrac{(256)(6)}{EI} - \dfrac{\tfrac{1}{2}(48)(6)(4)}{EI}$

$\qquad = 960 \dfrac{\text{kip-cu ft}}{EI}$ upward $= \dfrac{(960)(1{,}728)}{(30{,}000)(288)} = 0.192$ in. upward

11-4. The Conjugate-beam Method. A simple beam with some loading is shown in Fig. 11-11a. The conjugate beam for this situation, as shown in Fig. 11-11b, is this same simple beam loaded with the M/EI diagram for the actual loading on the beam as shown in Fig. 11-11a. It will be shown that the *positive shear at section C* (V'_C) *of the conjugate beam is equal to the clockwise rotation at C* (θ_C) *of the real beam, and the positive bending moment at C* (M'_C) *of the conjugate beam is equal to the downward deflection at C* (Δ_C) *of the real beam.*

Applying the moment-area theorems,

$$\theta_C = \theta_A - \left(\frac{M}{EI} \text{ area on } AC\right) = \frac{BB_1}{AB} - \left(\frac{M}{EI} \text{ area on } AC\right)$$

$$= \frac{\text{moment of } M/EI \text{ area on } AB \text{ about } B}{L} - \left(\frac{M}{EI} \text{ area on } AC\right)$$

and

$$\Delta_C = CC' = CC_1 - C_1C' = \theta_A(AC)$$

$$- \left(\text{moment of } \frac{M}{EI} \text{ area on } AC \text{ about } C\right)$$

$$= \frac{\text{moment of } M/EI \text{ area on } AB \text{ about } B}{L} (AC)$$

$$- \left(\text{moment of } \frac{M}{EI} \text{ area on } AC \text{ about } C\right)$$

The above two expressions for θ_C and Δ_C are actually the shear and bending moment at section C of the conjugate beam. Thus

$$\theta_C = V'_C \tag{11-6}$$
$$\Delta_C = M'_C \tag{11-7}$$
also
$$\theta_A = V'_A \tag{11-8}$$
$$\theta_B = V'_B \tag{11-9}$$

Although A and B in the above derivation are the two end points of a simple beam, they could have been any two points with zero deflection on a continuous elastic curve. In fact, in the general case, the conjugate-beam method may be applied between any two points with known deflections. In this event the rotation is measured from the direction of the straight line joining the two points on the elastic curve and the deflection perpendicular to the original straight axis of the member is measured from this same straight line.

Example 11-5. By the conjugate-beam method find the slope and deflection under the load in the beam shown in Fig. 11-12a.

SOLUTION. The M diagram and the modified M diagram on the basis of $I_c = 200$ in.⁴ are shown in Fig. 11-12b and c. Applying the conjugate-beam method to the elastic curve AB of Fig. 11-12c,

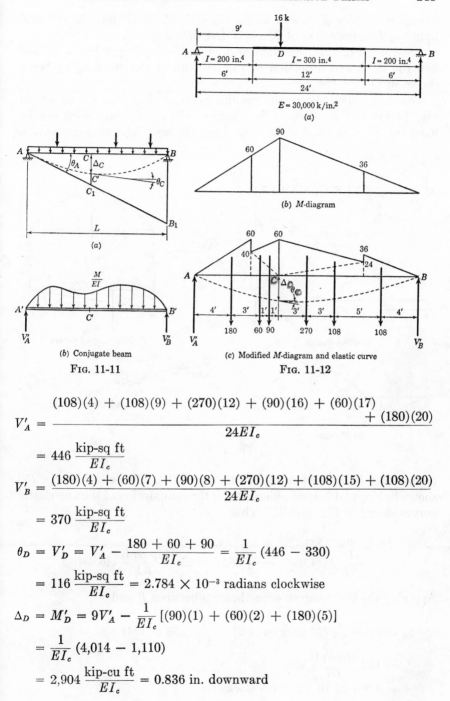

(b) M-diagram

(a)

(b) Conjugate beam

Fig. 11-11

(c) Modified M-diagram and elastic curve

Fig. 11-12

$$V'_A = \frac{(108)(4) + (108)(9) + (270)(12) + (90)(16) + (60)(17) + (180)(20)}{24EI_c}$$

$$= 446 \frac{\text{kip-sq ft}}{EI_c}$$

$$V'_B = \frac{(180)(4) + (60)(7) + (90)(8) + (270)(12) + (108)(15) + (108)(20)}{24EI_c}$$

$$= 370 \frac{\text{kip-sq ft}}{EI_c}$$

$$\theta_D = V'_D = V'_A - \frac{180 + 60 + 90}{EI_c} = \frac{1}{EI_c}(446 - 330)$$

$$= 116 \frac{\text{kip-sq ft}}{EI_c} = 2.784 \times 10^{-3} \text{ radians clockwise}$$

$$\Delta_D = M'_D = 9V'_A - \frac{1}{EI_c}[(90)(1) + (60)(2) + (180)(5)]$$

$$= \frac{1}{EI_c}(4{,}014 - 1{,}110)$$

$$= 2{,}904 \frac{\text{kip-cu ft}}{EI_c} = 0.836 \text{ in. downward}$$

The above values of θ_D and Δ_D (or V'_D and M'_D) could have been found by using DB, instead of AD, as the free body.

Example 11-6.   By the moment-area and/or conjugate-beam method find the slope and deflection at the free end of the overhanging beam shown in Fig. 11-13*a*.

SOLUTION.   The shear and bending-moment diagrams are shown in Fig. 11-13*b* and *c*.   Since I is constant, the M diagram need not be modified for slope and deflection computations.   The conjugate-beam

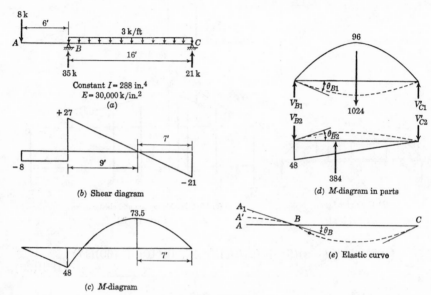

(*a*)

(*b*) Shear diagram

(*c*) *M*-diagram

(*d*) *M*-diagram in parts

(*e*) Elastic curve

FIG. 11-13

method will be applied to the elastic curve BC.   The elastic curve BC shown in Fig. 11-13*e* is considered to be the composition of the two elastic curves shown in Fig. 11-13*d*.   Thus

$$\theta_B = \theta_{B1} - \theta_{B2} = V'_{B1} - V'_{B2}$$
$$= \frac{1}{EI} [\tfrac{1}{2}(1,024) - \tfrac{2}{3}(384)] = 256 \frac{\text{kip-sq ft}}{EI} \text{ clockwise}$$

Applying the first moment-area theorem between B and A',

$$\theta_A = \theta_B - \frac{1}{EI} (M \text{ area on } AB) = \frac{1}{EI} (256 - 144)$$
$$= 112 \frac{\text{kip-sq ft}}{EI} \text{ clockwise}$$
$$= 1.867 \times 10^{-3} \text{ radians clockwise}$$

$$\Delta_A = A A' = A A_1 - A_1 A' = 6\theta_B - \frac{\frac{1}{2}(48)(6)(4)}{EI} = \frac{1,536 - 576}{EI}$$

$$= 960 \, \frac{\text{kip-cu ft}}{EI} \text{ upward}$$

$$= 0.192 \text{ in. upward}$$

A comparison of Examples 11-5 and 11-6 with Examples 11-3 and 11-4 indicates that the two solutions for each problem are really identical. The conception of the conjugate beam, when applied between two points of zero (or known) deflections, seems to give the required results more directly than the moment-area method; however, the latter deals physically with the elastic curve at almost every step, and this is sometimes an advantage.

11-5. The Unit-load Method. Let it be required to find the deflection Δ_C or the slope θ_C (Figs. 11-14b and 11-15b) at point C in a simple beam AB carrying two concentrated loads P_1 and P_2 as shown. In either

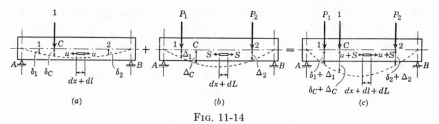

Fig. 11-14

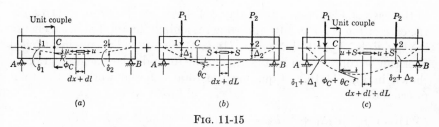

Fig. 11-15

Fig. 11-14b or 11-15b, P_1 and P_2 cause deflections Δ_1 and Δ_2 at points 1 and 2, and an internal tensile stress S in a typical fiber with a cross-sectional area dA and of an original length dx, but which is elongated by an amount dL. When P_1 and P_2 are gradually applied to the beam, by the law of conservation of energy, the total external work done on the beam must be equal to the total internal elastic energy stored in the beam, or

$$\frac{1}{2}P_1\Delta_1 + \frac{1}{2}P_2\Delta_2 = \Sigma\frac{1}{2}S \, dL \qquad (11\text{-}10)$$

A unit load or a unit couple applied at C will cause deflections δ_1 and δ_2 at points 1 and 2 and a deflection δ_C or rotation ϕ_C at point C as shown in

Figs. 11-14a or 11-15a. A typical fiber with an original length dx and a cross-sectional area dA will be subjected to an internal tensile stress u and an elongation of dl. Again, with gradual application of either the unit load or unit couple, the external work equals the internal energy. Accordingly,

$$\tfrac{1}{2}(1)(\delta_c) = \Sigma\tfrac{1}{2}u \, dl \qquad (11\text{-}11a)$$

or
$$\tfrac{1}{2}(1)(\phi_c) = \Sigma\tfrac{1}{2}u \, dl \qquad (11\text{-}11b)$$

Now if the unit load or the unit couple is first applied to the unloaded beam and the loads P_1 and P_2 are then added, the resulting condition will be as shown in Figs. 11-14c or 11-15c. The external work done on the beam due to the gradual application of the unit load or the unit couple is $\tfrac{1}{2}(1)(\delta_c)$ or $\tfrac{1}{2}(1)(\phi_c)$, while the internal energy is $\Sigma\tfrac{1}{2}u \, dl$.

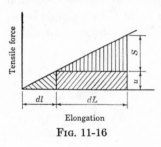

Because P_1 and P_2 go through the displacements Δ_1 and Δ_2 when the forces increase gradually from zero to P_1 and P_2, the external work done by P_1 and P_2 is $\tfrac{1}{2}P_1\Delta_1 + \tfrac{1}{2}P_2\Delta_2$. The external work done by the unit load or the unit couple, which is already on the beam, in going through the additional displacement Δ_C or rotation θ_C is $(1)(\Delta_c)$ or $(1)(\theta_c)$. The total external work is therefore

Elongation

Fig. 11-16

$$\tfrac{1}{2}(1)(\delta_c) + \tfrac{1}{2}P_1\Delta_1 + \tfrac{1}{2}P_2\Delta_2 + (1)(\Delta_c)$$

in the case of Fig. 11-14c, and $\tfrac{1}{2}(1)(\phi_c) + \tfrac{1}{2}P_1\Delta_1 + \tfrac{1}{2}P_2\Delta_2 + (1)(\theta_c)$ in the case of Fig. 11-15c. By the same reasoning, the total internal energy is $\Sigma\tfrac{1}{2}u \, dl + \Sigma\tfrac{1}{2}S \, dL + \Sigma u \, dL$. (In fact, the internal energy in a typical fiber can be represented by the area of the force-displacement diagram as shown in Fig. 11-16.) Equating the total external work to the total internal energy,

$$\tfrac{1}{2}(1)(\delta_c) + \tfrac{1}{2}P_1\Delta_1 + \tfrac{1}{2}P_2\Delta_2 + (1)(\Delta_c) = \Sigma\tfrac{1}{2}u \, dl + \Sigma\tfrac{1}{2}S \, dL + \Sigma u \, dL \qquad (11\text{-}12a)$$

or

$$\tfrac{1}{2}(1)(\phi_c) + \tfrac{1}{2}P_1\Delta_1 + \tfrac{1}{2}P_2\Delta_2 + (1)(\theta_c) = \Sigma\tfrac{1}{2}u \, dl + \Sigma\tfrac{1}{2}S \, dL + \Sigma u \, dL \qquad (11\text{-}12b)$$

Subtracting Eqs. (11-10) and (11-11) from Eq. (11-12),

$$(1)(\Delta_c) = \Sigma u \, dL \qquad (11\text{-}13a)$$

or
$$(1)(\theta_c) = \Sigma u \, dL \qquad (11\text{-}13b)$$

It must be pointed out that the value of u in Eq. (11-13a) is different from the value of u in Eq. (11-13b); one is the stress in a typical fiber due to the unit load, and the other is the stress in a typical fiber due to the unit couple.

Let M equal the bending moment in the beam due to P_1 and P_2 and m that due to the unit load or the unit couple. Then, in a typical fiber,

$$u = \frac{my}{I} \, dA$$

and

$$dL = \frac{My}{I} \frac{1}{E} \, dx$$

Substituting the above expressions for u and dL in Eq. (11-13),

$$(1)(\Delta_C) = \frac{my}{I} \, dA \, \frac{My}{EI} \, dx$$

due to unit load

$$= \int_0^L \int_0^A \frac{Mmy^2 \, dA}{EI^2} \, dx = \int_0^L \frac{Mm \, dx}{EI} \tag{11-14a}$$

or

$$(1)(\theta_C) = \int_0^L \frac{Mm \, dx}{EI} \tag{11-14b}$$

due to unit couple.

Note again that the m in Eq. (11-14a) is the bending moment due to the unit load and the m in Eq. (11-14b) is due to the unit couple.

The application of Eqs. (11-14a) and (11-14b) to the calculation of the deflections and slopes (or rotations) in statically determinate beams will be illustrated by the following examples.

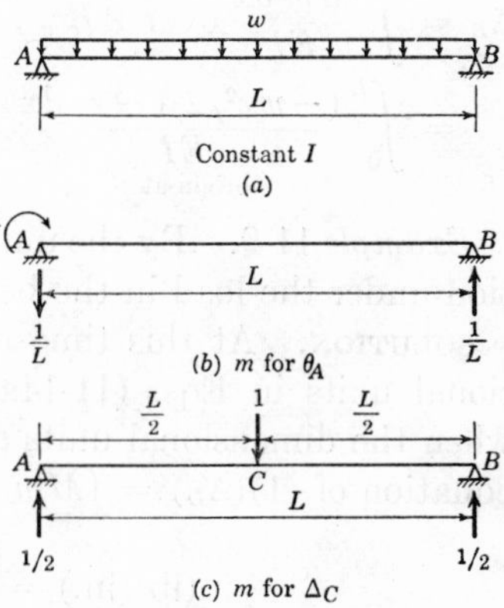

Example 11-7. Using the unit-load method, calculate the slopes of the elastic curve at A and B and the vertical deflection at the center of the beam shown in Fig. 11-17a.

SOLUTION

$$\theta_A = \int \frac{Mm \, dx}{EI} \text{ over } AB$$

$$= \int_0^L \frac{[(wL/2)x - (wx^2/2)][1 - (1/L)x] \, dx}{EI} \text{ with origin at } A$$

or

$$\theta_A = \int_0^L \frac{[(wL/2)x - (wx^2/2)][(1/L)x] \, dx}{EI} \text{ with origin at } B$$

$$= \frac{wL^3}{24EI} \text{ clockwise}$$

A positive result indicates that the direction of θ_A is the same as that of the unit couple, clockwise in this case.

Constant I
(a)

(b) m for θ_A

(c) m for Δ_C
Fig. 11-17

$$\Delta_C = \int \frac{Mm\,dx}{EI} \text{ over } AC + \int \frac{Mm\,dx}{EI} \text{ over } BC$$

$$= \underbrace{\int_0^{\frac{L}{2}} \frac{[(wL/2)x - (wx^2/2)](\tfrac{1}{2}x)\,dx}{EI}}_{\text{origin at } A} + \underbrace{\int_0^{\frac{L}{2}} \frac{[(wL/2)x - (wx^2/2)](\tfrac{1}{2}x)\,dx}{EI}}_{\text{origin at } B}$$

$$= 2 \int_0^L \frac{[(wL/2)x - (wx^2/2)](\tfrac{1}{2}x)\,dx}{EI} = \frac{5wL^4}{384EI} \text{ downward}$$

A positive result indicates that the direction of Δ_C is the same as that of the unit load, downward in this case.

Example 11-8. Use the unit-load method to calculate the slope and deflection at B of the cantilever beam shown in Fig. 11-18a.

SOLUTION

$$\theta_B = \int \frac{Mm\,dx}{EI} \text{ over } AB$$

$$= \underbrace{\int_0^L \frac{(-wx^2/2)(-1)\,dx}{EI}}_{\text{origin at } B} = \frac{wL^3}{6EI} \text{ clockwise}$$

(b) m for θ_B

$$\Delta_B = \int \frac{Mm\,dx}{EI} \text{ over } AB$$

$$= \underbrace{\int_0^L \frac{(-wx^2/2)(-x)\,dx}{EI}}_{\text{origin at } B} = \frac{wL^4}{8EI} \text{ downward}$$

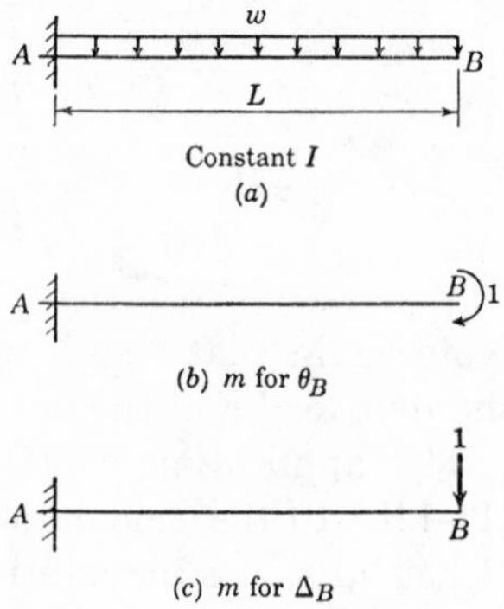

(c) m for Δ_B

FIG. 11-18

Example 11-9. By the unit-load method calculate the slope and deflection under the load in the beam shown in Fig. 11-19a.

SOLUTION. At this time it may be advisable to examine the dimensional units in Eqs. (11-14a) and (11-14b). In a numerical problem, when the dimensional units are like those in Fig. 11-19, the dimensional equation of $(1)(\Delta_D) = \int Mm\,dx/EI$ is

$$\text{(lb)(in.)} = \frac{\text{(kip-ft)(ft-lb)(ft)}}{\text{(ksi)(in.}^4)} \times 1,728$$

The dimensional equation of $(1)(\theta_D) = \int Mm\,dx/EI$ is

$$\text{(ft-lb)(radian)} = \frac{\text{(kip-ft)(ft-lb)(ft)}}{\text{(ksi)(in.}^4)} \times 144$$

Referring to Fig. 11-19,

$$\theta_D = \int \frac{Mm\,dx}{EI} \text{ on } AC + \int \frac{Mm\,dx}{EI} \text{ on } CD + \int \frac{Mm\,dx}{EI} \text{ on } DE$$
$$+ \int \frac{Mm\,dx}{EI} \text{ on } EB$$

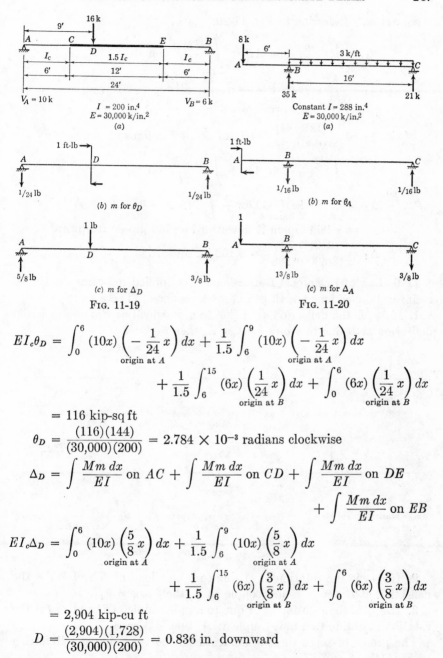

Fig. 11-19

Fig. 11-20

$$EI_c\theta_D = \int_0^6 (10x)\left(-\frac{1}{24}x\right)dx + \frac{1}{1.5}\int_6^9 (10x)\left(-\frac{1}{24}x\right)dx$$
$$\underset{\text{origin at }A}{} \qquad \underset{\text{origin at }A}{}$$

$$+ \frac{1}{1.5}\int_6^{15}(6x)\left(\frac{1}{24}x\right)dx + \int_0^6 (6x)\left(\frac{1}{24}x\right)dx$$
$$\underset{\text{origin at }B}{} \qquad \underset{\text{origin at }B}{}$$

$$= 116 \text{ kip-sq ft}$$

$$\theta_D = \frac{(116)(144)}{(30,000)(200)} = 2.784 \times 10^{-3} \text{ radians clockwise}$$

$$\Delta_D = \int \frac{Mm\,dx}{EI} \text{ on } AC + \int \frac{Mm\,dx}{EI} \text{ on } CD + \int \frac{Mm\,dx}{EI} \text{ on } DE$$

$$+ \int \frac{Mm\,dx}{EI} \text{ on } EB$$

$$EI_c\Delta_D = \int_0^6 (10x)\left(\frac{5}{8}x\right)dx + \frac{1}{1.5}\int_6^9 (10x)\left(\frac{5}{8}x\right)dx$$
$$\underset{\text{origin at }A}{} \qquad \underset{\text{origin at }A}{}$$

$$+ \frac{1}{1.5}\int_6^{15}(6x)\left(\frac{3}{8}x\right)dx + \int_0^6 (6x)\left(\frac{3}{8}x\right)dx$$
$$\underset{\text{origin at }B}{} \qquad \underset{\text{origin at }B}{}$$

$$= 2,904 \text{ kip-cu ft}$$

$$D = \frac{(2,904)(1,728)}{(30,000)(200)} = 0.836 \text{ in. downward}$$

Example 11-10.   Using the unit-load method calculate the slope and deflection at the free end of the overhanging beam shown in Fig. 11-20a.

SOLUTION.   Referring to Fig. 11-20,

$$\theta_A = \int \frac{Mm\,dx}{EI} \text{ on } AB + \int \frac{Mm\,dx}{EI} \text{ on } BC$$

$$EI\theta_A = \int_0^6 \underset{\text{origin at } A}{(-8x)(+1)\,dx} + \int_0^{16} \underset{\text{origin at } C}{(21x - \tfrac{3}{2}x^2)(\tfrac{1}{16}x)\,dx}$$
$$= 112 \text{ kip-sq ft}$$

$$\theta_A = \frac{(112)(144)}{(30,000)(288)} = 1.867 \times 10^{-3} \text{ radians clockwise}$$

$$\Delta_A = \int \frac{Mm\,dx}{EI} \text{ on } AB + \int \frac{Mm\,dx}{EI} \text{ on } BC$$

$$EI\Delta_A = \int_0^6 \underset{\text{origin at } A}{(-8x)(-x)\,dx} + \int_0^{16} \underset{\text{origin at } C}{(21x - \tfrac{3}{2}x^2)(-\tfrac{3}{8}x)\,dx}$$
$$= -960 \text{ kip-cu ft downward or } 960 \text{ kip-cu ft upward}$$

$$\Delta_A = \frac{(960)(1,728)}{(30,000)(288)} = 0.192 \text{ in. upward}$$

11-6. Law of Reciprocal Deflections.   As applied to beams the law of reciprocal deflections has three different versions.   These are:

1. If δ_{AB} is the deflection at A due to a unit load at B, and δ_{BA} is the deflection at B due to a unit load at A, then $\delta_{AB} = \delta_{BA}$.

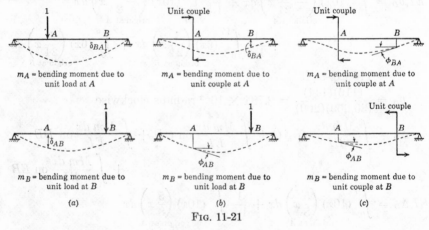

m_A = bending moment due to m_A = bending moment due to m_A = bending moment due to
       unit load at A unit couple at A unit couple at A

m_B = bending moment due to m_B = bending moment due to m_B = bending moment due to
       unit load at B unit load at B unit couple at B

(a)   (b)   (c)

FIG. 11-21

2. If ϕ_{AB} is the rotation at A due to a unit load at B and δ_{BA} is the deflection at B due to a unit couple at A, then $\phi_{AB} = \delta_{BA}$.

3. If ϕ_{AB} is the rotation at A due to a unit couple at B and ϕ_{BA} is the rotation at B due to a unit couple at A, then $\phi_{AB} = \phi_{BA}$.

The proofs of these three statements can be very simple.   Referring to Fig. 11-21a,

Let m_A = bending moment due to unit load at A
     m_B = bending moment due to unit load at B

Then
$$\delta_{AB} = \int \frac{Mm\,dx}{EI} = \int \frac{m_B m_A\,dx}{EI}$$

and
$$\delta_{BA} = \int \frac{Mm\,dx}{EI} = \int \frac{m_A m_B\,dx}{EI}$$

Therefore
$$\delta_{AB} = \delta_{BA}$$

Referring to Fig. 11-21b,

$$\delta_{BA} = \int \frac{Mm\,dx}{EI} = \int \frac{m_A m_B\,dx}{EI}$$

$$\phi_{AB} = \int \frac{Mm\,dx}{EI} = \int \frac{m_B m_A\,dx}{EI}$$

Therefore
$$\phi_{AB} = \delta_{BA}$$

Referring to Fig. 11-21c,

$$\phi_{BA} = \int \frac{Mm\,dx}{EI} = \int \frac{m_A m_B\,dx}{EI}$$

$$\phi_{AB} = \int \frac{Mm\,dx}{EI} = \int \frac{m_B m_A\,dx}{EI}$$

Therefore
$$\phi_{AB} = \phi_{BA}$$

The second statement is sometimes puzzling because one wonders how an angle in radians can be equal to a deflection in, for instance, feet. In fact, for $\phi_{AB} = \delta_{BA}$ to hold true, if the unit load at B is 1 kip, the unit couple at A must be 1 kip-ft.

Example 11-11. Referring to Fig. 11-22, verify all reciprocal relations after determining the deflection and rotation at B due to a 1-kip load or a 1-kip-ft couple at A, and the deflection and rotation at A due to a 1-kip load or a 1-kip-ft couple at B by the conjugate-beam method.

SOLUTION. Applying the conjugate-beam method to Fig. 11-22a,

$$EI\delta_{BA} = M'_B = (16)(3) - \tfrac{1}{2}(1)(3)(1) = 46.5 \text{ kip-cu ft}$$

$$\delta_{BA} = \frac{(46.5)(144)}{(30,000)(288)} = 0.775 \times 10^{-3} \text{ ft downward}$$

$$EI\phi_{BA} = V'_B = 16 - \tfrac{1}{2}(1)(3) = 14.5 \text{ kip-sq ft}$$

$$\phi_{BA} = \frac{(14.5)(144)}{(30,000)(288)} = {}^{725}\!/_3 \times 10^{-6} \text{ radians counterclockwise}$$

From Fig. 11-22b,

$$EI\delta'_{BA} = M'_B = (2)(3) - \tfrac{1}{2}(\tfrac{1}{6})(3)(1) = 5.75 \text{ kip-cu ft}$$

$$\delta'_{BA} = \frac{(5.75)(144)}{(30,000)(288)} = {}^{575}\!/_6 \times 10^{-6} \text{ ft downward}$$

$$EI\phi'_{BA} = V'_B = 2 - \tfrac{1}{2}(\tfrac{1}{6})(3) = 1.75 \text{ kip-sq ft}$$

$$\phi'_{BA} = \frac{(1.75)(144)}{(30,000)(288)} = {}^{175}\!/_6 \times 10^{-6} \text{ radians counterclockwise}$$

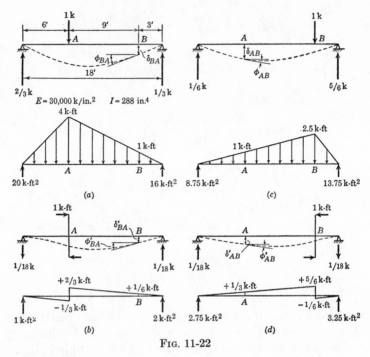

FIG. 11-22

From Fig. 11-22c,

$$EI\delta_{AB} = M'_A = (8.75)(6) - \tfrac{1}{2}(1)(6)(2) = 46.5 \text{ kip-sq ft}$$
$$\delta_{AB} = 0.775 \times 10^{-3} \text{ ft downward}$$
$$EI\phi_{AB} = V'_A = 8.75 - \tfrac{1}{2}(1)(6) = 5.75 \text{ kip-sq ft}$$
$$\phi_{AB} = {}^{575}\!\!/\!_6 \times 10^{-6} \text{ radians clockwise}$$

From Fig. 11-22d,

$$EI\delta'_{AB} = M'_A = (2.75)(6) - \tfrac{1}{2}(\tfrac{1}{3})(6)(2) = 14.5 \text{ kip-cu ft}$$
$$\delta'_{AB} = {}^{725}\!\!/\!_3 \times 10^{-6} \text{ ft downward}$$
$$EI\phi'_{AB} = V'_A = 2.75 - \tfrac{1}{2}(\tfrac{1}{3})(6) = 1.75 \text{ kip-sq ft}$$
$$\phi'_{AB} = {}^{175}\!\!/\!_6 \times 10^{-6} \text{ radians clockwise}$$

Thus the four reciprocal relations which are now verified are

$$\delta_{BA} = \delta_{AB} \qquad \delta'_{BA} = \phi_{AB} \qquad \phi_{BA} = \delta'_{AB} \quad \text{and} \quad \phi'_{BA} = \phi'_{AB}$$

11-7. Statically Indeterminate Beams with One Redundant. The analysis of statically indeterminate beams with one redundant by the method of consistent deformation will be illustrated by the following two examples.

Example 11-12. Analyze the statically indeterminate beam shown in Fig. 11-23a by the method of consistent deformation. Draw the shear and bending-moment diagrams and sketch the elastic curve.

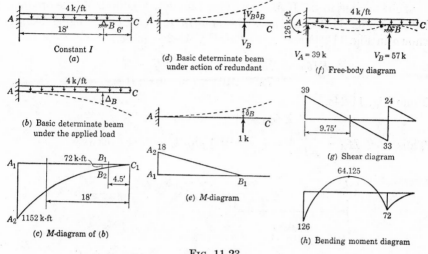

FIG. 11-23

FIRST SOLUTION. If V_B is chosen to be the redundant, the support at B is assumed to be removed and the basic determinate beam becomes a cantilever beam fixed at A only. The given beam of Fig. 11-23a is the sum of those in Fig. 11-23b and d. Equating Δ_B to $V_B \delta_B$,

$$V_B = \frac{\Delta_B}{\delta_B}$$

From Fig. 11-23c,

$$
\begin{aligned}
EI\Delta_B &= \text{moment of area } A_1 A_2 B_1 B_2 \text{ about } B \\
&= \text{moment of area } A_1 A_2 C_1 \text{ about } B \\
&\qquad - \text{moment of area } B_1 B_2 C_1 \text{ about } B \\
&= (\tfrac{1}{3})(1{,}152)(24)(12) - \tfrac{1}{3}(72)(6)(-1.5) \\
&= 110{,}592 + 216 = 110{,}808 \text{ kip-cu ft}
\end{aligned}
$$

From Fig. 11-23e,

$$
\begin{aligned}
EI\delta_B &= \text{moment of area } A_1 A_2 B_1 \text{ about } B \\
&= \tfrac{1}{2}(18)(18)(12) = 1{,}944 \text{ kip-cu ft}
\end{aligned}
$$

Thus $\qquad V_B = \dfrac{EI\Delta_B}{EI\delta_B} = \dfrac{110{,}808}{1{,}944} = 57 \text{ kips upward}$

By statics (Fig. 11-23f),

$$V_A = 96 - 57 = 39 \text{ kips upward}$$
$$M_A = (96)(12) - (57)(18) = 126 \text{ kip-ft counterclockwise}$$

The shear and bending-moment diagrams are shown in Fig. 11-23g and h, and the elastic curve is sketched in Fig. 11-23f.

SECOND SOLUTION. If M_A is assumed to be the redundant, the original beam in Fig. 11-24a becomes the sum of the two overhanging beams shown in Fig. 11-24b and d. From Fig. 11-24b and c,

$$EI\theta_A = EI\theta_{A1} - EI\theta_{A2} = V'_{A1} - V'_{A2}$$
$$= 972 - 216 = 756 \text{ kip-sq ft}$$

From Fig. 11-24e,

$$EI\phi_A = V'_A = 6 \text{ kip-sq ft}$$

Thus $\qquad M_A = \dfrac{EI\theta_A}{EI\phi_A} = \dfrac{756}{6} = 126 \text{ kip-ft counterclockwise}$

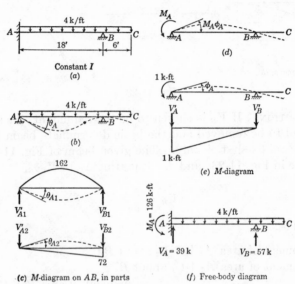

(a) Constant I

(b)

(c) M-diagram on AB, in parts

(d)

(e) M-diagram

(f) Free-body diagram

FIG. 11-24

By statics (Fig. 11-24f),

$$V_A = \frac{126 + 96(6)}{18} = 39 \text{ kips upward}$$

$$V_B = \frac{(96)(12) - 126}{18} = 57 \text{ kips upward}$$

Example 11-13. Analyze the statically indeterminate beam shown in Fig. 11-25a by the method of consistent deformation. Draw the shear and bending-moment diagrams and sketch the elastic curve.

FIRST SOLUTION. If V_B is chosen as the redundant, the given beam of Fig. 11-25a may be considered to be equivalent to the two simple beams shown in Fig. 11-25b and d. For zero deflection at B in Fig. 11-25a, Δ_B in Fig. 11-25b must be equal to $V_B\delta_B$ in Fig. 11-25d; thus

$$V_B = \frac{\Delta_B}{\delta_B}$$

Δ_B is conveniently calculated by the conjugate-beam method. Portion 4 of the bending-moment diagram in Fig. 11-25c is equal to the bending-moment diagram of a simple beam with span of 12 ft and subjected to a

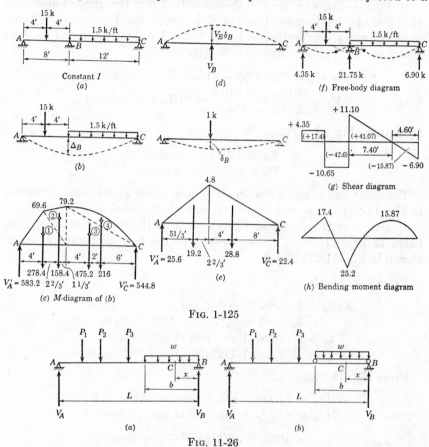

Fig. 1-125

Fig. 11-26

uniform load of 1.5 kips per ft. This can be proved by comparing the bending moments at point C of the beams shown in Fig. 11-26a and b. $(M_C$ in Fig. 11-26a$) - (M_C$ in Fig. 11-26b$)$

$$= \left(V_B x - \frac{wx^2}{2}\right) - \left[\left(V_B - \frac{wb}{2}\right)x\right] = \frac{wb}{2}x - \frac{wx^2}{2}$$

In the above equation, it is seen that the difference in the bending moments at C in Fig. 11-26a and b, which is the vertical intercept within, for instance, area of portion 4 in Fig. 11-25c, is equal to the bending moment in a simple beam with span equal to b and subjected to a uniform load. From Fig. 11-25c,

$$EI\Delta_B = M'_B = (583.2)(8) - (158.4)(\tfrac{4}{3}) - (278.4)(4)$$
$$= 3,340.8 \text{ kip-cu ft}$$

In Fig. 11-25d, δ_B is used in the sense that it is the upward deflection at B due to a 1-kip upward load at B; however, this is numerically equal to the downward deflection at B due to a 1-kip downward load at B, as shown in Fig. 11-25e. Referring to Fig. 11-25e,

$$EI\delta_B = M'_B = (25.6)(8) - (19.2)(2\tfrac{2}{3}) = 153.6 \text{ kip-cu ft}$$

Thus $\quad\quad V_B = \dfrac{EI\Delta_B}{EI\delta_B} = \dfrac{3,340.8}{153.6} = 21.75 \text{ kips upward}$

By statics (Fig. 11-25f),

$$V_A = 4.35 \text{ kips upward}$$
$$V_C = 6.90 \text{ kips upward}$$

The shear and bending-moment diagrams of the given beam are shown in Fig. 11-25g and h, and the elastic curve is sketched on Fig. 11-25f.

SECOND SOLUTION. When V_A is chosen as the redundant, the given beam of Fig. 11-27a is the equivalent of the two overhanging beams shown in Fig. 11-27b and d.

From Fig. 11-27c,

$$EI\theta_B = \tfrac{2}{3}(\text{area } 2) - \tfrac{1}{2}(\text{area } 1)$$
$$= \tfrac{2}{3}(360) - \tfrac{1}{2}(216) = 132 \text{ kip-sq ft}$$
$$EI\Delta_A = EI\theta_B(8) + \text{moment of area 3 about } A$$
$$= (132)(8) + (120)(6\tfrac{2}{3}) = 1,856 \text{ kip-cu ft}$$

From Fig. 11-27e,

$$EI\theta_B = \tfrac{2}{3}(\text{area } 4) = \tfrac{2}{3}(48) = 32 \text{ kip-sq ft}$$
$$EI\delta_A = 8EI\theta_B + \text{moment of area 5 about } A$$
$$= 8(32) + 32(5\tfrac{1}{3}) = \frac{1,280}{3} \text{ kip-cu ft}$$

Thus $\quad\quad V_A = \dfrac{EI\Delta_A}{EI\delta_A} = \dfrac{1,856}{1,280/3} = 4.35 \text{ kips upward}$

By statics,

$$V_B = 21.75 \text{ kips upward}$$
$$V_C = 6.90 \text{ kips upward}$$

11-8. Statically Indeterminate Beams with Two Redundants. The analysis of statically indeterminate beams with two redundants by the method of consistent deformation will be illustrated by the following two examples.

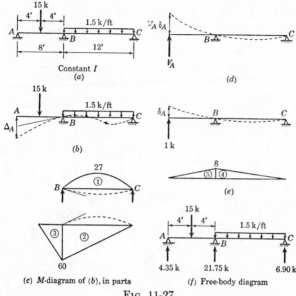

(c) *M*-diagram of (*b*), in parts   (*f*) Free-body diagram

FIG. 11-27

Example 11-14. Analyze the statically indeterminate beam shown in Fig. 11-28*a* by the method of consistent deformation. Draw the shear and bending-moment diagrams and sketch the elastic curve.

SOLUTION. The beam in Fig. 11-28*a* is statically indeterminate to the second degree, because it has four unknown reaction components M_A, V_A, V_B, and V_C, while statics provides only two independent equations of equilibrium for a coplanar-parallel-force system. When V_B and V_C are chosen as the redundants, the given beam in Fig. 11-28*a* becomes the composite of the three cantilever beams shown in Fig. 11-28*bcd*. The notation δ_{PQ} will be used to designate the deflection at P due to a unit load at Q. Thus the two equations for consistent deformation are

$$V_B\delta_{BB} + V_C\delta_{BC} = \Delta_B$$

and

$$V_B\delta_{CB} + V_C\delta_{CC} = \Delta_C$$

Values of Δ_B, Δ_C, δ_{BB}, $\delta_{BC} = \delta_{CB}$ and δ_{CC} will be found by the moment-area method. Referring to Fig. 11-28*b*,

$EI\Delta_B$ = moment of areas 1, 2, 3, and 4 about B

$= (624)(6\frac{2}{3}) + (360)(5\frac{1}{3}) + (360)(2\frac{2}{3}) + (216)(1\frac{1}{3})$

$= 7,328$ kip-cu ft

$EI\Delta_C$ = moment of areas 1, 2, 3, 4, and 5 about C

$= (624)(18\frac{2}{3}) + (360)(17\frac{1}{3}) + (360)(14\frac{2}{3}) + (216)(13\frac{1}{3})$

$+ \frac{1}{3}(108)(12)(9)$

$= 29,936$ kip-cu ft

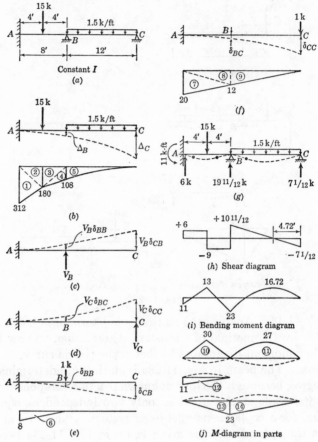

FIG. 11-28

Referring to Fig. 11-28e,

$$EI\delta_{BB} = \text{moment of area 6 about } B$$
$$= (32)(5\tfrac{1}{3}) = 170\tfrac{2}{3} \text{ kip-cu ft}$$
$$EI\delta_{CB} = \text{moment of area 6 about } C$$
$$= (32)(17\tfrac{1}{3}) = 554\tfrac{2}{3} \text{ kip-cu ft}$$

Referring to Fig. 11-28f,

$$EI\delta_{BC} = \text{moment of areas 7 and 8 about } B$$
$$= (80)(5\tfrac{1}{3}) + (48)(2\tfrac{2}{3}) = 554\tfrac{2}{3} \text{ kip-cu ft}$$
$$EI\delta_{CC} = \text{moment of areas 7, 8, and 9 about } C$$
$$= (80)(17\tfrac{1}{3}) + (48)(14\tfrac{2}{3}) + (72)(8) = 2,666\tfrac{2}{3} \text{ kip-cu ft}$$

Note that δ_{CB} and δ_{BC} have been computed independently, but, by the law of reciprocal deflections, they are naturally equal.

The two equations for consistent deformation become, numerically,

$$\frac{512}{3} V_B + \frac{1,664}{3} V_C = 7,328$$

$$\frac{1,664}{3} V_B + \frac{8,000}{3} V_C = 29,936$$

Solving,

$$V_B = 191\frac{1}{12} \text{ kips upward}$$
$$V_C = 7\frac{1}{12} \text{ kips upward}$$

By statics,

$$V_A = 33 - V_B - V_C = 6 \text{ kips upward}$$
$$M_A = (15)(4) + (18)(14) - 8V_B - 20V_C$$
$$= 11 \text{ kip-ft counterclockwise}$$

The free-body, shear, and bending-moment diagrams for the given beam are shown in Fig. 11-28ghi and the elastic curve is sketched in Fig. 11-28g.

A check on the correctness of the above solution could be made by choosing, say, V_A and M_A as the redundants. It is preferable, however, to check the consistency of the elastic curve by use of the bending-moment diagram in Fig. 11-28i. The bending-moment diagram in Fig. 11-28i may be resolved into parts as shown in Fig. 11-28j. Applying the conjugate-beam method to span AB,

$$EI\theta_A = \frac{1}{2}(\text{area } 10) - \frac{2}{3}(\text{area } 12) - \frac{1}{3}(\text{area } 13)$$
$$= 60 - \frac{2}{3}(44) - \frac{1}{3}(92) = 0$$
$$EI\theta_B = \frac{1}{2}(\text{area } 10) - \frac{1}{3}(\text{area } 12) - \frac{2}{3}(\text{area } 13)$$
$$= 60 - \frac{1}{3}(44) - \frac{2}{3}(92) = -16 \text{ or } 16 \text{ kip-sq ft clockwise}$$

Applying the conjugate-beam method to span BC,

$$EI\theta_B = \frac{1}{2}(\text{area } 11) - \frac{2}{3}(\text{area } 14) = \frac{1}{2}(216) - \frac{2}{3}(138)$$
$$= 16 \text{ kip-sq ft clockwise}$$

The fact that $\theta_A = 0$ and θ_B in span AB is equal to θ_B in span BC is assurance that two conditions of geometry have been satisfied in this beam which is statically indeterminate to the second degree. Thus the correctness of the solution is ensured.

Example 11-15. Analyze the statically indeterminate beam shown in Fig. 11-29a by the method of consistent deformation. Draw the shear and bending-moment diagrams and sketch the elastic curve.

SOLUTION. With four unknown reaction components and only two equations of statics, the given beam is statically indeterminate to the second degree. If M_A and M_B are chosen as the redundants, the given beam with fixed ends becomes the composite of the three simple beams

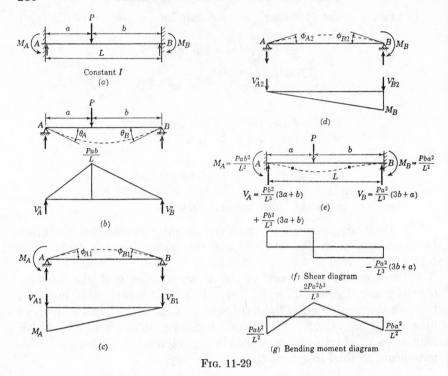

Fig. 11-29

shown in Fig. 11-29*bcd*. The two conditions for consistent deformation are

$$\theta_A = \phi_{A1} + \phi_{A2}$$
$$\theta_B = \phi_{B1} + \phi_{B2}$$

Applying the conjugate-beam method to Fig. 11-29*b*,

$$EI\theta_A = V'_A = \frac{1}{L}\left[\frac{Pab^2}{2L}\left(\frac{2}{3}b\right) + \frac{Pa^2b}{2L}\left(b + \frac{a}{3}\right)\right] = \frac{Pab}{6L}(a + 2b)$$

$$EI\theta_B = V'_B = \frac{1}{L}\left[\frac{Pab^2}{2L}\left(a + \frac{b}{3}\right) + \frac{Pa^2b}{2L}\left(\frac{2}{3}a\right)\right] = \frac{Pab}{6L}(2a + b)$$

From Fig. 11-29*c*,

$$EI\phi_{A1} = V'_{A1} = \frac{M_AL}{3}$$

$$EI\phi_{B1} = V'_{B1} = \frac{M_AL}{6}$$

From Fig. 11-29*d*,

$$EI\phi_{A2} = V'_{A2} = \frac{M_BL}{6}$$

$$EI\phi_{B2} = V'_{B2} = \frac{M_BL}{3}$$

Substituting,

$$\frac{M_A L}{3} + \frac{M_B L}{6} = \frac{Pab}{6L}(a + 2b)$$

$$\frac{M_A L}{6} + \frac{M_B L}{3} = \frac{Pab}{6L}(2a + b)$$

Solving,

$$M_A = \frac{Pab^2}{L^2} \quad \text{and} \quad M_B = \frac{Pba^2}{L^2}$$

The free-body, shear, and bending-moment diagrams, and the elastic curve are shown in Fig. 11-29efg.

11-9. Influence Diagrams of Statically Indeterminate Beams. Let it be required to construct the influence diagram for the reaction at B of the continuous beam ABC shown in Fig. 11-30a. To determine the

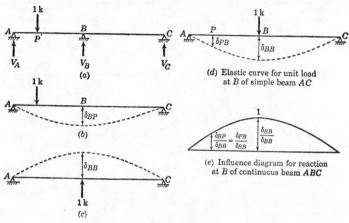

(d) Elastic curve for unit load
at B of simple beam AC

(e) Influence diagram for reaction
at B of continuous beam ABC

FIG. 11-30

influence diagram, it will be necessary to compute the values of V_B for various positions of the unit load on the beam. It would appear, then, that each ordinate of the influence diagram must be evaluated by a statically indeterminate analysis. For instance, for a unit load at point P in Fig. 11-30a, V_B is equal to δ_{BP} in Fig. 11-30b divided by δ_{BB} in Fig. 11-30c. In the influence diagram for V_B (Fig. 11-30e), this value of $V_B = \delta_{BP}/\delta_{BB}$ is the ordinate at P. However, by the law of reciprocal deflections, δ_{BP} is equal to δ_{PB}; thus the influence ordinate V_B at P is equal to δ_{PB}/δ_{BB}, in which the numerator is the deflection at P due to a unit load at B and the denominator is a constant. The elastic curve for a unit load at B as shown in Fig. 11-30d gives the values of δ_{PB} along the beam. The influence diagram for V_B is obtained by dividing all ordinates in the elastic curve by δ_{BB}.

This discussion again demonstrates the fact that influence diagrams are deflection diagrams. Thus, in this case, the influence diagram for V_B may be obtained by introducing a unit deflection at point B in the simple beam AC.

If it is required to construct the influence diagram for the reactions at B and C of the continuous beam $ABCD$ shown in Fig. 11-31a, it will be necessary to compute V_B and V_C for a unit load at any position such as P. Applying the two conditions for consistent deformation to Fig. 11-31bcd,

$$V_B \delta_{BB} + V_C \delta_{BC} = \delta_{BP}$$
$$V_B \delta_{CB} + V_C \delta_{CC} = \delta_{CP}$$

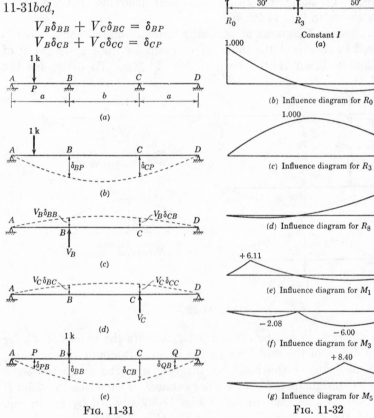

Fig. 11-31

Fig. 11-32

All the δ quantities in the above equations may be taken from the elastic curve for a unit load at B on a simple beam AD, because, by use of symmetry and the law of reciprocal deflections,

$$\delta_{BC} = \delta_{CB}$$
$$\delta_{CC} = \delta_{BB}$$
$$\delta_{BP} = \delta_{PB}$$
$$\delta_{CP} = \delta_{BQ} = \delta_{QB}$$

Note that points P and Q are symmetrical with respect to the center line of the beam. Thus, for a symmetrical three-span continuous beam, a single elastic curve as shown in Fig. 11-31e will supply all the necessary data for computing the values of V_B and V_C for various positions of the unit load.

Example 11-16. Construct the influence diagrams for R_0, R_3, R_8, M_1, M_3, and M_5 in the continuous beam shown in Fig. 11-32a.

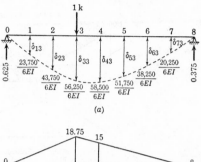

(a)

SOLUTION. Although any one of the three reactions can be made the redundant, for convenience R_3 will be chosen as the redundant. In Table 11-1 are shown the values of R_3, R_0, R_8, M_1, M_3, and M_5 due to a unit load at the successive points. The influence diagrams shown in Fig. 11-32 are plotted by using the values of the influence ordinates in Table 11-1.

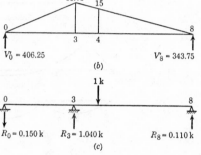

(b)

(c)

Fig. 11-33

To illustrate the procedure for making these calculations, the computations for the values of R_3, R_0, R_8, M_1, M_3, and M_5 due to a unit load at point 4 (Fig. 11-33c) will be shown. $EI\delta_{43}$ in Fig. 11-33a is equal to the bending moment at point 4 in the conjugate beam shown in Fig. 11-33b.

$$EI\delta_{43} = (343.75)(40) - \tfrac{1}{2}(15)(40)(4\tfrac{0}{3}) = 9{,}750 \text{ kip-cu ft}$$

TABLE 11-1. INFLUENCE TABLE

Load at	$R_3 = \dfrac{\delta_{P3}}{\delta_{33}}$	R_0	R_8	M_1	M_3	M_5
0	0	+1.000	0	0	0	0
1	+0.422	+0.611	−0.033	+6.11	−1.66	−1.00
2	+0.778	+0.264	−0.042	+2.64	−2.08	−1.25
3	+1.000	0	0	0	0	0
4	+1.040	−0.150	+0.110	−1.50	−4.50	+3.30
5	+0.920	−0.200	+0.280	−2.00	−6.00	+8.40
6	+0.680	−0.175	+0.495	−1.75	−5.25	+4.85
7	+0.360	−0.100	+0.740	−1.00	−3.00	+2.20
8	0	0	+1.000	0	0	0

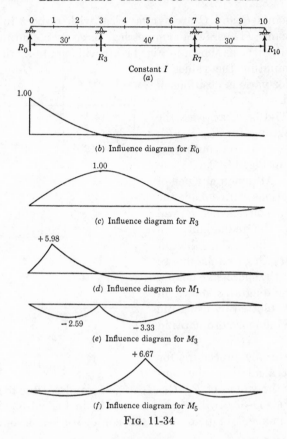

(b) Influence diagram for R_0

(c) Influence diagram for R_3

(d) Influence diagram for M_1

(e) Influence diagram for M_3

(f) Influence diagram for M_5

Fig. 11-34

Likewise, $EI\delta_{33}$ is equal to the bending moment at point 3.

$$EI\delta_{33} = (406.25)(30) - \tfrac{1}{2}(18.75)(30)(10) = 9,375 \text{ kip-cu ft}$$

Thus
$$R_3 = \frac{\delta_{43}}{\delta_{33}} = \frac{9,750}{9,375} = 1.040 \text{ kips upward}$$

By statics (Fig. 11-33c),

$$R_0 = 0.150 \text{ kip downward}$$
$$R_8 = 0.110 \text{ kip upward}$$

From the free-body diagram shown as Fig. 11-33c,

$$M_1 = -10R_0 = -1.50 \text{ kip-ft}$$
$$M_3 = -30R_0 = -4.50 \text{ kip-ft}$$
$$M_5 = +30R_8 = +3.30 \text{ kip-ft}$$

Other values in the influence table are similarly computed. The required influence diagrams are plotted in Fig. 11-32.

Example 11-17.   Construct the influence diagrams for R_0, R_3, M_1, M_3, and M_5 in the continuous beam shown in Fig. 11-34a.

SOLUTION

TABLE 11-2. INFLUENCE TABLE

Load at	R_0	R_3	M_1	M_3	M_5
0	+1.000	0	0	0	0
1	+0.598	+0.469	+5.98	−2.08	−0.74
2	+0.247	+0.836	+2.47	−2.59	−0.93
3	0	+1.000	0	0	0
4	−0.108	+0.896	−1.08	−3.25	+2.50
5	−0.111	+0.611	−1.11	−3.33	+6.67
6	−0.058	+0.271	−0.58	−1.75	+2.50
7	0	0	0	0	0
8	+0.025	−0.108	+0.25	+0.74	−0.93
9	+0.020	−0.086	+0.20	+0.59	−0.74
10	0	0	0	0	0

The values in Table 11-2 are computed, and the required influence diagrams are drawn as shown in Fig. 11-34.

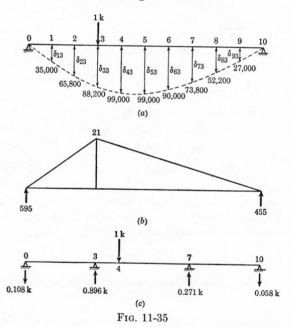

(a)

(b)

(c)

FIG. 11-35

For purpose of illustration, computations for values of R_0, R_3, M_1, M_3, and M_5 in the continuous beam due to a unit load at point 4 will be shown.   In this analysis, R_3 and R_7 are chosen as the redundants.   With a unit load at point 4, the two conditions for consistent deformation are

$$R_3\delta_{33} + R_7\delta_{37} = \delta_{34}$$
$$R_3\delta_{73} + R_7\delta_{77} = \delta_{74}$$

All δ quantities may be taken from the elastic curve of Fig. 11-35a. They have been computed from the conjugate beam shown in Fig. 11-35b. Thus

$$EI\delta_{77} = EI\delta_{33} = M_3' = (595)(30) - \tfrac{1}{2}(21)(30)(10)$$
$$= 88,200 \text{ kip-cu ft}$$
$$EI\delta_{37} = EI\delta_{73} = M_7' = (455)(30) - \tfrac{1}{2}(9)(30)(10)$$
$$= 73,800 \text{ kip-cu ft}$$
$$EI\delta_{34} = EI\delta_{43} = M_4' = (455)(60) - \tfrac{1}{2}(18)(60)(20)$$
$$= 99,000 \text{ kip-cu ft}$$
$$EI\delta_{74} = EI\delta_{36} = EI\delta_{63} = M_6' = (455)(40) - \tfrac{1}{2}(12)(40)(4\%)$$
$$= 90,000 \text{ kip-cu ft}$$

Substituting and solving,

$$R_3 = 0.8958 \text{ kip upward}$$
$$R_7 = 0.2708 \text{ kip upward}$$

By statics (Fig. 11-35c),

$$R_0 = 0.1083 \text{ kip downward}$$
$$R_{10} = 0.0583 \text{ kip downward}$$

From Fig. 11-35c,

$$M_1 = -10R_0 = -1.083 \text{ kip-ft}$$
$$M_3 = -30R_0 = -3.249 \text{ kip-ft}$$
$$M_5 = -50R_{10} + 20R_7 = +2.501 \text{ kip-ft}$$

The influence diagrams as shown in Fig. 11-34 will clearly indicate the spans which should be loaded with uniform live load for maximum effect. For instance, for maximum negative moment at point 3, the two adjacent spans should be loaded but the third span should remain unloaded.

PROBLEMS

11-1. By the moment-area method calculate the slopes of the elastic curve at A, B, and C and the vertical deflection at C of the simple beam.

11-2. By the moment-area method find the slope and deflection at the free end B of the cantilever beam AB.

11-3. By the moment-area method find the slope and deflection at the free end A of the cantilever beam AB.

11-4. By the moment-area method find the slopes at ends A and B and the vertical deflection at the center of the simple beam.

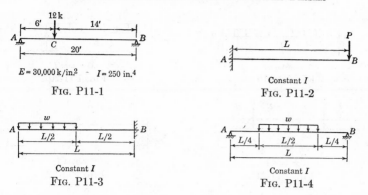

$E = 30,000\,\text{k/in.}^2 \quad I = 250\,\text{in.}^4$

FIG. P11-1

Constant I

FIG. P11-2

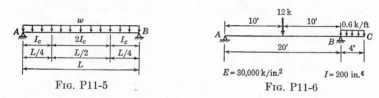

Constant I

FIG. P11-3

Constant I

FIG. P11-4

11-5. By the moment-area method find the slopes at ends A and B and the vertical deflection at the center of the simple beam.

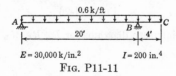

FIG. P11-5

$E = 30,000\,\text{k/in.}^2 \qquad I = 200\,\text{in.}^4$

FIG. P11-6

11-6. By the moment-area method calculate the slope and deflection at the free end C of the overhanging beam ABC.

11-7. Solve Prob. 11-1 by the conjugate-beam method. Determine also the location and amount of the maximum deflection.

11-8. Solve Prob. 11-4 by the conjugate-beam method.

11-9. Solve Prob. 11-5 by the conjugate-beam method.

11-10. Solve Prob. 11-6 by the moment-area and/or conjugate-beam method.

$E = 30,000\,\text{k/in.}^2 \qquad I = 200\,\text{in.}^4$

FIG. P11-11

11-11. By the moment-area and/or conjugate-beam method calculate the slope and deflection at the free end C of the overhanging beam ABC.

11-12. Solve Prob. 11-1 by the unit-load method.

11-13. Solve Prob. 11-2 by the unit-load method.

11-14. Solve Prob. 11-3 by the unit-load method.

11-15. Solve Prob. 11-4 by the unit-load method.

11-16. Solve Prob. 11-5 by the unit-load method.

11-17. Solve Prob. 11-6 by the unit-load method.

11-18. Solve Prob. 11-11 by the unit-load method.

11-19 to 11-24. Analyze the statically indeterminate beams shown by the method of consistent deformation. Draw the shear and bending-moment diagrams and sketch the elastic curve.

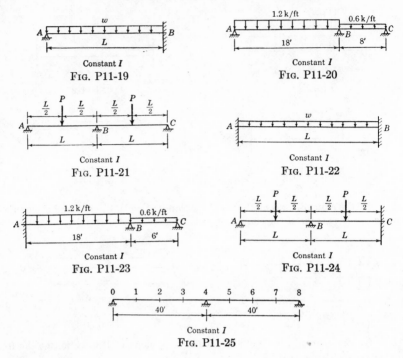

Constant *I*

FIG. P11-19

Constant *I*

FIG. P11-20

Constant *I*

FIG. P11-21

Constant *I*

FIG. P11-22

Constant *I*

FIG. P11-23

Constant *I*

FIG. P11-24

Constant *I*

FIG. P11-25

11-25. Construct the influence diagrams for R_0, R_4, M_2, and M_4 in the continuous beam shown.

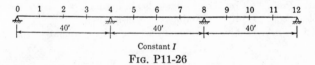

Constant *I*

FIG. P11-26

11-26. Construct the influence diagrams for R_0, R_4, M_2, M_4, and M_6 in the continuous beam shown.

CHAPTER 12

ANALYSIS OF STATICALLY INDETERMINATE
RIGID FRAMES

12-1. Statically Determinate vs. Statically Indeterminate Rigid Frames. A rigid frame has been defined as a structure in which the members are joined together by rigid connections, such as welded steel joints or monolithic connections in reinforced concrete. In contradistinction to a hinged joint, members are not free to rotate around a rigid joint when subjected to geometric deformation, or more specifically, the angle between the tangents to the elastic curves of any two adjacent members must remain constant. When deformations of rigid frames are evaluated quantitatively, this requirement is of paramount importance.

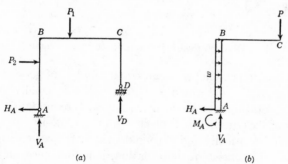

Fig. 12-1. Statically Determinate Rigid Frames.

Rigid frames may be statically determinate or statically indeterminate; in the case of single-story rigid frames, the degree of indeterminacy is equal to the number of unknown reaction components in excess of the three unknowns which are dependent on the three conditions of equilibrium for a general coplanar-force system. For instance, the two rigid frames shown in Fig. 12-1 are statically determinate because in either frame there are only three unknown reaction components, H_A, V_A, and V_D in Fig. 12-1a and M_A, H_A, and V_A in Fig. 12-1b. However, the rigid frames shown in Fig. 12-2 are statically indeterminate, the degree of

267

indeterminacy being equal to the number of unknown reaction components minus 3. Thus the rigid frame of Fig. 12-2a is statically indeterminate to the first degree; of Fig. 12-2b, to the second degree; of Fig. 12-2c, to the third degree; and of Fig. 12-2d, to the fifth degree.

As in the case of statically indeterminate beams, the basic method of analyzing statically indeterminate rigid frames is the method of consistent deformation. Other more convenient methods, such as the slope-deflection or the moment-distribution methods, may be used to analyze both statically indeterminate beams and rigid frames. The two last-mentioned methods will be treated in Chaps. 14 and 15. In the method of consistent deformation, a basic determinate structure is first derived from the original indeterminate structure by removing the redundant reaction components and treating them as unknown forces or moments acting on the basic structure. Then these unknown redundant forces

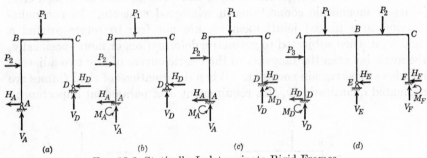

Fig. 12-2. Statically Indeterminate Rigid Frames.

or moments are determined from an equal number of geometrical conditions relating to the deformed structure. These usually require zero deflection or rotation at the location of each redundant. Numerical applications of this method will be shown in Art. 12-4.

In order that statically indeterminate rigid frames may be analyzed by the method of consistent deformation, it is first necessary to discuss the methods of finding deflections and rotations in a deformed statically determinate rigid frame. The two methods to be treated here are the unit-load method and the moment-area (including the conjugate-beam) method.

12-2. Deflections of Statically Determinate Rigid Frames: the Moment-area Method. In the preceding chapter the moment-area (including conjugate-beam) method was used to determine the deflections and slopes at various points in statically determinate beams. The moment-area or conjugate-beam theorems may be applied to each straight member in a rigid frame by treating the member as a beam or a member subjected to bending only. Actually the members in a rigid frame are subjected to combined bending and direct stress and must be treated as such in the design of the cross sections.   In general the axial deformation in the member due to the direct stress is almost always so small in comparison with the transverse deflection due to bending that direct stress deformation may be neglected when considering the geometric distortion of the whole rigid frame.   In accordance with this fundamental assumption, the lengths of all members in a rigid frame are assumed to remain unchanged during distortion.   Thus the moment-area method may be used to determine the transverse deflections of any member in a rigid frame.

Since any application of the moment-area or the conjugate-beam theorems involves the computation for some geometric property of the elastic curve, it is most desirable to indicate the result thus obtained on a sketch of the elastic curve.   To determine deflections or rotations in statically determinate rigid frames, the following steps are necessary:

1. Compute reaction components.   Draw a free body and the bending-moment diagram for each member.

2. Draw the M diagram on the rigid frame.   A modified M diagram is required if the members have different moments of inertia.

3. By visualization determine the most probable new locations of the joints after deformation.   Sketch the elastic curve.

4. Apply the moment-area or conjugate-beam theorems to evaluate any desired quantity on the elastic curve.

The above procedure will be used in the solutions of the following examples.

Example 12-1.   By the moment-area method determine the rotation, horizontal deflection, and vertical deflection of each joint in the rigid frame shown in Fig. 12-3a.

SOLUTION.   The reaction components H_A, V_A, and V_D are first computed by applying the equations of statics to Fig. 12-3a.

$$\Sigma M_A = 0: \qquad (10)(12) + (72)(12) = 24V_D$$
$$V_D = 41 \text{ kips}$$
$$\Sigma F_x = 0: \qquad H_A = 10 \text{ kips}$$
$$\Sigma F_y = 0: \qquad V_A = 72 - 41 = 31 \text{ kips}$$

CHECK.   By $\Sigma M_D = 0$,

$$(31)(24) + (10)(12) = (72)(12)$$
$$864 = 864$$

The free-body, shear, and bending-moment diagrams of all members are shown in Fig. 12-3bcd.   When all free bodies are in equilibrium, and all shear and bending-moment diagrams close, the correctness of the analysis indicated in Fig. 12-3*bcd* is certain.

The bending-moment diagram for each member in Fig. 12-3*bcd* is next drawn on a sketch of the rigid frame, as shown in Fig. 12-3*e*. If members vary in cross section, the modified M diagram of Fig. 12-3*f* may be obtained by dividing the ordinates in the M diagram of Fig. 12-3*e*

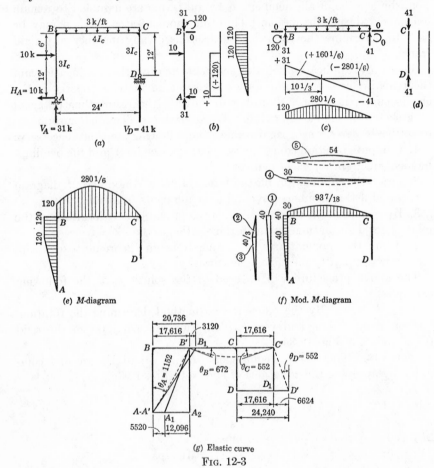

(a)

(b)

(c)

(d)

(e) M-diagram

(f) Mod. M-diagram

(g) Elastic curve

FIG. 12-3

by the number of I_c's in the respective members. The properties of the modified M diagram on member AB may be easily found as it is, or it may be decomposed into triangles 1 and 2 + 3 as shown in Fig. 12-3*f*. By referring to Fig. 12-3*b* it is seen that actually member AB may be considered to be a simple span subjected first to a 10-kip load without end moments and then to an end moment of 120 kip-ft. Triangle 1 in Fig. 12-3*f* is the modified M diagram due to the upper end moment only, while the triangle $2 + 3$ is that for the 10-kip load only. Similarly triangle 4 is the modified M diagram on member BC due to the left end moment only and parabola 5 is that due to the uniform load on a simple beam BC. Obviously it will be more convenient to use the properties of triangle 4 and parabola 5 than to use those of the combined modified M diagram as shown on member BC of Fig. 12-3f.

A sketch of the deformed structure is shown in Fig. 12-3g. It is noted that A and A' must coincide because of the hinge at A. D moves out horizontally to D'. Now, since the lengths of members AB and CD are assumed not to change (even though there are direct stresses in them), the deflected points B' and C' must remain on the same elevation as B and C. Moreover, BB' and CC' must be equal in order that the length of member BC remains constant. The elastic curve $B'C'$ must be concave on the top because the bending moment on member BC causes compression throughout the entire length. Similarly, elastic curve $A'B'$ is concave to the left and $C'D'$ is straight. Because the joints at B and C are assumed to be rigid, the tangents at these joints both rotate so that the angles at B' and C' are still right angles.

Upon completion of the above-described preliminaries, the actual computations for the rotations and deflections now become relatively simple. Applying the conjugate-beam method to member BC,

$$EI_c\theta_B = \tfrac{1}{2}(\text{area } 5) + \tfrac{2}{3}(\text{area } 4)$$
$$= \tfrac{1}{2}(\tfrac{2}{3})(54)(24) + \tfrac{2}{3}(\tfrac{1}{2})(30)(24) = 672 \text{ kip-sq ft}$$
$$\theta_B = 672 \frac{\text{kip-sq ft}}{EI_c} \text{ clockwise}$$
$$EI_c\theta_C = \tfrac{1}{2}(\text{area } 5) + \tfrac{1}{3}(\text{area } 4)$$
$$= \tfrac{1}{2}(864) + \tfrac{1}{3}(360) = 552 \text{ kip-sq ft}$$
$$\theta_C = 552 \frac{\text{kip-sq ft}}{EI_c} \text{ counterclockwise}$$

Applying the moment-area theorems to member AB,

$$EI_c\theta_A = EI\theta_B + (\text{area } 1 + \text{area } 2 + \text{area } 3)$$
$$= 672 + 360 + 40 + 80 = 1{,}152 \text{ kip-sq ft}$$
$$\theta_A = 1{,}152 \frac{\text{kip-sq ft}}{EI_c} \text{ clockwise}$$
$$BB' = BB_1 - B_1B'$$
$$= 18\theta_A - \frac{1}{EI_c} (\text{moment of areas 1, 2, and 3 about } B)$$
$$= \frac{1}{EI_c} [(18)(1{,}152) - (360)(6) - (40)(4) - (80)(10)]$$
$$= 17{,}616 \frac{\text{kip-cu ft}}{EI_c}$$

or $\quad BB' = AA_2 = AA_1 + A_1A_2$

$\qquad = \dfrac{1}{EI_c}$ (moment of areas 1, 2, and 3 about A) $+ 18\theta_B$

$\qquad = \dfrac{1}{EI_c}[(360)(12) + (80)(8) + (40)(14) + 18(672)]$

$\qquad = 17,616 \dfrac{\text{kip-cu ft}}{EI_c}$

$\quad CC' = BB' = 17,616 \dfrac{\text{kip-cu ft}}{EI_c}$

Referring to $C'D'$,

$\qquad \theta_D = \theta_C = 552 \dfrac{\text{kip-sq ft}}{EI_c}$ counterclockwise

$\quad DD' = DD_1 + D_1D' = CC' + 12\theta_C$

$\qquad = \dfrac{1}{EI_c}[17,616 + 12(552)] = 24,240 \dfrac{\text{kip-cu ft}}{EI_c}$

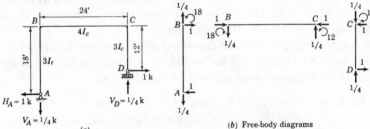

(a)

(b) Free-body diagrams

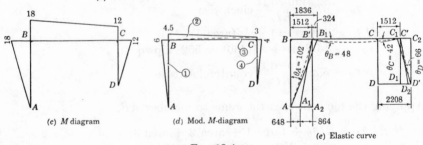

(c) M diagram

(d) Mod. M-diagram

(e) Elastic curve

FIG. 12-4

Example 12-2. By the moment-area method determine the rotation, horizontal deflection, and vertical deflection of each joint in the rigid frame shown in Fig. 12-4a.

SOLUTION. The reaction components due to the 1-kip horizontal load at D are found by applying the equations of statics to Fig. 12-4a. Thus

$$H_A = 1 \text{ kip to the left}$$
$$V_A = \tfrac{1}{4} \text{ kip downward}$$
and $\qquad V_D = \tfrac{1}{4} \text{ kip upward}$

The free-body diagrams of the individual members of the frame are shown in Fig. 12-4b.   The M and the modified M diagrams are shown in Fig. 12-4c and d, respectively.   The elastic curve $A'B'C'D'$ is sketched and shown in Fig. 12-4e.

Applying the conjugate-beam method to member BC,

$$EI_c\theta_B = \tfrac{2}{3}(\text{area 2}) + \tfrac{1}{3}(\text{area 3})$$
$$= \tfrac{2}{3}(54) + \tfrac{1}{3}(36) = 48 \text{ kip-sq ft clockwise}$$
$$EI_c\theta_C = \tfrac{1}{3}(\text{area 2}) + \tfrac{2}{3}(\text{area 3})$$
$$= \tfrac{1}{3}(54) + \tfrac{2}{3}(36) = 42 \text{ kip-sq ft counterclockwise}$$

Applying the moment-area theorems to member AB,

$$EI_c\theta_A = EI_c\theta_B + (\text{area 1})$$
$$= 48 + 54 = 102 \text{ kip-sq ft clockwise}$$
$$BB' = BB_1 - B_1B'$$
$$= 18\theta_A - \frac{1}{EI_c}(\text{moment of area 1 about } B)$$
$$EI_c(BB') = 18(102) - (54)(6) = 1{,}512 \text{ kip-cu ft}$$

or
$$BB' = AA_2 = AA_1 + A_1A_2$$
$$= \frac{1}{EI_c}(\text{moment of area 1 about } A) + 18\theta_B$$
$$EI_c(BB') = (54)(12) + 18(48) = 1{,}512 \text{ kip-cu ft}$$

Applying the moment area theorems to member CD,

$$EI_c\theta_D = EI_c\theta_C + (\text{area 4})$$
$$= 42 + 24 = 66 \text{ kip-sq ft counterclockwise}$$
$$DD' = DD_1 + D_1D_2 + D_2D'$$
$$= CC' + 12\theta_C + \frac{1}{EI_c}(\text{moment of area 4 about } D)$$
$$EI_c(DD') = 1{,}512 + 12(42) + (24)(8) = 2{,}208 \text{ kip-cu ft}$$

or
$$DD' = CC_2 = CC' + C'C_2 = CC' + (C_1C_2 - C_1C')$$
$$= CC' + 12\theta_D - \frac{1}{EI_c}(\text{moment of area 4 about } C)$$
$$EI_c(DD') = 1{,}512 + 12(66) - (24)(4) = 2{,}208 \text{ kip-cu ft}$$

Example 12-3.   By the moment-area method determine the rotation, horizontal deflection, and vertical deflection of each joint in the rigid frame shown in Fig. 12-5a.

SOLUTION.   By applying the three equations of statics to the free body shown in Fig. 12-5a, the reaction components are found to be $H_A = 10$ kips to the left, $V_A = 72$ kips upward, and $M_A = 984$ kip-ft counterclockwise.   The M and the modified M diagrams are shown in Fig. 12-5b and c, respectively.   By starting at the fixed end at A, the elastic curve $A'B'C'D'$ may be sketched as shown in Fig. 12-5d. The moment-area theorems are then applied successively to the individual members of the frame.

$$EI_c\theta_B = \text{area } 2 + \text{area } 3 = (288)(18) + \frac{1}{2}(40)(12)$$
$$= 5{,}184 + 240 = 5{,}424 \text{ kip-sq ft clockwise}$$
$$EI_c(BB') = \text{moment of areas 2 and 3 about } B$$
$$= (5{,}184)(9) + (240)(14) = 50{,}016 \text{ kip-cu ft}$$
$$EI_c\theta_C = EI_c\theta_B + (\text{area } 1) = 5{,}424 + 1{,}728$$
$$= 7{,}152 \text{ kip-sq ft clockwise}$$

$$C_1C' = C_1C_2 + C_2C' = 24\theta_B + \frac{1}{EI_c}\ (\text{moment of area 1 about } C)$$

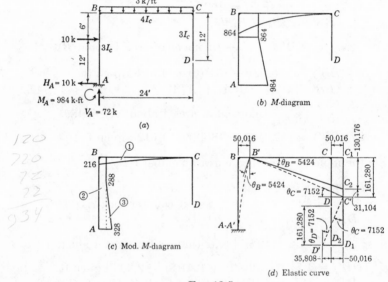

(a)

(b) M-diagram

(c) Mod. M-diagram

(d) Elastic curve

FIG. 12-5

$$EI(C_1C') = (24)(5{,}424) + (1{,}728)(18) = 161{,}280 \text{ kip-cu ft}$$
$$DD_2 = C_1C' = 161{,}280\ \frac{\text{kip-cu ft}}{EI_c}$$
$$\theta_D = \theta_C = 7{,}152\ \frac{\text{kip-sq ft}}{EI_c} \text{ clockwise}$$
$$D_2D' = D_1D' - D_1D_2 = 12\theta_C - CC_1$$
$$EI_c(D_2D') = 12(7{,}152) - 50{,}016 = 35{,}808 \text{ kip-cu ft}$$

Example 12-4. By the moment-area method determine the rotation, horizontal deflection, and vertical deflection of each joint in the rigid frame shown in Fig. 12-6a.

SOLUTION.   The reaction components at A due to the 1 kip-ft moment at D are found by the equations of statics to be $H_A = 0$, $V_A = 0$, and $M_A = 1$ kip-ft clockwise (Fig. 12-6a).   The M diagram, the modified M diagram, and a sketch of the elastic curve are shown in Fig. 12-6bcd.

$$EI_c\theta_B = \text{area } 1 = 6 \text{ kip-sq ft counterclockwise}$$
$$EI_c(BB') = \text{moment of area 1 about } B = (6)(9) = 54 \text{ kip-cu ft}$$
$$EI_c\theta_C = EI_c\theta_B + \text{area } 2 = 6 + 6 = 12 \text{ kip-sq ft counterclockwise}$$
$$CC_1 = BB'$$
$$C_1C' = C_1C_2 + C_2C' = 24\theta_B + \frac{1}{EI_c} \text{ (moment of area 2 about } C)$$

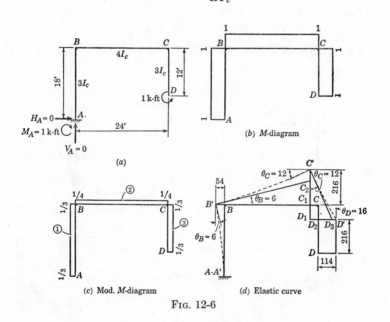

(a)

(b) M-diagram

(c) Mod. M-diagram

(d) Elastic curve

FIG. 12-6

$$EI_c(C_1C') = (24)(6) + (6)(12) = 216 \text{ kip-cu ft}$$
$$EI_c\theta_D = EI_c\theta_C + \text{area } 3 = 12 + 4 = 16 \text{ kip-sq ft counterclockwise}$$
$$D_2D = C_1C'$$
$$D_2D' = D_1D_3 + D_3D' - D_1D_2$$
$$= 12\theta_C + \frac{1}{EI_c} \text{ (moment of area 3 about } D) - \frac{54}{EI_c}$$
$$EI_c(D_2D') = (12)(12) + (4)(6) - 54 = 114 \text{ kip-cu ft}$$

Example 12-5.   By the moment-area method determine the rotation, horizontal deflection, and vertical deflection of each joint in the rigid frame shown in Fig. 12-7a.

SOLUTION. The reaction components at A due to the 1-kip horizontal load at D are found by statics to be $H_A = 1$ kip to the left, $V_A = 0$, and $M_A = 6$ kip-ft counterclockwise (Fig. 12-7a). The M diagram, the modified M diagram, and the sketched elastic curve are shown in Fig. 12-7bcd. Note that B and B' happen to coincide in this particular problem.

$$EI_c\theta_B = (\text{area } 2) - (\text{area } 1)$$
$$= 24 - 6 = 18 \text{ kip-sq ft counterclockwise}$$
$$EI_c(BB') = (\text{moment of area } 1 \text{ about } B) - (\text{moment of area } 2 \text{ about } B)$$
$$= (6)(16) - (24)(4) = 0$$
$$EI_c\theta_C = EI_c\theta_B + (\text{area } 3)$$
$$= 18 + 72 = 90 \text{ kip-sq ft counterclockwise}$$
$$CC' = CC_1 + C_1C' = 24\theta_B + \frac{1}{EI_c}(\text{moment of area } 3 \text{ about } C)$$

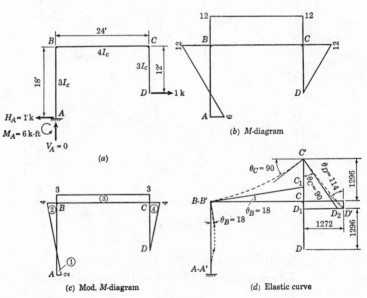

(a)

(b) M-diagram

(c) Mod. M-diagram

(d) Elastic curve

FIG. 12-7

$$EI_c(CC') = (24)(18) + (72)(12) = 1,296 \text{ kip-cu ft}$$
$$EI_c\theta_D = EI_c\theta_C + (\text{area } 4) = 90 + 24$$
$$= 114 \text{ kip-sq ft counterclockwise}$$
$$D_1D' = D_1D_2 + D_2D' = 12\theta_C + \frac{1}{EI_c}(\text{moment of area } 4 \text{ about } D)$$
$$EI_c(D_1D') = (12)(90) + (24)(8) = 1,272 \text{ kip-cu ft}$$

Example 12-6. By the moment-area method determine the rotation, horizontal deflection, and vertical deflection of each joint in the rigid frame shown in Fig. 12-8a.

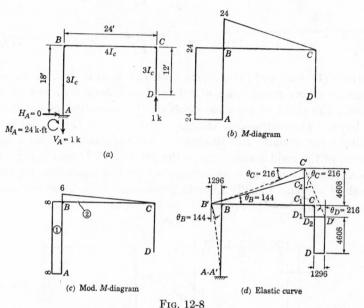

(a)

(b) *M*-diagram

(c) Mod. *M*-diagram

(d) Elastic curve

Fig. 12-8

SOLUTION. The reaction components at *A* due to the 1-kip vertical load at *D* are found by statics to be $H_A = 0$, $V_A = 1$ kip downward, and $M_A = 24$ kip-ft clockwise (Fig. 12-8a). The *M* diagram, the modified *M* diagram, and the sketched elastic curve are shown in Fig. 12-8bcd.

$$EI_c\theta_B = \text{area } 1 = 144 \text{ kip-sq ft counterclockwise}$$
$$EI_cBB' = \text{moment of area 1 about } B$$
$$= (144)(9) = 1{,}296 \text{ kip-cu ft}$$
$$EI_c\theta_C = EI_c\theta_B + (\text{area } 2) = 144 + 72 = 216 \text{ kip-sq ft}$$
$$C_1C' = C_1C_2 + C_2C'$$
$$= 24\theta_B + \frac{1}{EI_c} (\text{moment of area 2 about } C)$$
$$EI_c(C_1C') = (24)(144) + (72)(16) = 4{,}608 \text{ kip-cu ft}$$
$$\theta_D = \theta_C$$
$$D_2D' = D_1D' - D_1D_2 = 12\theta_C - D_1D_2$$
$$EI_c(D_2D') = (12)(216) - 1{,}296 = 1{,}296 \text{ kip-cu ft}$$

12-3. Deflections of Statically Determinate Rigid Frames: the Unit-load Method.

The derivation and application of the unit-load method to finding deflections and rotations in statically determinate beams were described in Art. 11-5.   The formulas previously derived are:

$$(1)(\Delta_C) = \int_0^L \frac{Mm \, dx}{EI} \qquad (11\text{-}14a)$$

and

$$(1)(\theta_C) = \int_0^L \frac{Mm \, dx}{EI} \qquad (11\text{-}14b)$$

Equations (11-14a) and (11-14b) may also be used to find the deflection or rotation at any point in a statically determinate rigid frame provided that only the effect of bending stress on the distortion of the frame is considered.   The expression on the right side of Eqs. (11-14a) and (11-14b) must include the summation or integration through all the members of the rigid frame.   Since the product of M and m is involved in these summations, it is important to use the same sign convention for both M and m.   Usually a bending moment which causes compression in the outer fibers of the section is considered to be positive.

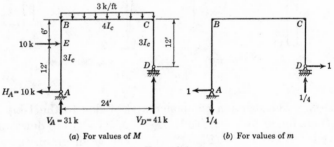

(a) For values of M (b) For values of m

Fig. 12-9

The following examples will illustrate the application of the unit-load method to the finding of deflections and slopes in statically determinate rigid frames.

Example 12-7.   By the unit-load method determine the horizontal deflection at the roller support of the rigid frame shown in Fig. 12-9a.

SOLUTION

Segment of frame	AE	EB	BC	CD
Origin...........	A	E	C	D
Limits..........	0 to 12	0 to 6	0 to 24	0 to 12
M..............	$10x$	120	$41x - \tfrac{3}{2}x^2$	0
m..............	x	$x + 12$	$12 + \tfrac{1}{4}x$	x
I..............	$3I_c$	$3I_c$	$4I_c$	$3I_c$

$$EI_c\Delta_H \text{ of } D = \tfrac{1}{3} \int_0^{12} (10x)(x)dx + \tfrac{1}{3} \int_0^6 (120)(x+12)dx$$

$$+ \tfrac{1}{4} \int_0^{24} (41x - \tfrac{3}{2}x^2)(12 + \tfrac{1}{4}x)dx$$

$$= \left[\frac{10x^3}{9}\right]_0^{12} + \left[20x^2 + 480x\right]_0^6$$

$$+ \left[\frac{123x^2}{2} - \frac{3x^3}{2} + \frac{41}{48}x^3 - \frac{3}{128}x^4\right]_0^{24}$$

$$= 1{,}920 + 3{,}600 + 18{,}720 = 24{,}240 \text{ kip-cu ft}$$

The fact that a positive result is obtained indicates that the direction of the horizontal deflection at D agrees with that of the unit load applied at D as shown in Fig. 12-9b (toward the right in this case). Thus

$$\Delta_H \text{ of } D = 24{,}240 \frac{\text{kip-cu ft}}{EI_c} \text{ to the right}$$

Example 12-8. By the unit-load method determine the horizontal deflection at the roller support of the rigid frame shown in Fig. 12-10a.

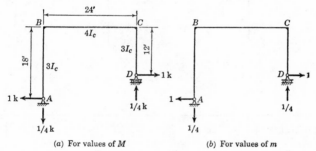

(a) For values of M (b) For values of m

Fig. 12-10

SOLUTION

Segment of frame	AB	BC	CD
Origin...........	A	B	D
Limits...........	0 to 18	0 to 24	0 to 12
M..............	x	$18 - \tfrac{1}{4}x$	x
m..............	x	$18 - \tfrac{1}{4}x$	x
I...............	$3I_c$	$4I_c$	$3I_c$

$$EI_c\Delta_H \text{ of } D = \tfrac{1}{3} \int_0^{18} x^2\, dx + \tfrac{1}{4} \int_0^{24} (18 - \tfrac{1}{4}x)^2 dx + \tfrac{1}{3} \int_0^{12} x^2\, dx$$

$$= [\tfrac{1}{9}x^3]_0^{18} + [81x - \tfrac{9}{8}x^2 + \tfrac{1}{192}x^3]_0^{24} + [\tfrac{1}{9}x^3]_0^{12}$$

$$= 648 + 1{,}368 + 192 = 2{,}208 \text{ kip-cu ft}$$

Thus $\qquad \Delta_H$ of $D = 2{,}208 \dfrac{\text{kip-cu ft}}{EI_c}$ to the right

Example 12-9. By the unit-load method determine the rotation, horizontal deflection, and vertical deflection at the free end of the rigid frame shown in Fig. 12-11a.

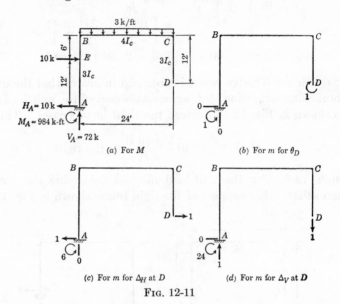

(a) For M

(b) For m for θ_D

(c) For m for Δ_H at D

(d) For m for Δ_V at D

FIG. 12-11

SOLUTION

Segment of frame	AE	EB	BC	CD
Origin..............	A	E	C	D
Limits..............	0 to 12	0 to 6	0 to 24	0 to 12
M..................	$-984 + 10x$	-864	$-\frac{3}{2}x^2$	0
m for θ_D............	-1	-1	-1	-1
m for Δ_H at D........	$-6 + x$	$-6 + (x + 12)$	$+12$	$+x$
m for Δ_V at D........	-24	-24	$-x$	0
I..................	$3I_c$	$3I_c$	$4I_c$	$3I_c$

$$EI_c\theta_D = \tfrac{1}{3}\int_0^{12}(-984 + 10x)(-1)\,dx + \tfrac{1}{3}\int_0^6 (-864)(-1)\,dx$$
$$+ \tfrac{1}{4}\int_0^{24}(-\tfrac{3}{2}x^2)(-1)\,dx$$
$$= [328x - \tfrac{5}{3}x^2]_0^{12} + [288x]_0^6 + [\tfrac{1}{8}x^3]_0^{24}$$
$$= 3{,}696 + 1{,}728 + 1{,}728 = 7{,}152 \text{ kip-sq ft}$$
$$\theta_D = 7{,}152 \frac{\text{kip-sq ft}}{EI_c} \text{ clockwise}$$

$$EI_c\Delta_H \text{ at } D = \tfrac{1}{3} \int_0^{12} (-984 + 10x)(-6 + x)dx$$

$$+ \tfrac{1}{3} \int_0^6 (-864)(x + 6)dx + \tfrac{1}{4} \int_0^{24} (-\tfrac{3}{2}x^2)(+12)dx$$

$$= \left[1{,}968x - 174x^2 + \frac{10x^3}{9} \right]_0^{12} + [-144x^2 - 1{,}728x]_0^6$$

$$+ [-\tfrac{3}{2}x^3]_0^2;$$

$$= 480 - 15{,}552 - 20{,}736 = -35{,}808$$

$$\Delta_H \text{ at } D = 35{,}808 \frac{\text{kip-cu ft}}{EI_c} \text{ to the left}$$

$$EI_c\Delta_V \text{ at } D = \tfrac{1}{3} \int_0^{12} (-984 + 10x)(-24)dx + \tfrac{1}{3} \int_0^6 (-864)(-24)dx$$

$$+ \tfrac{1}{4} \int_0^{24} (-\tfrac{3}{2}x^2)(-x)dx$$

$$= [7{,}872x - 40x^2]_0^{12} + [6{,}912x]_0^6 + [\tfrac{3}{3}{}_2 x^4]_0^{24}$$

$$= 88{,}704 + 41{,}472 + 31{,}104 = 161{,}280$$

$$\Delta_V \text{ at } D = 161{,}280 \frac{\text{kip-cu ft}}{EI_c} \text{ downward}$$

Example 12-10. By the unit-load method determine the rotation, horizontal deflection, and vertical deflection at the free end of the rigid frame shown in Fig. 12-12a.

SOLUTION

Segment of frame	AB	BC	CD
Origin.............	A	C	D
Limits.............	0 to 18	0 to 24	0 to 12
M................	$+1$	$+1$	$+1$
m for θ_D..........	$+1$	$+1$	$+1$
m for Δ_H at D......	$-6 + x$	$+12$	$+x$
m for Δ_V at D......	$+24$	$+x$	0
I................	$3I_c$	$4I_c$	$3I_c$

$$EI_c\theta_D = \tfrac{1}{3} \int_0^{18} (+1)^2 dx + \tfrac{1}{4} \int_0^{24} (+1)^2 dx + \tfrac{1}{3} \int_0^{12} (+1)^2 dx$$

$$= 16 \text{ kip-sq ft counterclockwise}$$

$$EI_c\Delta_H \text{ at } D = \tfrac{1}{3} \int_0^{18} (+1)(-6 + x)dx + \tfrac{1}{4} \int_0^{24} (+1)(+12)dx$$

$$+ \tfrac{1}{3} \int_0^{12} (+1)(+x)dx$$

$$= 114 \text{ kip-cu ft to the right}$$

$$EI_c\Delta_V \text{ at } D = \tfrac{1}{3} \int_0^{18} (+1)(+24)dx + \tfrac{1}{4} \int_0^{24} (+1)(+x)dx$$

$$+ \tfrac{1}{3} \int_0^{12} (+1)(0)dx$$

$$= 216 \text{ kip-cu ft upward}$$

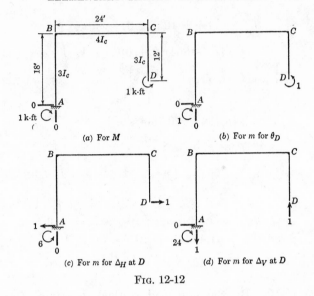

(a) For M (b) For m for θ_D

(c) For m for Δ_H at D (d) For m for Δ_V at D

FIG. 12-12

Example 12-11. By the unit-load method determine the rotation, horizontal deflection, and vertical deflection at the free end of the rigid frame shown in Fig. 12-13a.

SOLUTION

Segment of frame	AB	BC	CD
Origin............	A	C	D
Limits............	0 to 18	0 to 24	0 to 12
M................	$-6 + x$	$+12$	$+x$
m for θ_D..........	$+1$	$+1$	$+1$
m for Δ_H at D......	$-6 + x$	$+12$	$+x$
m for Δ_V at D......	$+24$	$+x$	0
I................	$3I_c$	$4I_c$	$3I_c$

$$EI_c\theta_D = \tfrac{1}{3} \int_0^{18} (-6 + x)(+1)dx + \tfrac{1}{4} \int_0^{24} (+12)(+1)dx$$
$$+ \tfrac{1}{3} \int_0^{12} (+x)(+1)dx$$

$$= 114 \text{ kip-sq ft counterclockwise}$$

$$EI_c\Delta_H \text{ at } D = \tfrac{1}{3} \int_0^{18} (-6 + x)^2 dx + \tfrac{1}{4} \int_0^{24} (+12)^2 dx$$
$$+ \tfrac{1}{3} \int_0^{12} (+x)^2 dx$$

$$= 1{,}272 \text{ kip-cu ft to the right}$$

$$EI_c\Delta_V \text{ at } D = \tfrac{1}{3}\int_0^{18}(-6+x)(+24)dx + \tfrac{1}{4}\int_0^{24}(+12)(+x)dx$$

$$+ \tfrac{1}{3}\int_0^{12}(+x)(0)dx$$

$$= 1{,}296 \text{ kip-cu ft upward}$$

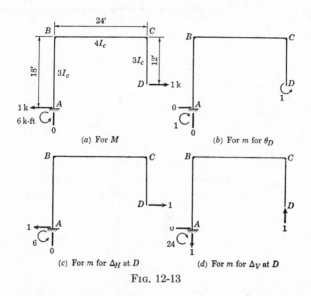

(a) For M (b) For m for θ_D

(c) For m for Δ_H at D (d) For m for Δ_V at D

FIG. 12-13

Example 12-12.   By the unit-load method determine the rotation, horizontal deflection, and vertical deflection at the free end of the rigid frame shown in Fig. 12-14a.

SOLUTION

Segment of frame	AB	BC	CD
Origin	A	C	D
Limits	0 to 18	0 to 24	0 to 12
M	$+24$	$+x$	0
m for θ_D	$+1$	$+1$	$+1$
m for Δ_H at D	$-6+x$	$+12$	$+x$
m for Δ_V at D	$+24$	$+x$	0
I	$3I_c$	$4I_c$	$3I_c$

$$EI_c\theta_D = \tfrac{1}{3}\int_0^{18}(+24)(+1)dx + \tfrac{1}{4}\int_0^{24}(+x)(+1)dx$$

$$+ \tfrac{1}{3}\int_0 (0)(+1)dx$$

$$= 216 \text{ kip-sq ft counterclockwise}$$

$$EI_c\Delta_H \text{ at } D = \tfrac{1}{3}\int_0^{18}(+24)(-6+x)dx + \tfrac{1}{4}\int_0^{24}(+x)(+12)dx$$
$$+ \tfrac{1}{3}\int_0^{12}(0)(+x)dx$$

$$= 1{,}296 \text{ kip-cu ft to the right}$$

$$EI_c\Delta_V \text{ at } D = \tfrac{1}{3}\int_0^{18}(+24)^2dx + \tfrac{1}{4}\int_0^{24}(+x)^2dx + \tfrac{1}{3}\int_0^{12}(0)^2dx$$

$$= 4{,}608 \text{ kip-cu ft upward}$$

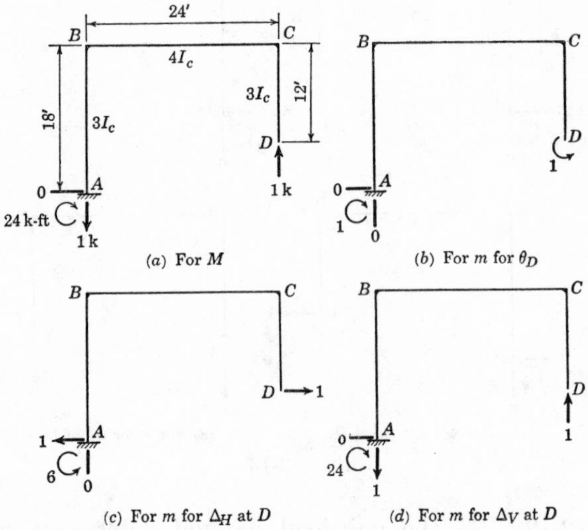

(a) For M   (b) For m for θ_D

(c) For m for Δ_H at D   (d) For m for Δ_V at D

Fig. 12-14

12-4. Analysis of Statically Indeterminate Rigid Frames by the Method of Consistent Deformation. The analysis of statically indeterminate rigid frames by the method of consistent deformation is quite similar to that of statically indeterminate beams. The degree of indeterminacy of the rigid frame is first observed. A basic determinate rigid frame is then derived from the given indeterminate rigid frame by removing the restraints (equal in number to the degree of indeterminacy) and treating the redundant reaction components as loads acting on the basic structure. The unknown redundant forces are determined from an equal number of conditions of consistent deformation, which usually require zero rotation or zero deflection at the location of each redundant. Once the redundants are known, the remaining reaction components may be determined by the laws of statics.

Example 12-13. Analyze the rigid frame shown in Fig. 12-15a by the method of consistent deformation. Draw the free-body, shear, and bending-moment diagrams for all members. Sketch the elastic curve of the deformed structure.

SOLUTION. A single-span rigid frame with two hinged supports is statically indeterminate to the first degree because there are four unknown reaction components H_A, V_A, H_D, and V_D as shown in Fig. 12-15a, but only three equations of statics. In this case the basic determinate structure may be obtained by replacing the hinged support at D by a roller support. Then the original structure in Fig. 12-15a is the equivalent of the two structures shown in Fig. 12-15b and c. In the present instance, the deflection Δ_D to the right and caused by the loading on the statically

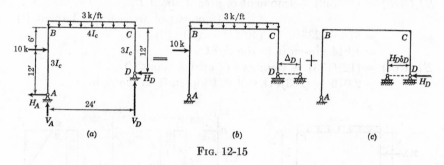

FIG. 12-15

determinate structure must be wholly counteracted by the redundant reaction H_D which must act to the left. To satisfy the condition of zero horizontal deflection at D, H_D must be equal to Δ_D/δ_D. From the result of either Example 12-1 or Example 12-7,

$$EI_c\Delta_D = 24{,}240 \text{ kip-cu ft to the right}$$

and from either Example 12-2 or Example 12-8,

$$EI_c\delta_D = 2{,}208 \text{ kip-cu ft to the left}$$

Therefore

$$H_D = \frac{EI_c\Delta_D}{EI_c\delta_D} = \frac{24{,}240}{2{,}208} = 10.98 \text{ kips to the left}$$

By statics (Fig. 12-16a),

$$H_A = 0.98 \text{ kip to the right}$$
$$V_A = 33.74 \text{ kips upward}$$
$$V_D = 38.26 \text{ kips upward}$$

The free-body, shear, and bending-moment diagrams for the three members are shown in Fig. 12-16bcd. The modified M diagram and its parts are shown in Fig. 12-16e. The elastic curve is shown in Fig. 12-16f. Numerical properties of the elastic curve may be computed as shown below.

$EI_c\theta_B = \frac{1}{2}(\text{area } 4) - \frac{2}{3}(\text{area } 5) - \frac{1}{3}(\text{area } 6)$
$\qquad = \frac{1}{2}(864) - \frac{2}{3}(233) - \frac{1}{3}(396) = 145$ kip-sq ft clockwise
$EI_c\theta_C = \frac{1}{2}(\text{area } 4) - \frac{1}{3}(\text{area } 5) - \frac{2}{3}(\text{area } 6)$
$\qquad = \frac{1}{2}(864) - \frac{1}{3}(233) - \frac{2}{3}(396)$
$\qquad = 90$ kip-sq ft counterclockwise
$EI_c\theta_A = 145 - (\text{area } 3) + (\text{area } 1) + (\text{area } 2)$
$\qquad = 145 - 233 + 40 + 80 = 32$ kip-sq ft clockwise
$EI_c\theta_D = (\text{area } 7) - 90 = 263 - 90 = 173$ kip-sq ft clockwise
$EI_c(BB') = (18)(32) + (\text{moment of area 3 about } B)$
$\qquad\qquad\qquad - (\text{moment of areas 1 and 2 about } B)$
$\qquad = 576 + 233(6) - (40)(4) - (80)(10)$
$\qquad = 1,014$ kip-cu ft to the right
$EI_c(CC') = (12)(173) - (\text{moment of area 7 about } C)$
$\qquad = 2,076 - (263)(4) = 1,024$ kip-cu ft to the right

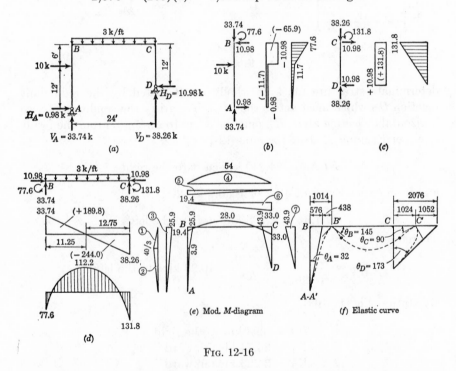

FIG. 12-16

The fact that all free-body diagrams (Fig. 12-16abcd) are in equilibrium shows that all equations of statics are satisfied, and the fact that BB' and CC' are found to be equal shows that the one important condition of geometry has been satisfied. When all the conditions of statics and of the geometry of deformation have been satisfied, the correctness of the solution is ensured.

Example 12-14.   Analyze the rigid frame shown in Fig. 12-17a by the method of consistent deformation.   Draw the free-body, shear, and bending-moment diagrams of all members.   Sketch the elastic curve of the deformed structure.

SOLUTION.   Because there are six unknown reaction components, and only three equations of statics, a single-span rigid frame with two fixed supports is statically indeterminate to the third degree.   Since any three of the six reaction components may be taken as the redundants, there are several different ways of choosing the basic determinate structure. If M_D, H_D, and V_D are chosen the redundants, the basic determinate structure is a cantilever structure with a fixed support at A and a free end at D.   The original statically indeterminate structure in Fig. 12-17a

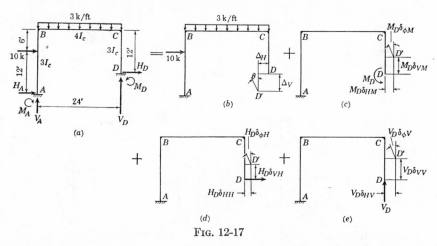

FIG. 12-17

then becomes the equivalent of the four determinate structures shown in Fig. 12-17bcde.

Let θ, Δ_H, and Δ_V be the rotation, horizontal deflection, and vertical deflection at D due to the applied loads on the basic determinate structure; $\delta_{\phi M}$, δ_{HM}, and δ_{VM} be those due to a 1 kip-ft counterclockwise moment at D; $\delta_{\phi H}$, δ_{HH}, and δ_{VH} be those due to a 1-kip horizontal load acting to the right at D; and $\delta_{\phi V}$, δ_{HV}, and δ_{VV} be those due to a 1-kip vertical load acting upward at D.   The three conditions of consistent deformation requiring zero rotation and zero horizontal and vertical deflections at D are:

$$\theta_D + M_D\delta_{\phi M} + H_D\delta_{\phi H} + V_D\delta_{\phi V} = 0$$
$$\Delta_H + M_D\delta_{HM} + H_D\delta_{HH} + V_D\delta_{HV} = 0$$
$$\Delta_V + M_D\delta_{VM} + H_D\delta_{VH} + V_D\delta_{VV} = 0$$

By assigning a positive sign to counterclockwise rotation, horizontal deflection to the right, and vertical deflection upward, the following values have been taken from the results of Examples 3 to 6, or Examples 9 to 12:

$$EI_c\theta_D = -7{,}152 \text{ kip-sq ft} \qquad EI_c\delta_{\phi M} = +16 \text{ kip-sq ft}$$

$$EI_c\Delta_H = -35{,}808 \text{ kip-cu ft} \qquad EI_c\delta_{HM} = +114 \text{ kip-cu ft}$$

$$EI_c\Delta_V = -161{,}280 \text{ kip-cu ft} \qquad EI_c\delta_{VM} = +216 \text{ kip-cu ft}$$

$$EI_c\delta_{\phi H} = +114 \text{ kip-sq ft} \qquad EI_c\delta_{\phi V} = +216 \text{ kip-sq ft}$$

$$EI_c\delta_{HH} = +1{,}272 \text{ kip-cu ft} \qquad EI_c\delta_{HV} = +1{,}296 \text{ kip-cu ft}$$

$$EI_c\delta_{VH} = +1{,}296 \text{ kip-cu ft} \qquad EI_c\delta_{VV} = +4{,}608 \text{ kip-cu ft}$$

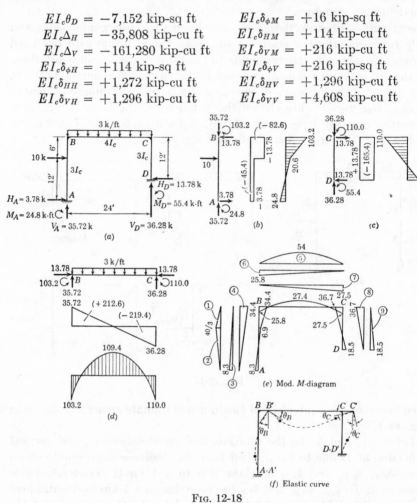

Fig. 12-18

In the above tabulation, it should be noted that

$$\delta_{\phi H} = \delta_{HM}$$

$$\delta_{\phi V} = \delta_{VM}$$

$$\delta_{HV} = \delta_{VH}$$

According to the law of reciprocal deflections, these relations must hold true. It should also be noted that in the moment-area method each value in the pair of δ's is independently determined; therefore, the fact that they must be equal serves as an excellent check. In the unit-load method, however, the expressions for both δ values are identical, and this does not afford an independent check.

Substituting the above deflection quantities in the equations of consistent deformation,

$$-7{,}152 + 16M_D + 114H_D + 216V_D = 0$$
$$-35{,}808 + 114M_D + 1{,}272H_D + 1{,}296V_D = 0$$
$$-161{,}280 + 216M_D + 1{,}296H_D + 4{,}608V_D = 0$$

Solving,

$$M_D = +55.4 \text{ kip-ft or } 55.4 \text{ kip-ft counterclockwise}$$
$$H_D = -13.78 \text{ kips or } 13.78 \text{ kips to the left}$$
$$V_D = +36.28 \text{ kips or } 36.28 \text{ kips upward}$$

Applying the laws of statics to Fig. 12-18a,

$$H_A = 3.78 \text{ kips to the right}$$
$$V_A = 35.72 \text{ kips upward}$$
$$M_A = 24.8 \text{ kip-ft clockwise}$$

The free-body, shear, and bending-moment diagrams of the three members of the frame are shown in Fig. 12-18bcd. The modified M diagram and its separate parts are shown in Fig. 12-18e. The elastic curve is shown in Fig. 12-18f. Numerical properties of the elastic curve may be computed as shown below.

$$
\begin{aligned}
EI_c\theta_B &= \tfrac{1}{2}(\text{area 5}) - \tfrac{2}{3}(\text{area 6}) - \tfrac{1}{3}(\text{area 7})\\
&= \tfrac{1}{2}(864) - \tfrac{2}{3}(309.6) - \tfrac{1}{3}(330.0)\\
&= 116 \text{ kip-sq ft clockwise}
\end{aligned}
$$

or
$$
\begin{aligned}
EI_c\theta_B &= (\text{area 4}) - (\text{area 3}) - (\text{areas 1 and 2})\\
&= 309.6 - 74.7 - 40 - 80\\
&= 115 \text{ kip-sq ft clockwise} \quad (check)
\end{aligned}
$$

$$
\begin{aligned}
EI_c\theta_C &= \tfrac{1}{2}(\text{area 5}) - \tfrac{1}{3}(\text{area 6}) - \tfrac{2}{3}(\text{area 7})\\
&= \tfrac{1}{2}(864) - \tfrac{1}{3}(309.6) - \tfrac{2}{3}(330.0)\\
&= 109 \text{ kip-sq ft counterclockwise}
\end{aligned}
$$

or
$$
\begin{aligned}
EI_c\theta_C &= (\text{area 8}) - (\text{area 9})\\
&= 220.2 - 111.0 = 109 \text{ kip-sq ft counterclockwise} \quad (check)
\end{aligned}
$$

$$
\begin{aligned}
EI_c(BB') &= (\text{moment of area 4 about } B)\\
&\qquad - (\text{moment of areas 1, 2, and 3 about } B)\\
&= (309.6)(6) - (40)(4) - (80)(10) - (74.7)(12)\\
&= 1 \text{ kip-cu ft to the right}
\end{aligned}
$$

$$
\begin{aligned}
EI_c(CC') &= (\text{moment of area 9 about } C) - (\text{moment of area 8 about } C)\\
&= (111.0)(8) - (220.2)(4)\\
&= 7 \text{ kip-cu ft to the right} \quad (check)
\end{aligned}
$$

Inasmuch as this rigid frame with two fixed supports is statically indeterminate to the third degree, three consistent checks must be expected in the computed results of the elastic curve. These are: (1) $EI_c\theta_B = 116$ kip-sq ft clockwise as computed for member BC should be equal to $EI_c\theta_B = 115$ kip-sq ft clockwise as computed for member AB; (2) $EI_c\theta_C = 109$ kip-sq ft counterclockwise as computed for member BC should be equal to $EI_c\theta_C = 109$ kip-sq ft counterclockwise as computed for member CD, and (3) $EI_c(BB') = 1$ kip-cu ft to the right should be equal to $EI_c(CC') = 7$ kip-cu ft to the right. This last equality seems absurd but it is nevertheless within the permissible limit of the error expected because, in this instance, either 1 or 7 is the difference in the last significant number of values with three or four significant figures.

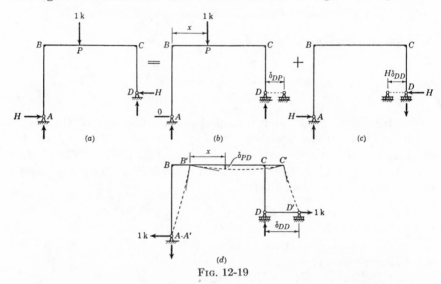

Fig. 12-19

12-5. Influence Diagrams for Statically Indeterminate Rigid Frames.

Statically indeterminate rigid frames may carry moving loads, especially on the horizontal member when this structure is used in bridges or to carry crane loads in industrial buildings. In these cases it may be necessary to construct influence diagrams for bending moments or other functions. With the aid of the influence diagrams, the critical loading conditions for maximum bending moments or other desired functions can usually be determined by inspection or by some cut-and-try method.

The discussion in this article will be limited to influence diagrams for a single-span rigid frame with two hinged supports. It will be seen that the use of the law of reciprocal deflections greatly simplifies the work involved in the computations for the ordinates in the influence diagram for the horizontal reaction. For instance, let it be required to find the horizontal reaction H due to a 1-kip vertical load at P in the rigid frame with two hinged supports shown in Fig. 12-19a. Applying the method of consistent deformation to Fig. 12-19abc,

$$H = \frac{\delta_{DP}}{\delta_{DD}}$$

in which δ_{DP} is the horizontal deflection at D due to a unit vertical load at P and δ_{DD} is the horizontal deflection at D due to a unit horizontal load

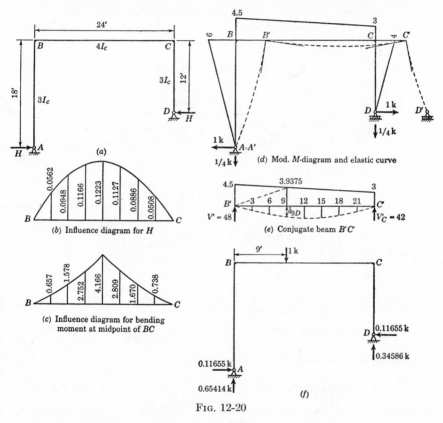

Fig. 12-20

at D. But, by the law of reciprocal deflections, δ_{DP} in Fig. 12-19b is equal to δ_{PD} in Fig. 12-19d. Substituting,

$$H \ (\text{Fig. 12-19}a) = \frac{\delta_{PD} \ (\text{Fig. 12-19}d)}{\delta_{DD} \ (\text{Fig. 12-19}d)}$$

Thus the elastic curve $B'C'$ in Fig. 12-19d is geometrically similar to the influence diagram for H. In fact, the influence ordinates for H may be obtained by dividing the ordinates to the elastic curve $B'C'$ by the constant δ_{DD}. This method will be used in the following illustrative example.

Example 12-15.   Construct the influence diagrams for the horizontal reaction and for the bending moment at the mid-point of the horizontal member in the rigid frame shown in Fig. 12-20a.

SOLUTION.   The required influence diagrams are shown in Fig. 12-20b and c.   To illustrate the procedure, the values in the influence table for a unit load at a point 9 ft from point B will be computed.   Applying the conjugate-beam method to the elastic curve $B'C'$ (Fig. 12-20e),

$$EI_c\delta_{9D} = (48)(9) - \tfrac{1}{2}(4.5)(9)(6) - \tfrac{1}{2}(3.9375)(9)(3)$$
$$= 257.34 \text{ kip-cu ft}$$

Referring to Examples 12-2 or 12-8,

$$EI_c\delta_{DD} \text{ (Fig. 12-20d)} = 2{,}208 \text{ kip-cu ft}$$

TABLE 12-1. INFLUENCE TABLE

Distance from B, ft	$\delta_{PD} \dfrac{\text{kip-cu ft}}{EI_c}$	$H = \dfrac{\delta_{PD}}{\delta_{DD}}$	Moment at mid-point of BC, ft
0	0	0	0
3	124.03125	0.05617	+0.657
6	209.25	0.09477	+1.578
9	257.34375	0.11655	+2.752
12	270	0.12228	+4.166
15	248.90625	0.11273	+2.809
18	195.75	0.08865	+1.670
21	112.21875	0.05082	+0.738
24	0	0	0

Thus

$$H \text{ (due to 1-kip load at 9 ft from } B) = \frac{257.34}{2{,}208} = 0.1166 \text{ kip}$$

Referring to Fig. 12-20f,

$$M \text{ (at mid-point of } BC) = (0.34586)(12) - (0.11655)(12)$$
$$= +2.752 \text{ kip-ft}$$

PROBLEMS

12-1 to 12-3. By the moment-area method determine the rotation, horizontal deflection, and vertical deflection of each joint in the rigid frame shown.

12-4. By the moment-area method determine the rotation, horizontal deflection, and vertical deflection at the free end of the rigid frame shown due to a 1-kip-ft moment acting counterclockwise at the free end.

12-5. By the moment-area method determine the rotation, horizontal deflection, and vertical deflection at the free end of the rigid frame shown due to a 1-kip horizontal load acting to the right at the free end.

12-6. By the moment-area method determine the rotation, horizontal deflection, and vertical deflection at the free end of the rigid frame shown due to a 1-kip vertical load acting upward at the free end.

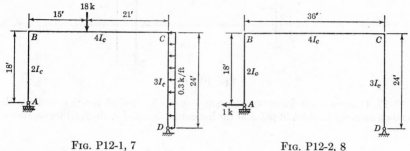

FIG. P12-1, 7   FIG. P12-2, 8

12-7 and 12-8. By the unit-load method determine the horizontal deflection at the roller support of the rigid frame shown.

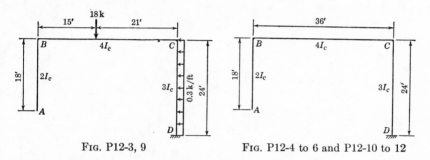

FIG. P12-3, 9   FIG. P12-4 to 6 and P12-10 to 12

12-9. By the unit-load method determine the rotation, horizontal deflection, and vertical deflection at the free end of the rigid frame shown.

12-10 to 12-12. Solve Probs. 12-4 to 12-6 by the unit-load method.

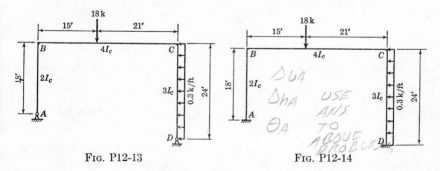

FIG. P12-13   FIG. P12-14

12-13 and 12-14. Analyze the rigid frame shown by the method of consistent deformation. Draw the free-body, shear, and bending-moment diagrams for all members. Sketch the elastic curve of the deformed structure.

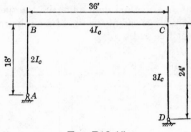

FIG. P12-15

12-15. Construct the influence diagrams for the horizontal reaction and for the bending moment at the mid-point of the horizontal member in the rigid frame shown.

CHAPTER 13

ANALYSIS OF STATICALLY INDETERMINATE TRUSSES

13-1. Statically Determinate vs. Statically Indeterminate Trusses. A truss is a structure in which all members are usually considered to be connected by smooth pins, and subjected to loads applied only at the joints. All members in a truss are thus *two-force* members. A truss is completely analyzed when the kind and amount of direct stress in each member are determined.

A truss is statically determinate if it can be completely analyzed by the laws of statics alone. A truss may be statically indeterminate because it has either external redundant reactions or internal redundant members or both, the degree of indeterminacy being equal to the combined number of redundant reactions and members. If a truss is subjected to a general coplanar-force system, the number of external redundants is equal to the total number of external reaction components minus 3. The number of internal redundant members present in a truss is equal to the total number of members minus $(2j - 3)$ in which j is the total number of joints. This statement may be proved by first establishing that there are two conditions of statics for each joint as a free body, or a total of $2j$ conditions to match $2j$ unknowns, three of which are the three external reaction components, thus leaving

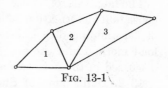

FIG. 13-1

$2j - 3$ conditions of statics available for determining the stresses in the internal members. Another method of [arriving at the same conclusion is to postulate that a truss is internally stable if it consists of a series of triangles as shown in Fig. 13-1. The first triangle is made up of three joints and three members; each successive triangle requires two additional members but only one additional joint. Thus, if m is the number of members in the truss and j is the number of joints, $(m - 3) = 2(j - 3)$, or $m = 2j - 3$.

The truss shown in Fig. 13-2a is statically determinate because it just has three unknown external reaction components and

$$2j - 3 = 2(16) - 3 = 29 \text{ members}$$

295

This is also obvious because the truss diagram is a compilation of simple triangles.   The truss shown in Fig. 13-2b is statically indeterminate to the second degree because it has two external redundant reaction components, although it has just the right number of internal members, $m = 2j - 3 = 2(28) - 3 = 53$.   The truss shown in Fig. 13-2c is statically indeterminate to the third degree because it has one external and two internal redundants.   The truss of Fig. 13-2d is statically inde-- terminate to the third degree because it has three internal redundants.

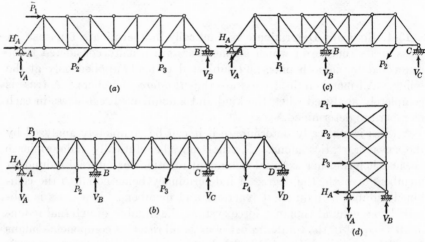

Fig. 13-2

The analysis of statically indeterminate trusses by the method of consistent deformation requires first of all a knowledge of the methods of determining deflections of statically determinate trusses.   The unit-load method and the graphical (Williot-Mohr diagram) method will be discussed in the next two articles.

13-2. Deflections of Statically Determinate Trusses: the Unit-load Method.   Incidental to developing the unit-load method of determining the deflections of statically determinate beams, one formula takes the form

$$(1)(\Delta_C) = \Sigma u \, dL \qquad (11\text{-}13a)$$

This formula may be applied to the calculation of deflections of trusses. In a truss there is a finite number of members in which u, the stress due to the unit load, is constant for each member and dL is the change in the length of the member due to the applied loads.   Applying Hooke's law,

$$dL = \frac{SL}{AE} \qquad (13\text{-}1)$$

wherein S is the stress in the member due to the applied loads, L and A are, respectively, the length and cross-sectional area of the member, and E is the modulus of elasticity of the material. Substituting (13-1) in (11-13a),

$$\Delta_C = \sum \frac{SuL}{AE} \qquad (13\text{-}2)$$

Example 13-1.   Determine the horizontal and vertical deflections of joint L_2 of the truss shown in Fig. 13-3a by the unit-load method.

SOLUTION.   The stress S in each truss member (kind and amount) is computed as indicated in Fig. 13-3b.   The horizontal and vertical unit loads are separately applied and the resulting u stresses calculated as

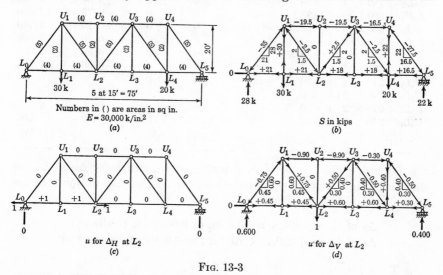

Numbers in ( ) are areas in sq in.
$E = 30,000$ k/in.^{2}
(a)

S in kips
(b)

u for Δ_H at L_2
(c)

u for Δ_V at L_2
(d)

FIG. 13-3

shown in Fig. 13-3c and d, respectively. The algebraic methods of joints or sections may be freely used; by showing the horizontal and vertical components of the stresses in the inclined members, the equilibrium of the concurrent-force system at each joint may be easily checked by inspection.

The required computations are shown in Table 13-1.   Thus

$$\Delta_H \text{ of } L_2 = +63 \times 10^{-3} \text{ in. or } 0.063 \text{ in. to the right}$$
and $$\Delta_V \text{ of } L_2 = +244.7 \times 10^{-3} \text{ in. or } 0.2447 \text{ in. downward}$$

It is to be noted that a positive result means that the actual direction of the deflection or movement agrees with that of the assumed unit load.

Example 13-2.   Determine the horizontal and vertical deflections of joint L_3 of the truss shown in Fig. 13-4a by the unit-load method.

TABLE 13-1

Member	L, in.	A, sq in.	S, kips	$\Delta L = \dfrac{SL}{AE}$, 10^{-3} in.	u for Δ_H at L_2	Δ_H at L_2 $= \Sigma u\,\Delta L$, 10^{-3} in.	u for Δ_V at L_2	Δ_V at L_2 $= \Sigma u\,\Delta L$, 10^{-3} in.
U_1U_2	180	4	-19.5	-29.25	0	0	-0.90	$+26.325$
U_2U_3	180	4	-19.5	-29.25	0	0	-0.90	$+26.325$
U_3U_4	180	4	-16.5	-24.75	0	0	-0.30	$+7.425$
L_0L_1	180	4	$+21$	$+31.5$	$+1$	$+31.5$	$+0.45$	$+14.175$
L_1L_2	180	4	$+21$	$+31.5$	$+1$	$+31.5$	$+0.45$	$+14.175$
L_2L_3	180	4	$+18$	$+27$	0	0	$+0.60$	$+16.200$
L_3L_4	180	4	$+18$	$+27$	0	0	$+0.60$	$+16.200$
L_4L_5	180	4	$+16.5$	$+24.75$	0	0	$+0.30$	$+7.425$
L_0U_1	300	5	-35	-70	0	0	-0.75	$+52.500$
U_1L_2	300	5	-2.5	-5	0	0	$+0.75$	-3.750
L_2U_3	300	5	$+2.5$	$+5$	0	0	$+0.50$	$+2.500$
U_3L_4	300	5	-2.5	-5	0	0	-0.50	$+2.500$
U_4L_5	300	5	-27.5	-55	0	0	-0.50	$+27.500$
U_1L_1	240	2	$+30$	$+120$	0	0	0	0
U_2L_2	240	2	0	0	0	0	0	0
U_3L_3	240	2	0	0	0	0	0	0
U_4L_4	240	2	$+22$	$+88$	0	0	$+0.40$	$+35.200$
Σ						$+63$		$+244.70$

SOLUTION. The values of S, u for Δ_H at L_3, and u for Δ_V at L_3, are shown in Fig. 13-4bcd. Computations for Δ_H and Δ_V at joint L_3 are shown in Table 13-2. Thus

$$\Delta_H \text{ of } L_3 = -64.8 \times 10^{-3} \text{ in. or } 0.0648 \text{ in. to the left}$$
$$\Delta_V \text{ of } L_3 = +320.7 \times 10^{-3} \text{ in. or } 0.3207 \text{ in. downward}$$

TABLE 13-2

Member	L, in.	A, sq in.	S, kips	$\Delta L = \dfrac{SL}{AE}$, 10^{-3} in.	u for Δ_H at L_3	Δ_H at L_3 $= \Sigma u\,\Delta L$, 10^{-3} in.	u for Δ_V at L_3	Δ_V at L_3 $= \Sigma u\,\Delta L$, 10^{-3} in.
U_0U_1	72	4	$+36$	$+21.6$	0	......	$+1.50$	$+32.4$
U_1U_2	72	4	$+18$	$+10.8$	0	......	$+0.75$	$+8.1$
L_0L_1	72	4	-54	-32.4	$+1$	-32.4	-2.25	$+72.9$
L_1L_2	72	4	-36	-21.6	$+1$	-21.6	-1.50	$+32.4$
L_2L_3	72	4	-18	-10.8	$+1$	-10.8	-0.75	$+8.1$
U_0L_1	120	5	$+30$	$+24$	0	......	$+1.25$	$+30.0$
U_1L_2	120	5	$+30$	$+24$	0	......	$+1.25$	$+30.0$
U_2L_3	120	5	$+30$	$+24$	0	......	$+1.25$	$+30.0$
U_1L_1	96	2	-24	-38.4	0	......	-1	$+38.4$
U_2L_2	96	2	-24	-38.4	0	......	-1	$+38.4$
Σ						-64.8		$+320.7$

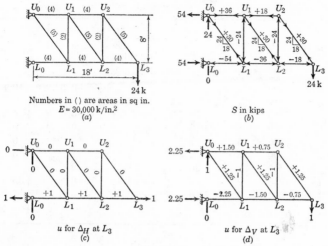

Numbers in ( ) are areas in sq in.
$E = 30,000$ k/in.^{2}
(a)

S in kips
(b)

u for Δ_H at L_3
(c)

u for Δ_V at L_3
(d)

FIG. 13-4

Example 13-3. Determine the relative movement in the direction L_1U_2 between the joints L_1 and U_2 of the truss shown in Fig. 13-5a by the unit-load method.

SOLUTION. The separating relative movement between the joints L_1 and U_2, in the direction L_1U_2, is equal to the sum of the two absolute movements in the direction L_1U_2 of joints L_1 and U_2 (the movement upward to the right of joint U_2 and the movement downward to the left

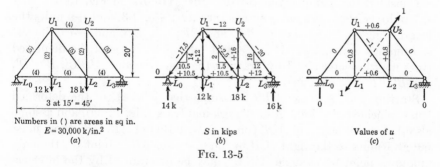

Numbers in ( ) are areas in sq in.
$E = 30,000$ k/in.^{2}
(a)

S in kips
(b)

Values of u
(c)

FIG. 13-5

of joint L_1). Each of the two absolute movements may be found by applying a single unit load at a time, but by applying a pair of unit loads at L_1 and U_2 as shown in Fig. 13-5c, the combined relative movement may be found directly by $\Sigma u(\Delta L)$, in which u is the stress in any truss member due to the pair of unit loads. As shown in Table 13-3, this relative movement is $+83.25 \times 10^{-3}$ in., or 0.08325 in. away from each other.

13-3. Deflections of Statically Determinate Trusses: the Graphical Method. Since a truss is an assembly of triangles, the sides of which

TABLE 13-3

Member	L, in.	A, sq in.	S, kips	$L = \dfrac{SL}{AE}$, 10^{-3} in.	u	$\Delta = \Sigma u(\Delta L)$, 10^{-3} in.
U_1U_2	180	4	-12	-18	$+0.6$	-10.8
L_0L_1	180	4	$+10.5$	$+15.75$	0	0
L_1L_2	180	4	$+10.5$	$+15.75$	$+0.6$	$+\ 9.45$
L_2L_3	180	4	$+12$	$+18$	0	0
L_0U_1	300	5	-17.5	-35	0	0
U_1L_2	300	5	$+\ 2.5$	$+\ 5$	-1	$-\ 5$
U_2L_3	300	5	-20	-40	0	0
U_1L_1	240	2	$+12$	$+48$	$+0.8$	$+38.4$
U_2L_2	240	2	$+16$	$+64$	$+0.8$	$+51.2$
Σ						$+83.25$

remain straight while undergoing changes in length, it can be surmised that the shape of the deformed truss may be graphically determined by using the new lengths of members as the sides of the component triangles. But the changed lengths of the members are only a little longer or shorter than the original lengths, a fact which makes the deformed truss almost coincide with the original truss when ordinary scales are used. This difficulty may be avoided by using two different scales when plotting the original lengths L and the changes in length ΔL.

Consider, for example, the Warren truss $ABCDE$ in Fig. 13-6a. Let $+7$ units be the lengthening in member AB; $+6$, $+3$, and $+9$ units the lengthenings in members BC, BD, and BE; -8, -5, and -4 units the shortenings in members AC, CD, and DE. The shape of the deformed truss may be determined by drawing the new triangles $A'B'C'$, $B'C'D'$, and $B'D'E'$ in succession.

Starting with triangle $A'B'C'$, assume that joint A' of the deformed truss coincides with joint A of the original truss (Fig. 13-6a), and assume also that the direction of $A'B'$ coincides with that of AB. B', then, must fall at 7 units to the right of B in the direction AB. Joint C', the only unknown point on triangle $A'B'C'$, may be determined by the intersection of two arcs, using A' and B' as centers and the lengths of $A'C'$ and $B'C'$ as radii. This is performed by the following procedure: From C measure $CC_1 = 8$ units toward A (or A') because member AC is shortened by 8 units. AC_1 is then the new length of member AC. Now, instead of drawing the arc with A' as center and $A'C_1$ as radius, which can be done only if CC_1 is plotted on the same scale as that for AC, a perpendicular to AC is drawn at C_1. This is an approximation which is permissible because the deformations are very small in comparison with

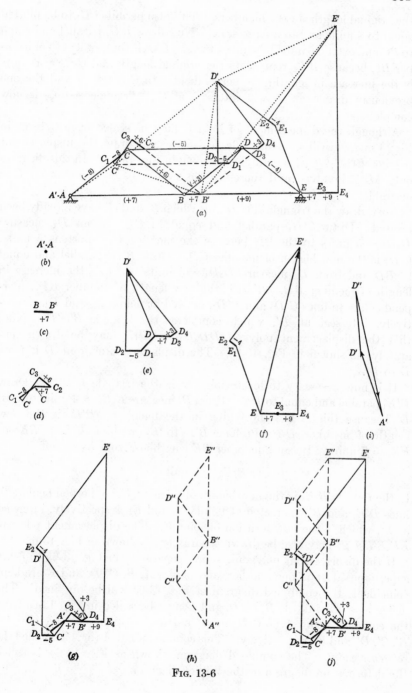

Fig. 13-6

the original lengths of the members. This also permits CC_1 to be plotted equal to 8 units and to a large scale. Now draw $B'C_2$ parallel and equal to BC and extend 6 units beyond C_2 to C_3. $B'C_3$ is the new length of member BC because $B'C_2$ represents the original length and $C_2C_3 = 6$ units is the increase in length. A perpendicular to C_3B' at C_3 and the one previously drawn at C_1 will intersect at C'. Triangle $A'B'C'$ is now completed.

Although the displacement of joint C may be scaled from C to C' in Fig. 13-6a, it will be preferable to isolate and draw the displacement polygon $CC_2C_3C'C_1C$ separately as shown in Fig. 13-6d. In this diagram only ΔL is involved and therefore the scale may be made as large as desirable.

Next, draw the triangle $B'C'D'$, for which B' and C' have already been located. Draw $C'D_1$ parallel and equal to CD. From D_1 measure $D_1D_2 = 5$ units to the left because the member CD shortens 5 units. $C'D_2$ is the new length of member CD. Draw $B'D_3$ parallel and equal to BD and from D_3 measure $D_3D_4 = 3$ units which is the increase in length of member BD. $B'D_4$ is the new length of member BD. Perpendiculars to lines $C'D_2$ and $B'D_4$, erected at points D_2 and D_4, respectively, intersect at D', which completes the triangle $B'C'D'$. Note that the displacement polygon $DD_3D_4D'D_2D_1D$ may be drawn separately, as shown in Fig. 13-6e. The displacement of joint D is from D to D'.

It is now necessary to locate joint E' in the triangle $B'D'E'$. Draw $D'E_1$ parallel and equal to DE. From E_1 measure $E_1E_2 = 4$ units toward D' because this is the shortening in the member. $D'E_2$ is the new length of member DE. Prolong $B'E$ to E_3, making $B'E_3 = BE$, or $EE_3 = 7$ units. Because member BE lengthens, from E_3 lay off

$$E_3E_4 = 9 \text{ units}$$

to the right. $B'E_4$ is the new length of member BE. Perpendiculars to lines $D'E_2$ and $B'E_4$, erected at points E_2 and E_4, respectively, intersect at E', which is the new location of joint E. The displacement polygon $EE_4E'E_2E_1E$ may also be drawn separately, as shown in Fig. 13-6f.

If the displacement polygons, shown separately in Fig. 13-6b to f, are superimposed upon one another, and points A, B, C, D, and E are kept coincident, the combined diagram of Fig. 13-6g will be obtained. The deflections of joints A, B, C, D, and E may be scaled in Fig. 13-6g from the common point A (B,C,D, or E) outward to points A' ($AA' = 0$), B', C', D', and E', respectively. This common point A (or A') is called the *reference point*. The combined diagram, shown as Fig. 13-6g, is called the deformation diagram or the Williot diagram.

The preceding explanation of the Williot diagram has been given in detail to ensure a basic understanding of the analysis. In practice, however, this construction may be greatly simplified. After choosing joint A as the reference point and member AB as the reference member, start with A' (Fig. 13-6g) and then, because the lengthening in AB is 7 units, locate point B' 7 units to the right of A. Because joint C moves 8 units downward to the left relative to joint A, from A' lay off $A'C_1 = 8$ units downward to the left and parallel to the direction of member AC. Since joint C moves upward to the left relative to joint B, from B' measure $B'C_3 = 6$ units upward to the left and parallel to the direction of member BC. The two perpendiculars to the directions of members AC and BC, erected at points C_1 and C_3, respectively, intersect at C'. From C' draw $C'D_2 = 5$ units horizontally to the left; and from B' measure $B'D_4 = 3$ units upward to the right in the direction of member BD. The two perpendiculars drawn at D_2 and D_4 intersect at D'. From D' lay off $D'E_2 = 4$ units upward to the left in the direction of member DE, and from B' draw $B'E_4 = 9$ units horizontally to the right. The two perpendiculars drawn at E_2 and E_4 intersect at E'. While studying this paragraph the reader is advised to make an independent sketch of the Williot diagram as shown in Fig. 13-6g.

The deformed truss now assumes the form $A'B'C'D'E'$ as shown in Fig. 13-6a, and the movement of each joint may be measured in Fig. 13-6g from the reference point to the single prime point in question. Obviously it is necessary to rotate the deformed truss $A'B'C'D'E'$ of Fig. 13-6a through a clockwise angle, in radians, of $E'E_4$ divided by the span of the truss so as to bring the point E' to the same level as the hinge at A and to coincide with point E_4. The additional displacements of all joints due to this rotation may be found as follows: In Fig. 13-6g, a vertical line through the hinge A (also called A' or A'') and a horizontal line through E' are made to intersect at E'' (see Fig. 13-6j); then the vertical distance $E''A''$ is the movement of joint E due to the above-mentioned rotation. With $A''E''$ as a base (Fig. 13-6h), $A''B''C''D''E''$ is drawn similar to the original truss $ABCDE$. In this case, the rotation diagram $A''B''C''D''E''$ should be drawn on the left of the base $A''E''$ because the criterion is that $A''B''C''D''E''$ can be rotated 90° to a position parallel and similar to the original truss, $ABCDE$.

It can be shown that the movements of joints B, C, D, E due to this rotation may be scaled in Fig. 13-6h from the double-prime point in question to the hinge A (A' or A''). For instance, it is required to prove that $D''A''$ in Fig. 13-6h equals line DA in Fig. 13-6a times the angle of rotation and that $D''A''$ in Fig. 13-6h is perpendicular to line DA in Fig. 13-6a.

PROOF.   Since triangle $D''A''E''$ (Fig. 13-6h) is similar to triangle DAE (Fig. 13-6a), $E''A''$ is perpendicular to EA, and

$$\text{angle } D''A''E'' = \text{angle } DAE$$

$D''A''$ is perpendicular to DA.   Also,

$$\frac{D''A''}{DA} = \frac{E''A''}{EA}$$

but $\dfrac{E''A''}{EA}$ = angle of rotation

Therefore, $\dfrac{D''A''}{DA}$ = angle of rotation

or $D''A'' = (DA)$ times the angle of rotation

If Fig. 13-6h is superimposed on Fig. 13-6g, Fig. 13-6j is obtained. The total movement of each joint is the vector sum of the "rotation movement" from the double-prime point to the hinge A (A' or A'') and the "deformation movement" from the reference point (in this case it happens that the hinge is chosen as the reference point) to the single-prime point.   (The vector diagram for joint D is shown as Fig. 13-6i.)

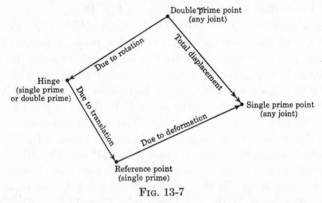

FIG. 13-7

More simply, the total movement may be measured directly from the double-prime point to the single-prime point in Fig. 13-6j.   Thus, in the graphical solution for the magnitude and direction of the deflection of each joint in a truss, a diagram similar to Fig. 13-6j is *all that is necessary*, and the deflection is always measured from a double-prime point to the corresponding single-prime point.   Figure 13-6g is known as the Williot diagram, Fig. 13-6h as the Mohr diagram, and Fig. 13-6j as the Williot-Mohr diagram.

If some joint other than the hinge had been chosen as the reference point, then owing to deformation only the hinge would have moved from the reference point to the single-prime point for the hinge.   It is necessary, therefore, in addition to the deformation and rotation effects, to translate the deformed truss through the displacement from the single-prime point for the hinge back to the reference point. Figure 13-7 shows the combination of all three vectors, but again, the total movement may be measured directly from the double-prime point to the single-prime point.

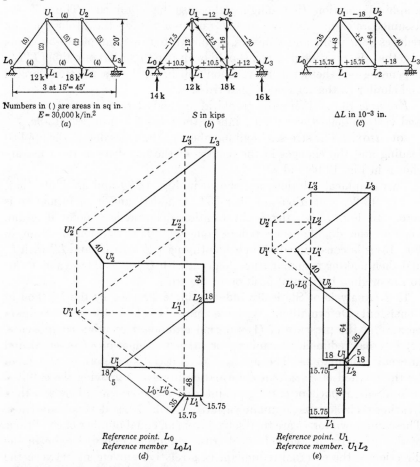

Note. Deflection is measured from the double prime point to the single prime point

FIG. 13-8

Example 13-4. Using the graphical method, determine the horizontal and vertical deflections of all joints of the truss shown in Fig. 13-8a.

SOLUTION. The stresses in all members of the truss due to the applied loading and the corresponding changes in the lengths of the members are shown in Fig. 13-8b and c.

Two graphical solutions are shown in Fig. 13-8d and e. It is to be noted that, had the same scale been used in both solutions, that shown in Fig. 13-8e would take much less space than does the solution in Fig. 13-8d. A good choice of reference point and the selection of a reference member with comparatively little rotation will require a relatively small rotation or correction diagram and therefore will yield a more compact graphical solution than might otherwise be obtained. Thus, if the assumed reference member or line does not rotate, the Mohr diagram reduces to a point and vanishes, as illustrated in Example 13-5. Note also that the rotation diagram should be drawn to the left or right of $L_0''L_3''$, depending on whether a 90° rotation will bring it to a position parallel and similar to that of the original truss.

Example 13-5. Using the graphical method, determine the horizontal and vertical deflections of all joints of the truss shown in Fig. 13-9a.

SOLUTION. The stresses in all members of the truss due to the applied loading and the changes in the lengths of the members of the truss are shown in Fig. 13-9b and c.

Two graphical solutions are shown in Fig. 13-9d and e. Note that, in Fig. 13-9c, a fictitious member U_0L_0 is added, and its deformation is zero. In Fig. 13-9d, the Mohr rotation diagram is a point diagram, viz., all the double-prime points coincide. The rotation diagram in Fig. 13-9e is constructed by first identifying U_0'' with U_0' and L_0'' with L_0' and then making a similar truss diagram with $U_0''L_0''$ as a base and at 90° to the original horizontal position of the truss.

13-4. Analysis of Statically Indeterminate Trusses by the Method of Consistent Deformation. A truss may be statically indeterminate because of the presence of (1) external redundant reaction components, (2) internal redundant members, or (3) a combination of external and internal redundants. The analysis of statically indeterminate trusses by the method of consistent deformation involves removing the external redundant supports or cutting the internal redundant members and then treating their actions as unknown forces on the basic determinate truss. These unknown forces are then solved from an equal number of conditions of the geometry of the deformed truss, requiring zero deflection in the direction of the external redundant or a relative movement between the cut ends of the internal redundant equal to the deformation of the redundant member. Because of the limited scope of this text, the following examples will deal only with trusses statically indeterminate to the first degree. One of these has an external redundant reaction and the other involves an internal redundant member.

Example 13-6. Analyze the statically indeterminate truss shown in Fig. 13-10a by the method of consistent deformation.

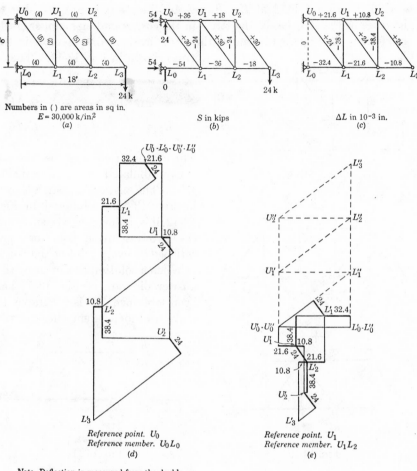

Numbers in ( ) are areas in sq in.
$E = 30,000\,\text{k/in.}^2$
(a)

S in kips
(b)

ΔL in 10^{-3} in.
(c)

Reference point. U_0
Reference member. $U_0 L_0$
(d)

Reference point. U_1
Reference member. $U_1 L_2$
(e)

Note. Deflection is measured from the double
prime point to the single prime point

Fig. 13-9

SOLUTION.   Any one of the three vertical reactions may be chosen as
the redundant force.   If R_2 is assumed to be redundant, the truss in
Fig. 13-10a is the equivalent of the two trusses shown in Fig. 13-10b and c.
For zero deflection at L_2,

$$R_2 = \frac{\Delta_2}{\delta_2}$$

in which Δ_2 is the downward deflection of L_2 due to the applied loads in
the basic determinate truss, and δ_2 is the upward deflection at L_2 due to a
1-kip upward load at L_2 in the basic determinate truss.   However, as
shown in Fig. 13-11, δ_2 may also be the downward deflection at L_2 due to

a 1-kip downward load at L_2. Let S be the stress in any member of the truss in Fig. 13-10a, S' the stress in the corresponding member of the simple truss in Fig. 13-10b, and u the stress in this same member in the truss shown in Fig. 13-11. Then

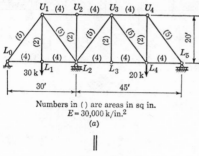

Numbers in ( ) are areas in sq in.
$E = 30,000$ k/in.^{2}
(a)

$\parallel$

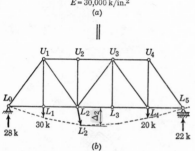

(b)

$$\Delta_2 = \sum \frac{S'uL}{AE} \qquad \delta_2 = \sum \frac{u^2L}{AE}$$

$$R_2 = \frac{\Delta_2}{\delta_2} \qquad S = S' - R_2 u$$

The computations may be conveniently tabulated as shown in Table 13-4. The answer diagram is shown in Fig. 13-12. As noted in Fig. 13-12, it is important to be sure that the two resolution equations are satisfied at every joint of the truss.

When a solution as shown by the answer diagram of Fig. 13-12 has been obtained, it is desirable to make an independent check of the

$+$

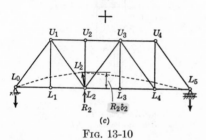

(c)
FIG. 13-10

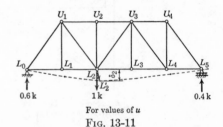

For values of u
FIG. 13-11

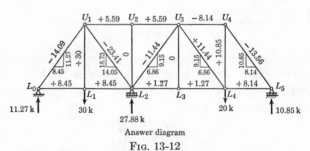

Answer diagram
FIG. 13-12

calculated stresses. One procedure is to assume either R_0 or R_5 as the redundant and then make another independent solution of the problem. It may be more convenient, however, to use a basic determinate structure different from the one used in the initial analysis and then check

TABLE 13-4

Member	L, in.	A, sq in.	S', kips	u	$\dfrac{S'uL}{AE}$, 10^{-3} in.	$\dfrac{u^2L}{AE}$, 10^{-3} in.	$-R_2u$, kips	$S = S' - R_2u$, kips
U_1U_2	180	4	-19.5	-0.90	$+\ 26.325$	$+1.215$	$+25.09$	$+\ 5.59$
U_2U_3	180	4	-19.5	-0.90	$+\ 26.325$	$+1.215$	$+25.09$	$+\ 5.59$
U_3U_4	180	4	-16.5	-0.30	$+\ 7.425$	$+0.135$	$+\ 8.36$	$-\ 8.14$
L_0L_1	180	4	$+21$	$+0.45$	$+\ 14.175$	$+0.30375$	-12.55	$+\ 8.45$
L_1L_2	180	4	$+21$	$+0.45$	$+\ 14.175$	$+0.30375$	-12.55	$+\ 8.45$
L_2L_3	180	4	$+18$	$+0.60$	$+\ 16.200$	$+0.540$	-16.73	$+\ 1.27$
L_3L_4	180	4	$+18$	$+0.60$	$+\ 16.200$	$+0.540$	-16.73	$+\ 1.27$
L_4L_5	180	4	$+16.5$	$+0.30$	$+\ 7.425$	$+0.135$	$-\ 8.36$	$+\ 8.14$
L_0U_1	300	5	-35	-0.75	$+\ 52.500$	$+1.125$	$+20.91$	-14.09
U_1L_2	300	5	$-\ 2.5$	$+0.75$	$-\ 3.750$	$+1.125$	-20.91	-23.41
L_2U_3	300	5	$+\ 2.5$	$+0.50$	$+\ 2.500$	$+0.500$	-13.94	-11.44
U_3L_4	300	5	$-\ 2.5$	-0.50	$+\ 2.500$	$+0.500$	$+13.94$	$+11.44$
U_4L_5	300	5	-27.5	-0.50	$+\ 27.500$	$+0.500$	$+13.94$	-13.56
U_1L_1	240	2	$+30$	0	0	0	0	$+30$
U_2L_2	240	2	0	0	0	0	0	0
U_3L_3	240	2	0	0	0	0	0	0
U_4L_4	240	2	$+22$	$+0.40$	$+\ 35.200$	$+0.640$	-11.15	$+10.85$
R_0	...	..	$+28$	$+0.60$	....	....	-16.73	$+11.27$
R_2	...	..	0	-1	....	....	$+27.88$	$+27.88$
R_5	...	..	$+22$	$+0.40$	....	....	-11.15	$+10.85$
Σ					$+244.700$ $(= \Delta_2)$	$+8.7775$ $(= \delta_2)$		

a geometrical condition. For instance, a truss supported at L_0 and L_2 and subjected to the loads of 30, 20, and 10.85 kips at L_0, L_2, and L_5, respectively, as shown in Fig. 13-13a may be used. Note that the load at L_5 is the previously calculated reaction but that it is now considered

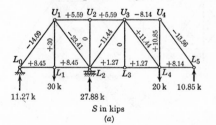

FIG. 13-13

to be an upward load acting at L_5. The condition of consistent deformation is that the deflection of this truss at L_5 should equal zero. Values of u are shown in Fig. 13-13b, and computations for the deflection at L_5 are arranged in Table 13-5. Thus, when a solution satisfies the necessary conditions of both statics and geometry, its correctness is ensured.

TABLE 13-5

Member	L, in.	A, sq in.	S, kips	$\Delta L = \dfrac{SL}{AE}$, 10^{-3} in.	u	$u\,\Delta L$, 10^{-3} in.
U_1U_2	180	4	$+\ 5.59$	$+\ \ 8.39$	$+2.25$	$+18.88$
U_2U_3	180	4	$+\ 5.59$	$+\ \ 8.39$	$+2.25$	$+18.88$
U_3U_4	180	4	$-\ 8.14$	$-\ 12.21$	$+0.75$	$-\ 9.16$
L_0L_1	180	4	$+\ 8.45$	$+\ 12.67$	-1.125	-14.25
L_1L_2	180	4	$+\ 8.45$	$+\ 12.67$	-1.125	-14.25
L_2L_3	180	4	$+\ 1.27$	$+\ \ 1.90$	-1.5	$-\ 2.85$
L_3L_4	180	4	$+\ 1.27$	$+\ \ 1.90$	-1.5	$-\ 2.85$
L_4L_5	180	4	$+\ 8.14$	$-\ 12.21$	-0.75	$-\ 9.16$
L_0U_1	300	5	-14.09	$-\ 28.18$	$+1.875$	-52.84
U_1L_2	300	5	-23.41	$-\ 46.82$	-1.875	$+87.79$
L_2U_3	300	5	-11.44	$-\ 22.88$	-1.25	$+28.60$
U_3L_4	300	5	$+11.44$	$+\ 22.88$	$+1.25$	$+28.60$
U_4L_5	300	5	-13.56	$-\ 27.12$	$+1.25$	-33.90
U_1L_1	240	2	$+30$	$+120$	0	0
U_2L_2	240	2	0	0	0	0
U_3L_3	240	2	0	0	0	0
U_4L_4	240	2	$+10.85$	$+\ 43.40$	-1	-43.40
Σ						$+\ 0.09 \approx 0$

Example 13-7.   Using the method of consistent deformation, analyze the statically indeterminate truss shown in Fig. 13-14a.

SOLUTION.   The given truss has no external redundant, but internally it has one redundant member ($m = 10$, $j = 6$). Any one of the six members around or within the middle panel may be cut without impair-

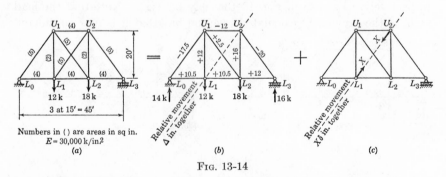

Numbers in ( ) are areas in sq in.
$E = 30{,}000$ k/in^2

(a)   (b)   (c)

FIG. 13-14

ing the static stability of the truss.   Normally, however, one of the two diagonals is taken as the redundant because obviously there is one extra or unnecessary diagonal in the middle panel.   If member L_1U_2 is cut, the given truss may be replaced by the two basic determinate trusses shown in Fig. 13-14b and c.   One of these carries the applied loads and the other is subjected to the pair of forces X and X, the unknown tensile stress in the redundant member.   Let

Δ = relative movement together between L_1 and U_2 due to the applied loads

δ = relative movement together between L_1 and U_2 due to a pair of 1-kip loads at joints L_1 and U_2 and acting in the direction of the cut member (Fig. 13-15)

Then the condition for consistent deformation becomes

$$-(\Delta + X\delta) = + \frac{XL}{AE}$$

or

$$X = - \frac{\Delta}{\delta + (L/AE)}$$

because actually joints L_1 and U_2 should move apart to accommodate an elongation of XL/AE in the redundant member.   Let S' be the stress

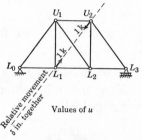

Values of u

Fig. 13-15

in any member of the truss shown in Fig. 13-14b, u be that due to a pair of 1-kip loads at joints L_1 and U_2 as shown in Fig. 13-15, and S be the resultant stress in the member, then

$$\Delta = \sum \frac{S'uL}{AE} \qquad \delta = \sum \frac{u^2L}{AE} \qquad S = S' + uX$$

Note that the above two summations do not include the redundant member.

From Table 13-6, it is seen that

$$\Delta = \sum \frac{S'uL}{AE} = -83.25 \times 10^{-3} \text{ in.}$$

and

$$\delta = \sum \frac{u^2L}{AE} = +8.20 \times 10^{-3} \text{ in.}$$

TABLE 13-6

Member	L, in.	A, sq in.	S', kips	u (Fig. 13-15)	$\dfrac{S'uL}{AE}$	$\dfrac{u^2L}{AE}$	$(u)(X)$	$S = S' + uX$	u (Fig. 13-17)	$\dfrac{SuL}{AE}$
(1)	(2)	(3)	(4)	(5)	(6)	(7)	(8)	(9)	(10)	(11)
U_1U_2	180	4	-12	-0.6	$+10.8$	$+0.54$	-4.90	-16.90	$+0.6$	-15.21
L_0L_1	180	4	$+10.5$	0	0	0	0	$+10.50$	0	0
L_1L_2	180	4	$+10.5$	-0.6	-9.45	$+0.54$	-4.90	$+5.60$	$+0.6$	$+5.04$
L_2L_3	180	4	$+12$	0	0	0	0	$+12.00$	0	0
L_0U_1	300	5	-17.5	0	0	0	0	-17.50	0	0
U_1L_2	300	5	$+2.5$	$+1.0$	$+5$	$+2.00$	$+8.16$	$+10.66$	0	0
L_1U_2	300	5	....	....	....	....	$+8.16$	$+8.16$	-1	-16.32
U_2L_3	300	5	-20	0	0	0	0	-20.00	0	0
U_1L_1	240	2	$+12$	-0.8	-38.4	$+2.56$	-6.53	$+5.47$	$+0.8$	$+17.50$
U_2L_2	240	2	$+16$	-0.8	-51.2	$+2.56$	-6.53	-9.47	$+0.8$	$+30.31$
Σ					-83.25 ($=\Delta$)	$+8.20$ ($=\delta$)				$+21.32$

Therefore

$$X = -\frac{(-83.25)}{(+8.20) + \dfrac{300}{5 \times 30}} = +\frac{83.25}{10.20} = +8.16 \text{ kips}$$

With the stress in the redundant known, the total stress S in any member is found from $S = S' + uX$, as shown in column 9 of Table 13-6. The answer diagram is shown in Fig. 13-16.

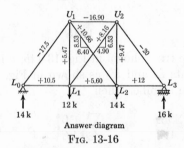

Answer diagram

FIG. 13-16

The conditions of static equilibrium may be checked by making sure that the two resolution equations are satisfied at every joint as shown in Fig. 13-16. The geometrical condition is checked by calculating the relative movement between joints U_1 and L_2 of the truss in Fig. 13-17a, which is the truss shown in Fig. 13-16 with member U_1L_2 cut and its action replaced by a pair of forces of 10.66 kips each. As shown in Table 13-6 (columns 10 and 11), joints U_1 and L_2 move 21.32×10^{-3} in. apart, which is just equal to the elongation caused by a tension of 10.66 kips

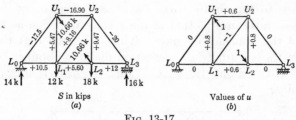

FIG. 13-17

in member U_1L_2.   The correctness of the solution is ensured because the conditions of statics and geometry are now both satisfied.

13-5. Influence Diagrams for Statically Indeterminate Trusses.   A truss supported at more than two points is called a continuous truss and is, of course, statically indeterminate.   In the analysis of statically indeterminate trusses carrying a system of moving loads, influence diagrams for the reactions or stresses in the truss members must be constructed in order to determine critical loading positions.   The discussion in this article will be limited to the determination of the influence diagram for the intermediate reaction on a two-span continuous truss.

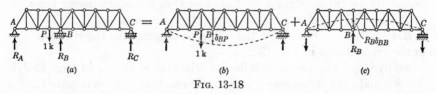

FIG. 13-18

Let it be required to construct the influence diagram for R_B of the truss shown in Fig. 13-18a.   In the development of this influence diagram, it will be recalled that the value of R_B due to a 1-kip load at P must be plotted directly under P, as shown in Fig. 13-20.   According to the method of consistent deformation (Fig. 13-18), the value of R_B due to a 1-kip load at P is

$$R_B = \frac{\delta_{BP}}{\delta_{BB}}$$

By the law of reciprocal deflections,

$$\delta_{BP} = \delta_{PB}$$

Therefore
$$R_B = \frac{\delta_{PB}}{\delta_{BB}}$$

Values of δ_{PB} and δ_{BB} may be taken directly from Fig. 13-19b, which is the elastic curve of the simple truss AC due to a unit load at B.   Thus the required influence diagram which is shown in Fig. 13-20b is similar to the elastic curve of Fig. 13-19b.   The scale has been changed because the ordinates to the elastic curve have been divided by the constant δ_{BB} and these ratios are plotted in the reverse direction as the positive ordinates of the influence diagram. Thus it is seen that the influence diagram for R_B is actually the deflection diagram obtained by applying a load at B of the simple truss AC so that the deflection caused at B is equal to unity.

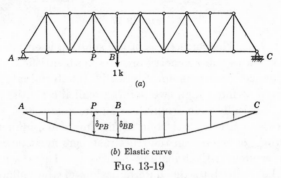

(a)

(b) Elastic curve

Fig. 13-19

Similarly, influence diagrams for R_A or R_C may be obtained independently by introducing a load at A or C of the overhanging truss BC or AB so that the deflection at A or C is equal to unity.

Example 13-8. Construct the influence diagram for the reaction at L_2 in the truss shown in Fig. 13-21a.

SOLUTION. The required influence diagram is shown in Fig. 13-21b. To calculate the influence ordinates, the reaction at L_2 is removed and a

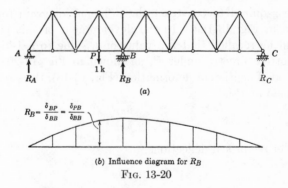

(a)

$$R_B = \frac{\delta_{BP}}{\delta_{BB}} = \frac{\delta_{PB}}{\delta_{BB}}$$

(b) Influence diagram for R_B

Fig. 13-20

load of 1 kip is placed at L_2 as shown in Fig. 13-22a. Then by any convenient method, either algebraic or graphic, determine the vertical deflections of the lower-chord panel points of the truss L_0L_5 and plot the elastic curve as shown in Fig. 13-22b.

Inasmuch as the influence ordinate at L_2 should be unity, divide the ordinates to the deflection curve in Fig. 13-22b by 8.7775 and plot these

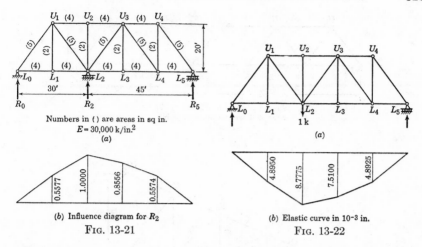

Numbers in ( ) are areas in sq in.
$E = 30,000 \, k/in.^2$
(a)

(b) Influence diagram for R_2

Fig. 13-21

(a)

(b) Elastic curve in 10^{-3} in.

Fig. 13-22

ratios above the base line as shown in Fig. 13-21b. This is the required influence diagram.

PROBLEMS

13-1. Determine the horizontal and vertical deflections of joint L_3 of the truss shown by the unit-load method.

13-2. Determine the horizontal and vertical deflections of joint L_5 of the truss shown by the unit-load method.

13-3. Determine the horizontal and vertical deflections of joints L_0 and L_2 of the truss shown by the unit-load method.

13-4. Determine the relative movement in the direction U_2L_3 between the joints U_2 and L_3 of the truss shown by the unit-load method.

13-5. Determine the relative movement in the direction L_2U_3 between the joints L_2 and U_3 of the truss shown by the unit-load method.

13-6 to 13-8. Determine the horizontal and vertical deflections of all joints of the truss shown by the graphical method.

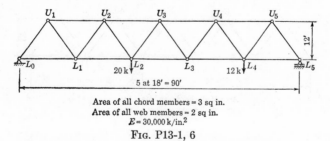

Area of all chord members = 3 sq in.
Area of all web members = 2 sq in.
$E = 30,000 \, k/in.^2$

Fig. P13-1, 6

13-9. Determine the relative movement in the direction U_2L_3 between the joints U_2 and L_3 of the truss shown by the graphical method.

13-10. Determine the relative movement in the direction L_2U_3 between the joints L_2 and U_3 of the truss shown by the graphical method.

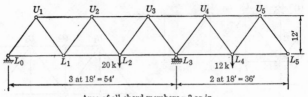

Area of all chord members = 3 sq in.
Area of all web members = 2 sq in.
$E = 30,000\ \text{k/in.}^2$

FIG. P13-2, 7

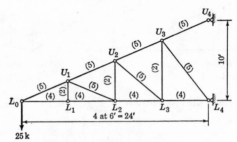

Numbers in ( ) are areas in sq in.
$E = 30,000\ \text{k/in.}^2$

FIG. P13-3, 8

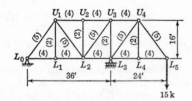

Numbers in ( ) are areas in sq in.
$E = 30,000\ \text{k/in.}^2$

FIG. P13-4, 9

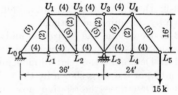

Numbers in ( ) are areas in sq in.
$E = 30,000\ \text{k/in.}^2$

FIG. P13-5, 10

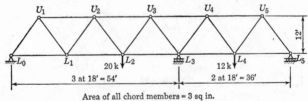

Area of all chord members = 3 sq in.
Area of all web members = 2 sq in.
$E = 30,000\ \text{k/in.}^2$

FIG. P13-11, 12

13-11. Using the reaction at L_3 as the redundant, analyze the statically indeterminate truss shown by the method of consistent deformation.

13-12. Using the reaction at L_5 as the redundant, analyze the statically indeterminate truss shown by the method of consistent deformation.

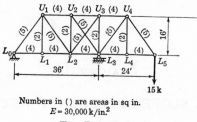

Numbers in ( ) are areas in sq in.
$E = 30,000 \, \text{k/in.}^2$

FIG. P13-13, 14

13-13. Using member U_2L_3 as the redundant, analyze the statically indeterminate truss shown by the method of consistent deformation.

13-14. Using member L_2U_3 as the redundant, analyze the statically indeterminate truss shown by the method of consistent deformation.

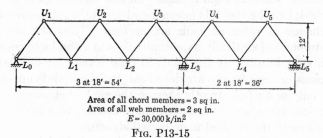

Area of all chord members = 3 sq in.
Area of all web members = 2 sq in.
$E = 30,000 \, \text{k/in.}^2$

FIG. P13-15

13-15. Construct the influence diagrams for the reactions at L_0, L_3, and L_5 in the truss shown.

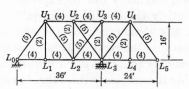

Numbers in ( ) are areas in sq in.
$E = 30,000 \, \text{k/in.}^2$

FIG. P13-16

13-16. Construct the influence diagrams for the stresses in members U_2L_3 and L_2U_3 in the truss shown.

CHAPTER 14

THE SLOPE-DEFLECTION METHOD

14-1. General Description of the Slope-deflection Method. The slope-deflection method may be used to analyze all types of statically indeterminate beams or rigid frames. In this method all joints are considered rigid; i.e., the angles between members at the joints are considered not to change in value as loads are applied. Thus the joints at the interior supports of statically indeterminate beams can be considered 180° rigid joints; and ordinarily the joints in rigid frames are 90° rigid joints. When beams or rigid frames are deformed, the rigid joints are considered to rotate only as a whole; in other terms, the angles between the tangents to the various branches of the elastic curve meeting at a joint remain the same as those in the original undeformed structure.

In the slope-deflection method the rotations of the joints are treated as unknowns. It will be shown later that for any one member bounded by two joints the end moments can be expressed in terms of the end rotations. But, to satisfy the condition of equilibrium, the sum of the end moments which any joint exerts on the ends of members meeting there must be zero, because the rigid joint in question is subjected to the sum of these end moments (only reversed in direction). This equation of equilibrium furnishes the necessary condition to cope with the unknown rotation of the joint, and when these unknown joint rotations are found, the end moments can be computed from the slope-deflection equations which will be derived in the next article.

The discussion in the preceding paragraph will be clarified by the following example: suppose that it is required to analyze the rigid frame loaded as shown in Fig. 14-1a. This frame is statically indeterminate to the sixth degree. The method of consistent deformation could be used, but the amount of work involved would make that method too laborious. Because the frame is kept from horizontal movement by its connection at A and vertical movement is prevented by the fixed bases at D and E, and since axial deformation of the members is usually neglected, all joints of this frame must remain in their original locations. (The cases in which some joints may change position when the frame is deformed will be taken up later.) *Clockwise* joint rotations, as shown in

Fig. 14-1a, are considered to be *positive*. The free-body diagrams of all members are shown in Fig. 14-1b. At any one end of each member, there are three reaction components: direct pull or thrust, end shear, and end moment. The end moment which acts at end A of member AB is denoted as M_{AB}; that at end B of member AB, as M_{BA}. *Counterclockwise* end moments acting on the members are considered to be *positive*, positive end moments being shown in Fig. 14-1b. It is possible, by the use of the slope-deflection equations to be derived in the next article, to express the end moments of each member in terms of the end rotations and the loading which acts on the member. Thus the eight end moments

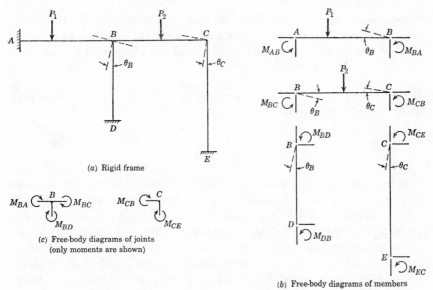

(a) Rigid frame

(c) Free-body diagrams of joints
(only moments are shown)

(b) Free-body diagrams of members

FIG. 14-1

in this problem may be expressed in terms of the two unknown joint rotations. The free-body diagrams of all joints are shown in Fig. 14-1c. Of course, the action of the member on the joint consists of a force in the direction of the axis of the member, a force perpendicular to this axis, and a moment, each being the opposite of the action of the joint on the member. In Fig. 14-1c, only the moments are shown. These moments are drawn in their *positive* direction, which is *clockwise*. For equilibrium, summation of all moments acting at each joint must be zero. Thus

Joint condition at B: $M_{BA} + M_{BC} + M_{BD} = 0$
Joint condition at C: $M_{CB} + M_{CE} = 0$

The above two equations are necessary and sufficient to determine the values of θ_B and θ_C. All end moments can then be found by substituting the known joint rotations in the *slope-deflection equations*. By the principles of statics the direct stress and shear and bending-moment diagrams for each member may be found.

It has been repeatedly pointed out that the analysis of statically indeterminate structures must satisfy both *statics* and *geometry*. In the slope-deflection method of analyzing rigid frames, the conditions required of the geometry of the deformed structure, which are those of the rigidity of the joints, are satisfied at the outset by calling the joint rotation one single unknown at each joint. Thus the conditions of statics, requiring that the sum of moments acting on each joint be zero, are used to solve for the joint rotations.

14-2. Derivation of the Slope-deflection Equations. In the slope-deflection equations, the end moments acting at the ends of a member are expressed in terms of the end rotations and the loading on the member.

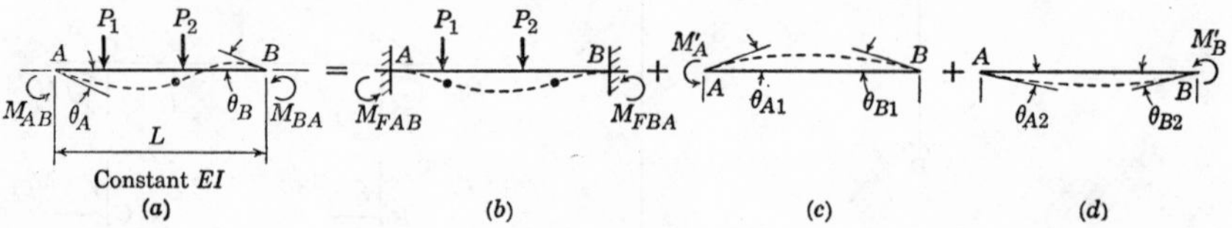

FIG. 14-2

Thus for span AB shown in Fig. 14-2a it is required to express M_{AB} and M_{BA} in terms of the end rotations θ_A and θ_B and the applied loading P_1 and P_2. Note that the end moments are shown as counterclockwise (positive) and the end rotations are shown as clockwise (positive). Now, with the applied loading on the member, the fixed end moments M_{FAB} and M_{FBA} (both shown as counterclockwise) are required to hold the tangents fixed at the ends (Fig. 14-2b). The additional end moments M'_A and M'_B should be such as to cause rotations of θ_A and θ_B, respectively. If θ_{A1} and θ_{B1} are the end rotations caused by M'_A and θ_{A2} and θ_{B2} by M'_B (Fig. 14-2c and d), the conditions required by geometry are

$$\theta_A = -\theta_{A1} + \theta_{A2}$$
$$\theta_B = +\theta_{B1} - \theta_{B2} \tag{14-1}$$

By superposition,

$$M_{AB} = M_{FAB} + M'_A$$
$$M_{BA} = M_{FBA} + M'_B \tag{14-2}$$

By the conjugate-beam method,

$$\theta_{A1} = \frac{M'_A L}{3EI} \qquad \theta_{B1} = \frac{M'_A L}{6EI}$$
$$\theta_{A2} = \frac{M'_B L}{6EI} \qquad \theta_{B2} = \frac{M'_B L}{3EI} \tag{14-3}$$

Substituting Eqs. (14-3) in Eqs. (14-1),

$$\theta_A = -\frac{M'_A L}{3EI} + \frac{M'_B L}{6EI}$$

$$\theta_B = +\frac{M'_A L}{6EI} - \frac{M'_B L}{3EI}$$

(14-4)

Solving Eqs. (14-4) for M'_A and M'_B,

$$M'_A = +\frac{2EI}{L}(-2\theta_A - \theta_B)$$

$$M'_B = +\frac{2EI}{L}(-2\theta_B - \theta_A)$$

(14-5)

let $K = \frac{4EI}{L}$

$M_{AB} = M_{FAB} + \frac{K}{2}(2\theta_A + \theta_B)$

$M_{BA} = M_{FBA} - \frac{K}{2}(\theta_A + 2\theta_B)$

Substituting Eqs. (14-5) in Eqs. (14-2),

$$M_{AB} = M_{FAB} + \frac{2EI}{L}(-2\theta_A - \theta_B)$$

$$M_{BA} = M_{FBA} + \frac{2EI}{L}(-2\theta_B - \theta_A)$$

(14-6)

Equations (14-6) are the slope-deflection equations which express the end moments in terms of the end rotations and the applied loading. Note again that counterclockwise moments acting at the ends of the member (M_{AB}, M_{BA}, M_{FAB}, and M_{FBA}) are positive and clockwise rotations (θ_A and θ_B) are positive.

The slope-deflection equations as shown in Eqs. (14-6) express the end moments in terms of the end rotations and the applied loading. If, in addition to the applied loading, the end joints are subjected to unequal movements in the direction perpendicular to the axis of the member, additional fixed end moments M'_{FAB} and M'_{FBA} (Fig. 14-3c) are induced to act on the member in order to keep the tangents at the ends fixed. Then M'_A and M'_B should be such as to cause rotations of θ_A and θ_B, respectively. As in the preceding case, the conditions required of geometry are

$$\theta_A = -\theta_{A1} + \theta_{A2}$$

$$\theta_B = +\theta_{B1} - \theta_{B2}$$

(14-1)

By superposition,

$$M_{AB} = M_{FAB} + M'_{FAB} + M'_A$$

$$M_{BA} = M_{FBA} + M'_{FBA} + M'_B$$

(14-7)

From Eqs. (14-5).

$$M'_A = +\frac{2EI}{L}(-2\theta_A - \theta_B)$$

$$M'_B = +\frac{2EI}{L}(-2\theta_B - \theta_A)$$

(14-5)

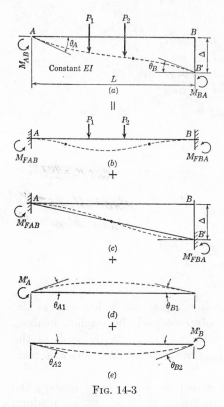

FIG. 14-3

The moment-area method will be used to determine M'_{FAB} and M'_{FBA} (Fig. 14-4). Note that R, the angle measured from the original direction of member AB to the line joining the displaced joints, is *positive* when *clockwise*. Note also that

$$R = \frac{\Delta}{L}$$

By the first moment-area theorem,

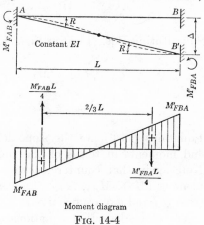

Moment diagram

FIG. 14-4

Change of slope between tangents at A and B'

$$= \text{area of } \frac{M}{EI} \text{ diagram between } A \text{ and } B' = 0$$

or

$$M'_{FAB} = M'_{FBA}$$

Deflection of B' from tangent at $A = \dfrac{1}{EI}\left[\dfrac{M'_{FAB}(L)}{4}\right]\left(\dfrac{2}{3}L\right)$

$$= \frac{M'_{FAB}L^2}{6EI} = \Delta$$

or

$$M'_{FAB} = M'_{FBA} = \frac{6EI\Delta}{L^2} = \frac{6EIR}{L} \tag{14-8}$$

Substituting Eqs. (14-5) and (14-8) in Eqs. (14-7),

$$M_{AB} = M_{FAB} + \frac{2EI}{L}(-2\theta_A - \theta_B + 3R)$$

$$M_{BA} = M_{FBA} + \frac{2EI}{L}(-2\theta_B - \theta_A + 3R) \tag{14-9}$$

Equations (14-9) are the general slope-deflection equations which express the end moments in terms of the end rotations, the applied loading, and the angle R between the line joining the deflected end joints and the original direction of the member. If the angle R is zero, Eqs. (14-9) reduce to Eqs. (14-6).

14-3. Application of the Slope-deflection Method to the Analysis of Statically Indeterminate Beams. The procedure for analyzing statically

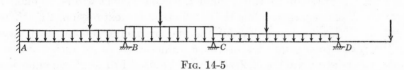

FIG. 14-5

indeterminate beams by the slope-deflection method is as follows (see (Fig. 14-5):

1. Determine the fixed-end moments due to the applied loads at the ends of each span, using the formulas shown in Fig. 14-6.

2. Determine the known value of R for each span ($R = 0$ in most cases unless there are unequal settlements at the supports).

3. Express all end moments in terms of the fixed-end moments due to the applied loads, the known values of R, and the unknown joint rotations by using the slope-deflection equations.

4. Establish simultaneous equations with the rotations at the supports as unknowns by applying the conditions that the sum of the end moments

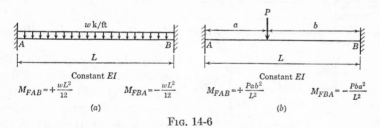

FIG. 14-6

acting on the ends of the two members meeting at each support is zero (except at the fixed end).

5. Solve for the rotations at all supports.

6. Substitute the rotations back into the slope-deflection equations, and compute the end moments.

7. Determine all reactions, draw shear and bending-moment diagrams, and sketch the elastic curve.

Example 14-1. Analyze the beam shown in Fig. 14-7a by the slope-deflection method. Draw shear and bending-moment diagrams. Sketch the elastic curve.

SOLUTION. In the slope-deflection equations

$$M_{AB} = M_{FAB} + \frac{2EI}{L}(-2\theta_A - \theta_B)$$

$$M_{BA} = M_{FBA} + \frac{2EI}{L}(-2\theta_B - \theta_A)$$

the coefficient $2EI/L$ is different for each span. If the $2EI/L$ values for all spans are made N times smaller, the only effect will be to make all θ values N times larger, while the products of the expressions $2EI/L$ and $(-2\theta_{near} - \theta_{far})$, or the values of the end moments, remain unchanged. If the absolute magnitudes of the θ values are not of direct interest, then, the relative values of $2EI/L$ may be used in the above equations. If the relative values of I/L are called the relative stiffness K, the slope-deflection equations become

$$M_{AB} = M_{FAB} + K_{AB}(-2\theta_A - \theta_B)$$
$$M_{BA} = M_{FBA} + K_{AB}(-2\theta_B - \theta_A) \qquad (14\text{-}10)$$

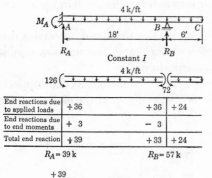

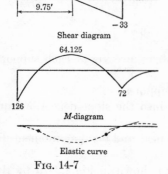

Fig. 14-7

Relative stiffness. Since there is only one span (or member) in this beam, the stiffness of the member may be regarded as unity.

Fixed-end moments.

$$M_{FAB} = +\frac{4(18)^2}{12} = +108 \text{ kip-ft}$$

$$M_{FBA} = -\frac{4(18)^2}{12} = -108 \text{ kip-ft}$$

Slope-deflection equations. By using the modified slope-deflection equations (14-10) and noting that θ_A is equal to zero, the following expressions for the end moments are obtained:

$$M_{AB} = +108 + 1(-2\theta_A - \theta_B)$$
$$= +108 - \theta_B$$
$$M_{BA} = -108 + 1(-2\theta_B - \theta_A)$$
$$= -108 - 2\theta_B$$

Joint condition

$$M_{BA} = -72 \text{ kip-ft}$$

Note: End moment acting on member BA at B is 72 kip-ft clockwise, or $M_{BA} = -72$ kip-ft.

Therefore,

$$-108 - 2\theta_B = -72$$
$$\theta_B = -18$$

Substituting,

$$M_{AB} = +108 - \theta_B = +108 - (-18) = +126 \text{ kip-ft}$$
$$M_{BA} = -108 - 2\theta_B = -108 - 2(-18) = -72 \text{ kip-ft} \quad (check)$$

It is to be noted that, in the expression $\theta_B = -18$, the minus sign indicates that the rotation at B is *counterclockwise*, but the value 18 is only relative, the true value of the rotation being $18/(2EI/L)$.

The end moments determined above are shown to act at the ends of span AB (Fig. 14-7). Note again that positive end moments are counterclockwise when acting on the member. Thus M_{AB} is 126 kip-ft counterclockwise and M_{BA} is 72 kip-ft clockwise. The end reactions are computed as the summation of those due to the applied loads and those due to the end moments, as shown in Fig. 14-7. The shear and bending-moment diagrams and the elastic curve are also shown in Fig. 14-7.

Example 14-2. Analyze the continuous beam shown in Fig. 14-8 by the slope-deflection method. Draw shear and bending-moment diagrams. Sketch the elastic curve.

SOLUTION. The values of the relative stiffness and the fixed-end moments are computed and shown in Fig. 14-8. Extreme care must be exercised in determining these values because the subsequent computation, even though its own correctness can be checked and thus assured, depends nevertheless on these preliminary quantities.

Relative stiffness

AB	$\dfrac{I}{8}(24)$	3
BC	$\dfrac{I}{12}(24)$	2

Fixed-end moments

$$M_{FAB} = +\frac{(15)(8)}{8} = +15 \text{ kip-ft}$$

$$M_{FBA} = -15 \text{ kip-ft}$$

$$M_{FBC} = +\frac{(1.5)(12)^2}{12} = +18 \text{ kip-ft}$$

$$M_{FCB} = -18 \text{ kip-ft}$$

Slope-deflection equations. The modified slope-deflection equations

$$M_{AB} = M_{FAB} + K_{AB}(-2\theta_A - \theta_B)$$
$$M_{BA} = M_{FBA} + K_{AB}(-2\theta_B - \theta_A)$$

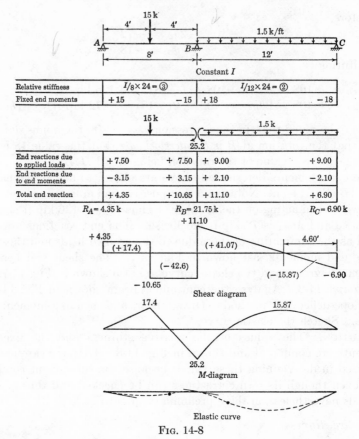

Relative stiffness	$I/8 \times 24 = ③$		$I/12 \times 24 = ②$	
Fixed end moments	$+15$	-15	$+18$	-18

End reactions due to applied loads	$+7.50$	$+7.50$	$+9.00$	$+9.00$
End reactions due to end moments	-3.15	$+3.15$	$+2.10$	-2.10
Total end reaction	$+4.35$	$+10.65$	$+11.10$	$+6.90$

$R_A = 4.35$ k　　　　　$R_B = 21.75$ k　　　　　$R_C = 6.90$ k

Shear diagram

M-diagram

Elastic curve

Fig. 14-8

will be used.

$$M_{AB} = +15 + 3(-2\theta_A - \theta_B) = +15 - 6\theta_A - 3\theta_B$$
$$M_{BA} = -15 + 3(-2\theta_B - \theta_A) = -15 - 6\theta_B - 3\theta_A$$
$$M_{BC} = +18 + 2(-2\theta_B - \theta_C) = +18 - 4\theta_B - 2\theta_C$$
$$M_{CB} = -18 + 2(-2\theta_C - \theta_B) = -18 - 4\theta_C - 2\theta_B$$

Joint conditions

Joint A:　　　　　$M_{AB} = 0$　　$-6\theta_A - 3\theta_B = -15$　　　　　(a)

Joint B:　　$M_{BA} + M_{BC} = 0$　　$-3\theta_A - 10\theta_B - 2\theta_C = -3$　　　　　(b)

Joint C:　　　　　$M_{CB} = 0$　　$-2\theta_B - 4\theta_C = +18$　　　　　(c)

Subtracting twice (b) from (a),

$$+17\theta_B + 4\theta_C = -9 \qquad\qquad (d)$$

Adding 8.5(c) to (d),

$$-30\theta_C = +144$$
$$\theta_C = -4.8 \qquad\qquad (e)$$

Substituting (e) in (d),

$$+17\theta_B - 19.2 = -9$$
$$\theta_B = +0.6 \qquad\qquad\qquad (f)$$

Substituting (f) in (a),

$$-6\theta_A - 1.8 = -15$$
$$\theta_A = +2.2 \qquad\qquad\qquad (g)$$

Computation of end moments

$$M_{AB} = +15 - 6\theta_A - 3\theta_B = +15 - 6(+2.2) - 3(+0.6) = 0$$
$$M_{BA} = -15 - 6\theta_B - 3\theta_A = -15 - 6(+0.6) - 3(+2.2)$$
$$= -25.2 \text{ kip-ft}$$
$$M_{BC} = +18 - 4\theta_B - 2\theta_C = +18 - 4(+0.6) - 2(-4.8)$$
$$= +25.2 \text{ kip-ft}$$
$$M_{CB} = -18 - 4\theta_C - 2\theta_B = -18 - 4(-4.8) - 2(+0.6) = 0$$

Note that the three joint conditions $M_{AB} = 0$, $M_{BA} + M_{BC} = 0$, and $M_{CB} = 0$ are satisfied.

The computations for the reactions, the shear and bending-moment diagrams, and the elastic curve are all shown in Fig. 14-8.

EXAMPLE 14-3. Analyze the continuous beam shown in Fig. 14-9 by the slope-deflection method. Draw shear and bending-moment diagrams. Sketch the elastic curve.

SOLUTION. The values of the relative stiffness and the fixed-end moments are computed and shown in Fig. 14-9.

Slope-deflection equations. In this problem, θ_A is zero because the beam is fixed at A.

$$M_{AB} = +15 + 3(-2\theta_A - \theta_B) = +15 - 3\theta_B$$
$$M_{BA} = -15 + 3(-2\theta_B - \theta_A) = -15 - 6\theta_B$$
$$M_{BC} = +18 + 2(-2\theta_B - \theta_C) = +18 - 4\theta_B - 2\theta_C$$
$$M_{CB} = -18 + 2(-2\theta_C - \theta_B) = -18 - 4\theta_C - 2\theta_B$$

Joint conditions

Joint B: $\qquad M_{BA} + M_{BC} = 0 \qquad -10\theta_B - 2\theta_C = -3 \qquad$ (a)
Joint C: $\qquad M_{CB} = 0 \qquad -2\theta_B - 4\theta_C = +18 \qquad$ (b)

Subtracting twice (a) from (b),

$$+18\theta_B = +24$$
$$\theta_B = +1.333 \qquad\qquad\qquad (c)$$

Substituting (c) in (b),

$$-2(+1.333) - 4\theta_C = +18$$
$$\theta_C = -5.167 \qquad\qquad\qquad (d)$$

Computation of end moments

$$M_{AB} = +15 - 3\theta_B = +15 - 3(+1.333) = +11.00 \text{ kip-ft}$$
$$M_{BA} = -15 - 6\theta_B = -15 - 6(+1.333) = -23.00 \text{ kip-ft}$$
$$M_{BC} = +18 - 4\theta_B - 2\theta_C = -18 - 4(+1.333) - 2(-5.167)$$
$$= +23.00 \text{ kip-ft}$$
$$M_{CB} = -18 - 4\theta_C - 2\theta_B = -18 - 4(-5.167) - 2(+1.333) = 0$$

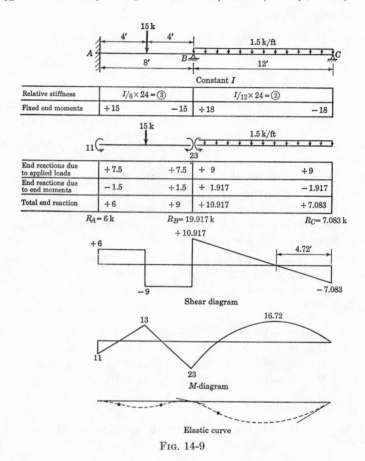

Fig. 14-9

Note that the joint conditions $M_{BA} + M_{BC} = 0$ and $M_{CB} = 0$ are satisfied. The reactions, shear and bending-moment diagrams, and the elastic curve are shown in Fig. 14-9.

Example 14-4. Analyze the continuous beam shown in Fig. 14-10 owing to the effect of a $\frac{1}{2}$-in. settlement at support B by the slope-deflection method. Calculate the reactions and draw the shear and bending-moment diagrams. Sketch the elastic curve.

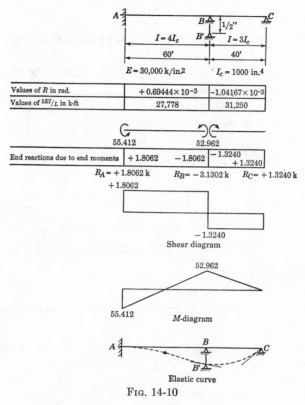

Fig. 14-10

SOLUTION. The general slope-deflection equations

$$M_{AB} = M_{FAB} + \frac{2EI}{L}(-2\theta_A - \theta_B + 3R)$$

$$M_{BA} = M_{FBA} + \frac{2EI}{L}(-2\theta_B - \theta_A + 3R)$$

will be used. In this example there are no fixed-end moments due to applied loadings, but there are R values for spans AB and BC. Inasmuch as the absolute values of R in radians are known, the absolute values of $2EI/L$ in kip-feet *must* be used.

Span AB: $R = +\dfrac{0.500 \text{ in.}}{720 \text{ in.}} = +0.69444 \times 10^{-3}$ radians

$$\frac{2EI}{L} = \frac{(2)(30,000)(4,000)}{(144)(60)} = 27,778 \text{ kip-ft}$$

Span BC: $R = -\dfrac{0.500 \text{ in.}}{480 \text{ in.}} = -1.04167 \times 10^{-3}$ radians

$$\frac{2EI}{L} = \frac{(2)(30,000)(3,000)}{(144)(40)} = 31,250 \text{ kip-ft}$$

Slope-deflection equations

$$M_{AB} = 0 + 27,778[-2\theta_A - \theta_B + 3(+0.69444 \times 10^{-3})]$$
$$= -27,778\theta_B + 57.862$$
$$M_{BA} = 0 + 27,778[-2\theta_B - \theta_A + 3(+0.69444 \times 10^{-3})]$$
$$= -55,556\theta_B + 57.862$$
$$M_{BC} = 0 + 31,250[-2\theta_B - \theta_C + 3(-1.04167 \times 10^{-3})]$$
$$= -62,500\theta_B - 31,250\theta_C - 97.656$$
$$M_{CB} = 0 + 31,250[-2\theta_C - \theta_B + 3(-1.04167 \times 10^{-3})]$$
$$= -62,500\theta_C - 31,250\theta_B - 97.656$$

Joint conditions

Joint B:

$$M_{BA} + M_{BC} = 0 \qquad -118,056\theta_B - 31,250\theta_C = +39.794 \qquad (a)$$

Joint C: $\quad M_{CB} = 0 \qquad -31,250\theta_B - 62,500\theta_C = +97.656 \qquad (b)$

Subtracting twice (a) from (b),

$$+204,862\theta_B = +18.068$$
$$\theta_B = +0.088196 \times 10^{-3} \text{ radians} \qquad (c)$$

Substituting (c) in (b),

$$-31,250(+0.0866196 \times 10^{-3}) - 62,500\theta_C = +97.656$$
$$\theta_C = -1.60659 \times 10^{-3} \text{ radians} \qquad (d)$$

Computation of end moments

$$M_{AB} = -27,778(+0.088196 \times 10^{-3}) + 57.862 = +55.412 \text{ kip-ft}$$
$$M_{BA} = -55,556(+0.088196 \times 10^{-3}) + 57.862 = +52.962 \text{ kip-ft}$$
$$M_{BC} = -62,500(+0.088196 \times 10^{-3}) - 31,250(-1.60659 \times 10^{-3})$$
$$\qquad\qquad - 97.656$$
$$= -52.962 \text{ kip-ft}$$
$$M_{CB} = -62,500(-1.60659 \times 10^{-3}) - 31,250(+0.088196 \times 10^{-3})$$
$$\qquad\qquad - 97.656$$
$$= 0$$

The reactions, shear and bending-moment diagrams, and the elastic curve are shown in Fig. 14-10.

A quick check of the above solution may be made by applying the moment-area method to the elastic curve shown in Fig. 14-10; thus

$$BB' = \text{deflection of } B' \text{ from the tangent at } A$$
$$= \frac{1,728}{(30,000)(4,000)} [(55.412)(^{60}\!/_2)(40) - (52.962)(^{60}\!/_2)(20)]$$
$$= 0.500 \text{ in.} \qquad (check)$$

Deflection of C from the tangent at A

$$= \frac{1,728}{(30,000)(4,000)} \, [(55.412)(^{60}\!\!/_2)(80) - (52.962)(^{60}\!\!/_2)(60)]$$

$$- \frac{1,728}{(30,000)(3,000)} \, [(52.962)(^{40}\!\!/_2)(^{80}\!\!/_3)]$$

$$= 0 \quad (check)$$

14-4. Application of the Slope-deflection Method to the Analysis of Statically Indeterminate Rigid Frames. *Case* 1 *Without Joint Movements.* The modified slope-deflection equations

$$M_{AB} = M_{FAB} + K_{AB}(+2\theta_A + \theta_B)$$
$$M_{BA} = M_{FBA} + K_{AB}(+2\theta_B + \theta_A)$$

may be used to analyze statically indeterminate rigid frames wherein all joints remain fixed in location during deformation. Again the axial deformation in the members due to direct stress is neglected in applying

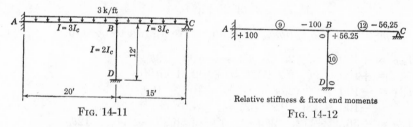

Relative stiffness & fixed end moments

FIG. 14-11   FIG. 14-12

the conditions of consistent deformation to the analysis; but direct stresses, together with shears and bending moments, must be considered in the design of sections. The conditions for consistent deformation are those of the rigidity of joints, or the angle between any two tangents to the elastic curves meeting at one joint must remain the same as that in the original undeformed structure. In the slope-deflection method the rotation at each joint is considered as the unknown, while the condition corresponding to this unknown is one of statics; i.e., the sum of the moments, as expressed by the slope-deflection equations, acting on the joint is equal to zero. Thus there are always as many conditions of statics as unknown rotations. After the latter are solved, all end moments may be found from the slope-deflection equations. With all end moments known, the direct stresses, shears, and bending moments in all members are found by applying the principles of statics to the individual members.

Example 14-5. Analyze the rigid frame shown in Fig. 14-11 by the slope-deflection method. Find the direct stresses, shears, and bending moments in all members. Sketch the deformed structure.

SOLUTION. Joints A, B, C, and D must all remain fixed in location; thus R is zero for all members.

Relative stiffness and fixed-end moments (Fig. 14-12)

AB	$\dfrac{3I_C}{20}\,(60)$	9
BC	$\left(\dfrac{3I_C}{15} = \dfrac{I_C}{5}\right)(60)$	12
BD	$\left(\dfrac{2I_C}{12} = \dfrac{I_C}{6}\right)(60)$	10

Span AB: $\qquad M_{FAB} = +\dfrac{3(20)^2}{12} = +100$ kip-ft

$\qquad\qquad\qquad M_{FBA} = -100$ kip-ft

Span BC: $\qquad M_{FBC} = +\dfrac{3(15)^2}{12} = +56.25$ kip-ft

$\qquad\qquad\qquad M_{FCB} = -56.25$ kip-ft

Slope-deflection equations. θ_A and θ_D are known to be zero.

$M_{AB} = +100 + 9(-2\theta_A - \theta_B) = +100 - 9\theta_B$
$\qquad\quad = +100 - 9(-0.27902) = +102.51$ kip-ft
$M_{BA} = -100 + 9(-2\theta_B - \theta_A) = -100 - 18\theta_B$
$\qquad\quad = -100 - 18(-0.27902) = -94.98$ kip-ft
$M_{BC} = +56.25 + 12(-2\theta_B - \theta_C) = +56.25 - 24\theta_B - 12\theta_C$
$\qquad\quad = +56.25 - 24(-0.27902) - 12(-2.20425)$
$\qquad\quad = +89.40$ kip-ft
$M_{CB} = -56.25 + 12(-2\theta_C - \theta_B) = -56.25 - 24\theta_C - 12\theta_B$
$\qquad\quad = -56.25 - 24(-2.20425) - 12(-0.27902) = 0$
$M_{BD} = 0 + 10(-2\theta_B - \theta_D) = -20(-0.27902) = +5.58$ kip-ft
$M_{DB} = 0 + 10(-2\theta_D - \theta_B) = -10\theta_B = -10(-0.27902)$
$\qquad\qquad\qquad\qquad\qquad\qquad\qquad\qquad\qquad\qquad = +2.79$ kip-ft

Joint conditions

$\qquad M_{BA} + M_{BC} + M_{BD} = 0 \qquad -62\theta_B - 12\theta_C = +43.75 \qquad$ (a)
$\qquad\qquad\quad M_{CB} = 0 \qquad -12\theta_B - 24\theta_C = +56.25 \qquad$ (b)

Subtracting twice (a) from (b),

$\qquad\qquad\qquad +112\theta_B = -31.25$
$\qquad\qquad\qquad\qquad \theta_B = -0.27902 \qquad\qquad\qquad$ (c)

Substituting (c) in (b),

$\qquad\qquad -12(-0.27902) - 24\theta_C = +56.25$
$\qquad\qquad\qquad\qquad\qquad \theta_C = -2.20425 \qquad\qquad$ (d)

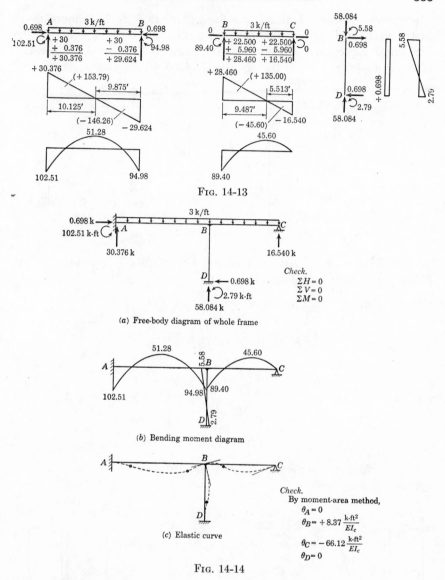

FIG. 14-13

(a) Free-body diagram of whole frame

Check.
$\Sigma H = 0$
$\Sigma V = 0$
$\Sigma M = 0$

(b) Bending moment diagram

(c) Elastic curve

Check.
By moment-area method,
$\theta_A = 0$
$\theta_B = +8.37 \dfrac{\text{k-ft}^2}{EI_c}$
$\theta_C = -66.12 \dfrac{\text{k-ft}^2}{EI_c}$
$\theta_D = 0$

FIG. 14-14

The free-body, shear, and bending-moment diagrams of the individual members are shown in Fig. 14-13. The free-body diagram, bending-moment diagram, and the elastic curve of the whole frame are shown in Fig. 14-14.

Example 14-6. Analyze the rigid frame shown in Fig. 14-15 by the slope-deflection method. Draw shear and bending-moment diagrams. Sketch the deformed structure.

SOLUTION. Joints A and D are fixed. Joints B and C cannot move in the vertical direction, but each may shift the same distance in the horizontal direction. In the present example, however, on account of the symmetry, both in the properties of the frame and in the applied

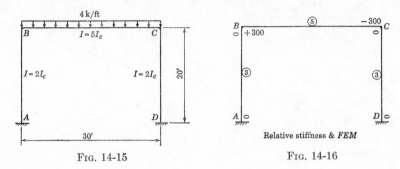

FIG. 14-15   FIG. 14-16

loading, joints B and C will not have any horizontal displacement. Thus R is equal to zero for all members.

Relative stiffness and fixed-end moments (Fig. 14-16)

AB, CD	$\left(\dfrac{2I_C}{20} = \dfrac{I_C}{10}\right)(30)$	3
BC	$\left(\dfrac{5I_C}{30} = \dfrac{I_C}{6}\right)(30)$	5

$$M_{FBC} = +\frac{4(30)^2}{12} = +300 \text{ kip-ft}$$

$$M_{FCB} = -300 \text{ kip-ft}$$

Slope-deflection equations. θ_A and θ_D are known to be zero. By symmetry, $\theta_C = -\theta_B$.

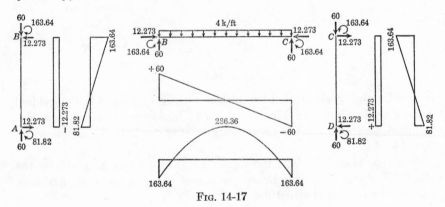

FIG. 14-17

$M_{AB} = 0 + 3(-2\theta_A - \theta_B) = -3\theta_B = -3(+27.273) = -81.82$ kip-ft
$M_{BA} = 0 + 3(-2\theta_B - \theta_A) = -6\theta_B = -6(+27.273) = -163.64$ kip-ft
$M_{BC} = +300 + 5(-2\theta_B - \theta_C) = +300 - 5\theta_B$
$\qquad = +300 - 5(+27.273) = +163.64$ kip-ft
$M_{CB} = -300 + 5(-2\theta_C - \theta_B) = -300 + 5\theta_B$
$\qquad = -300 + 5(+27.273) = -163.64$ kip-ft
$M_{CD} = 0 + 3(-2\theta_C - \theta_D) = +6\theta_B$
$\qquad = +6(+27.273) = +163.64$ kip-ft
$M_{DC} = 0 + 3(-2\theta_D - \theta_C) = +3\theta_B$
$\qquad = +3(+27.273) = +81.82$ kip-ft

Joint condition

Joint B: $\qquad\qquad M_{BA} + M_{BC} = 0$
$\qquad\qquad\qquad\qquad -11\theta_B + 300 = 0$
$\qquad\qquad\qquad\qquad\qquad \theta_B = +27.273$

The free-body, shear, and bending-moment diagrams of the individual members are shown in Fig. 14-17. The free-body diagram, bending-moment diagram, and the elastic curve of the whole frame are shown in Fig. 14-18.

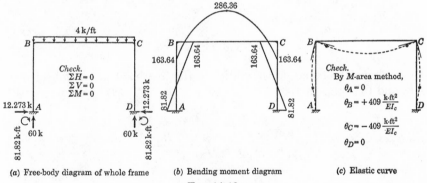

(a) Free-body diagram of whole frame   (b) Bending moment diagram   (c) Elastic curve

FIG. 14-18

14-5. Application of the Slope-deflection Method to the Analysis of Statically Indeterminate Rigid Frames. *Case 2 With Joint Movements.* When loads are applied to statically indeterminate rigid frames, there are cases in which some joints move unknown distances, although usually in known directions. Take, for instance, the rigid frame of Fig. 14-19a; joints D, E, and F are fixed; but joints A, B, and C may all move equal distances in the horizontal direction. This horizontal movement is generally called *sidesway*. Assume that the amount of sidesway is Δ to the right; then

$$R_{AD} = \frac{\Delta}{H_1} \qquad R_{BE} = \frac{\Delta}{H_2} \qquad R_{CF} = \frac{\Delta}{H_3}$$

Thus the slope-deflection equations (14-9)

$$M_{AB} = M_{FAB} + \frac{2EI}{L} \left(-2\theta_A - \theta_B + 3R \right)$$

$$M_{BA} = M_{FBA} + \frac{2EI}{L} \left(-2\theta_B - \theta_A + 3R \right)$$

(14-9)

must be used for members AD, BE, and CF. It is necessary, then, to seek another condition to cope with the unknown amount of sidesway Δ.

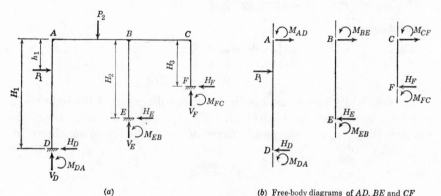

(a)

(b) Free-body diagrams of AD, BE and CF

Fig. 14-19

By applying the equations of statics to the free bodies of members AD, BE, and CF (Fig. 14-19b),

$$H_D = + \frac{P_1 h_1}{H_1} + \frac{M_{AD} + M_{DA}}{H_1}$$

$$H_E = \frac{M_{BE} + M_{EB}}{H_2}$$

$$H_F = \frac{M_{CF} + M_{FC}}{H_3}$$

Applying the equation of statics $\Sigma H = 0$ to the whole frame shown in Fig. 14-19a,

$$+P_1 - H_D - H_E - H_F = 0 \tag{14-11}$$

Equation (14-11) is generally called the *shear* equation or the *bent* equation. It furnishes the extra condition corresponding to the additional unknown Δ.

The application of the slope-deflection method to the analysis of single-span, one-story, statically indeterminate rigid frames in which some joints are displaced during deformation will be illustrated by the following examples.

Example 14-7. Analyze the rigid frame shown in Fig. 14-20*a* by the slope-deflection method. Draw shear and bending-moment diagrams. Sketch the deformed structure.

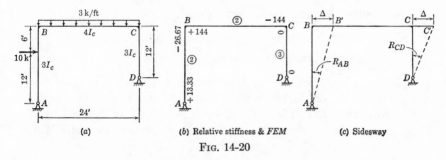

(a)　　　　　(b) Relative stiffness & *FEM*　　　　　(c) Sidesway

FIG. 14-20

SOLUTION. *Relative stiffness and fixed-end moments* (Fig. 14-20*b*)

AB	$\left(\dfrac{3I_C}{18} = \dfrac{I_C}{6}\right)(12)$	2
BC	$\left(\dfrac{4I_C}{24} = \dfrac{I_C}{6}\right)(12)$	2
CD	$\left(\dfrac{3I_C}{12} = \dfrac{I_C}{4}\right)(12)$	3

Span AB:　$M_{FAB} = +\dfrac{(10)(12)(6)^2}{(18)^2} = +13.33$ kip-ft

$M_{FBA} = -\dfrac{(10)(6)(12)^2}{(18)^2} = -26.67$ kip-ft

Span BC:　$M_{FBC} = +\dfrac{3(24)^2}{12} = +144$ kip-ft

$M_{FCB} = -144$ kip-ft

Relative values of R (Fig. 14-20*c*)

R_{AB}	$\dfrac{\Delta}{18}(36)$	$2R$
R_{BC}	0	0
R_{CD}	$\dfrac{\Delta}{12}(36)$	$3R$

If $R_{AB} = 2R$, then $R_{CD} = 3R$.

Slope-deflection equations.   The slope-deflection equations

$$M_{AB} = M_{FAB} + \frac{2EI}{L}(-2\theta_A - \theta_B + 3R)$$
$$M_{BA} = M_{FBA} + \frac{2EI}{L}(-2\theta_B - \theta_A + 3R)$$

(14-9)

may be modified to take the following form:

$$M_{AB} = M_{FAB} + K_{AB}(-2\theta_A - \theta_B + R_{\text{rel}})$$
$$M_{BA} = M_{FBA} + K_{AB}(-2\theta_B - \theta_A + R_{\text{rel}})$$

(14-12)

in which K_{AB} is the relative stiffness and R_{rel} is the relative size of the angle between the original axis and the straight line joining the deflected ends.   Note that $2EI/L$ is replaced by K_{AB} and $3R$ is replaced by R_{rel}. These changes do not affect the values of the end moments at all.   Equations (14-12) will be used in the present example.

$$
\begin{aligned}
M_{AB} &= +13.33 + 2(-2\theta_A - \theta_B + 2R) \\
       &= +13.33 - 4\theta_A - 2\theta_B + 4R \\
M_{BA} &= -26.67 + 2(-2\theta_B - \theta_A + 2R) \\
       &= -26.67 - 4\theta_B - 2\theta_A + 4R \\
M_{BC} &= +144 + 2(-2\theta_B - \theta_C) = +144 - 4\theta_B - 2\theta_C \\
M_{CB} &= +144 + 2(-2\theta_C - \theta_B) = -144 - 4\theta_C - 2\theta_B \\
M_{CD} &= 0 + 3(-2\theta_C - \theta_D + 3R) = -6\theta_C - 3\theta_D + 9R \\
M_{DC} &= 0 + 3(-2\theta_D - \theta_C + 3R) = -6\theta_D - 3\theta_C + 9R
\end{aligned}
$$

Joint conditions

$M_{AB} = 0$	$-4\theta_A - 2\theta_B + 4R = -13.33$	(a)
$M_{BA} + M_{BC} = 0$	$-2\theta_A - 8\theta_B - 2\theta_C + 4R = -117.33$	(b)
$M_{CB} + M_{CD} = 0$	$-2\theta_B - 10\theta_C - 3\theta_D + 9R = +144$	(c)
$M_{DC} = 0$	$-3\theta_C - 6\theta_D + 9R = 0$	(d)

Shear condition (Fig. 14-21).   Substituting

$$H_A = \frac{10}{3} + \frac{M_{AB} + M_{BA}}{18}$$

and

$$H_D = \frac{M_{CD} + M_{DC}}{12}$$

in the shear equation $+10 - H_A - H_D = 0$,

$$+10 - \left(\frac{10}{3} + \frac{M_{AB} + M_{BA}}{18}\right) - \frac{M_{CD} + M_{DC}}{12} = 0$$
$$+360 - 120 - 2(M_{AB} + M_{BA}) - 3(M_{CD} + M_{DC}) = 0$$
$$12\theta_A + 12\theta_B + 27\theta_C + 27\theta_D - 70R = -266.67 \quad \text{(e)}$$

Solving the five simultaneous equations (a) to (e),

$$\theta_A = +5.369$$
$$\theta_B = +24.174$$
$$\theta_C = -15.152$$
$$\theta_D = +28.761$$
$$R = +14.123$$

In deriving the shear equation (e), M_{AB} and M_{DC} could have been called zero; the equation then obtained will be actually the combination of Eqs. (a), (d), and (e) shown above.

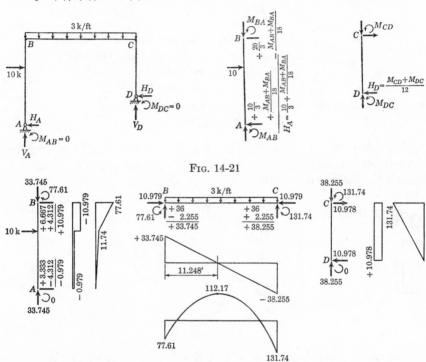

FIG. 14-21

FIG. 14-22

Computation of end moments

$$M_{AB} = +13.33 - 4(+5.369) - 2(+24.174) + 4(+14.123) = 0$$
$$M_{BA} = -26.67 - 4(+24.174) - 2(+5.369) + 4(+14.123)$$
$$= -77.61 \text{ kip-ft}$$
$$M_{BC} = +144 - 4(+24.174) - 2(-15.152) = +77.61 \text{ kip-ft}$$
$$M_{CB} = -144 - 4(-15.152) - 2(+24.174) = -131.74 \text{ kip-ft}$$
$$M_{CD} = -6(-15.152) - 3(+28.761) + 9(+14.123)$$
$$= +131.74 \text{ kip-ft}$$
$$M_{DC} = -6(+28.761) - 3(-15.152) + 9(+14.123) = 0$$

The free-body, shear, and bending-moment diagrams of the individual members are shown in Fig. 14-22. The free-body diagram, bending-moment diagram, and elastic curve of the whole frame are shown in Fig. 14-23.

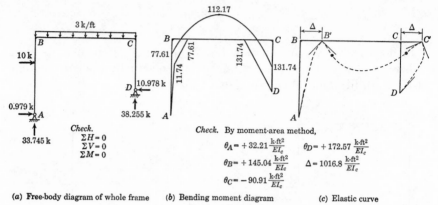

(a) Free-body diagram of whole frame   (b) Bending moment diagram   (c) Elastic curve

Check.
$\Sigma H = 0$
$\Sigma V = 0$
$\Sigma M = 0$

Check. By moment-area method,

$\theta_A = +32.21 \dfrac{\text{k-ft}^2}{EI_c}$ $\theta_D = +172.57 \dfrac{\text{k-ft}^2}{EI_c}$

$\theta_B = +145.04 \dfrac{\text{k-ft}^2}{EI_c}$ $\Delta = 1016.8 \dfrac{\text{k-ft}^2}{EI_c}$

$\theta_C = -90.91 \dfrac{\text{k-ft}^2}{EI_c}$

FIG. 14-23

Example 14-8. Analyze the rigid frame shown in Fig. 14-24a by the slope-deflection method. Draw shear and bending-moment diagrams. Sketch the deformed structure.

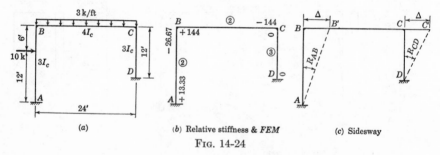

(a)   (b) Relative stiffness & *FEM*   (c) Sidesway

FIG. 14-24

SOLUTION. *Relative stiffness and fixed-end moments* (Fig. 14-24b)

AB	$\left(\dfrac{3I_C}{18} = \dfrac{I_C}{6}\right)(12)$	2
BC	$\left(\dfrac{4I_C}{24} = \dfrac{I_C}{6}\right)(12)$	3
CD	$\left(\dfrac{3I_C}{12} = \dfrac{I_C}{4}\right)(12)$	3

Span AB: $M_{FAB} = +\dfrac{10(12)(6)^2}{(18)^2} = +13.33$ kip-ft

$$M_{FBA} = -\frac{10(6)(12)^2}{(18)^2} = -26.67 \text{ kip-ft}$$

Span BC: $$M_{FBC} = +\frac{3(24)^2}{12} = +144 \text{ kip-ft}$$

$$M_{FCB} = -144 \text{ kip-ft}$$

Relative values of R (Fig. 14-24c)

R_{AB}	$\dfrac{\Delta}{18}$ (36)	$2R$
R_{BC}	0	0
R_{CD}	$\dfrac{\Delta}{12}$ (36)	$3R$

If $R_{AB} = 2R$, then $R_{CD} = 3R$.

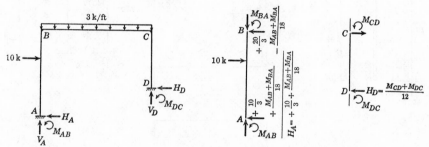

Fig. 14-25

Slope-deflection equations.   The slope-deflection equations

$$M_{AB} = M_{FAB} + K_{AB}(-2\theta_A - \theta_B + R_{rel})$$
$$M_{BA} = M_{FBA} + K_{AB}(-2\theta_B - \theta_A + R_{rel})$$

will be used. θ_A and θ_D are known to be zero.

$$M_{AB} = +13.33 + 2(-2\theta_A - \theta_B + 2R) = +13.33 - 2\theta_B + 4R$$
$$M_{BA} = -26.67 + 2(-2\theta_B - \theta_A + 2R) = -26.67 - 4\theta_B + 4R$$
$$M_{BC} = +144 + 2(-2\theta_B - \theta_C) = +144 - 4\theta_B - 2\theta_C$$
$$M_{CB} = -144 + 2(-2\theta_C - \theta_B) = -144 - 4\theta_C - 2\theta_B$$
$$M_{CD} = 0 + 3(-2\theta_C - \theta_D + 3R) = -6\theta_C + 9R$$
$$M_{DC} = 0 + 3(-2\theta_D - \theta_C + 3R) = -3\theta_C + 9R$$

Two joint conditions and one shear condition (Fig. 14-25)

Joint B: $M_{BA} + M_{BC} = 0$ $-8\theta_B - 2\theta_C + 4R = -117.33$ (a)
Joint C: $M_{CB} + M_{CD} = 0$ $-2\theta_B - 10\theta_C + 9R = +144$ (b)
Shear:

$$+10 - H_A - H_D = 0 \qquad -12\theta_B - 27\theta_C + 70R = +266.67 \quad (c)$$

Solving the three simultaneous equations (a) to (c),

$$\theta_B = +19.260$$
$$\theta_C = -18.154$$
$$R = +0.109$$

Computation of end moments

$$M_{AB} = +13.33 - 2(+19.260) + 4(+0.109) = -24.75 \text{ kip-ft}$$
$$M_{BA} = -26.67 - 4(+19.260) + 4(+0.109) = -103.27 \text{ kip-ft}$$
$$M_{BC} = +144 - 4(+19.260) - 2(-18.154) = +103.27 \text{ kip-ft}$$
$$M_{CB} = -144 - 4(-18.154) - 2(+19.260) = -109.90 \text{ kip-ft}$$
$$M_{CD} = -6(-18.154) + 9(+0.109) = +109.90 \text{ kip-ft}$$
$$M_{DC} = -3(-18.154) + 9(+0.109) = +55.44 \text{ kip-ft}$$

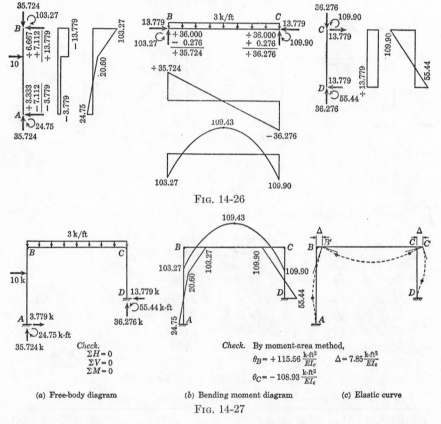

Fig. 14-26

(a) Free-body diagram   (b) Bending moment diagram   (c) Elastic curve

Fig. 14-27

The free-body, shear, and bending-moment diagrams of the individual members are shown in Fig. 14-26. The free-body diagram, bending-moment diagram, and elastic curve of the whole frame are shown in Fig. 14-27.

PROBLEMS

14-1 to 14-22. Analyze the statically indeterminate beams or rigid frames shown by the slope-deflection method. Draw shear and bending-moment diagrams. Sketch the deformed structure.

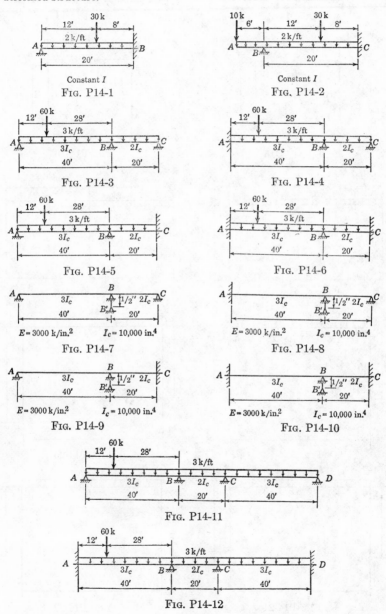

Constant I

FIG. P14-1

Constant I

FIG. P14-2

FIG. P14-3

FIG. P14-4

FIG. P14-5

FIG. P14-6

$E = 3000$ k/in.^{2} $I_c = 10,000$ in.^{4}

FIG. P14-7

$E = 3000$ k/in.^{2} $I_c = 10,000$ in.^{4}

FIG. P14-8

$E = 3000$ k/in.^{2} $I_c = 10,000$ in.^{4}

FIG. P14-9

$E = 3000$ k/in.^{2} $I_c = 10,000$ in.^{4}

FIG. P14-10

FIG. P14-11

FIG. P14-12

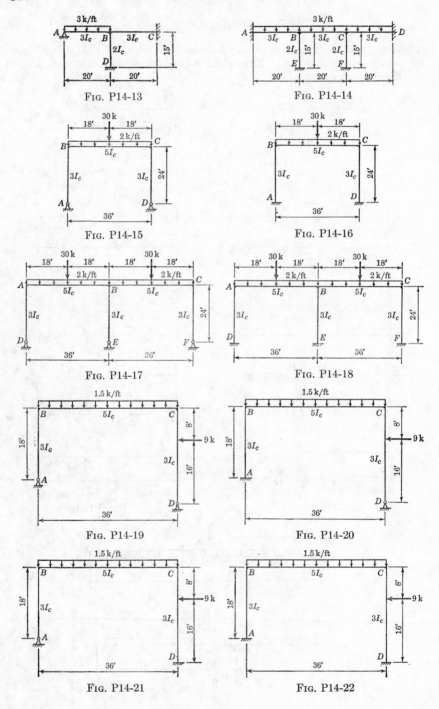

Fig. P14-13

Fig. P14-14

Fig. P14-15

Fig. P14-16

Fig. P14-17

Fig. P14-18

Fig. P14-19

Fig. P14-20

Fig. P14-21

Fig. P14-22

CHAPTER 15

THE MOMENT-DISTRIBUTION METHOD

15-1. General Description of the Moment-distribution Method. The moment-distribution method may be used to analyze all types of statically indeterminate beams or rigid frames. Essentially it consists in solving the simultaneous equations in the slope-deflection method by successive approximations. In developing the method, it will be helpful to consider the following problem: If a clockwise moment of M_A kip-ft is applied at the simple support of a straight member of constant cross section simply supported at one end and fixed at the other end, find the rotation θ_A at the simple support and the moment M_B at the fixed end (Fig. 15-1). The method of consistent deformation will be used. The condition of geometry required is, in this case,

$$\theta_B = 0 \quad \text{or} \quad \theta_{B1} = \theta_{B2} \quad (15\text{-}1)$$

By the conjugate-beam method,

$$\theta_{B1} = \frac{M_A L}{6EI} \qquad \theta_{B2} = \frac{M_B L}{3EI} \quad (15\text{-}2)$$

Substituting Eq. (15-2) in Eq. (15-1),

$$\frac{M_A L}{6EI} = \frac{M_B L}{3EI} \qquad M_B = \tfrac{1}{2}M_A \quad (15\text{-}3)$$

Also,

$$\theta_A = \theta_{A1} - \theta_{A2} = \frac{M_A L}{3EI} - \frac{M_B L}{6EI} = \frac{M_A L}{3EI} - \frac{(\tfrac{1}{2}M_A)L}{6EI}$$

$$= \frac{L}{4EI}\,M_A \tag{15-4}$$

Solving for M_A in Eq. (15-4),

$$M_A = \frac{4EI}{L}\,\theta_A \tag{15-5}$$

345

Thus, for a span AB which is simply supported at A and fixed at B, a clockwise rotation of θ_A may be effected by applying a clockwise moment of $M_A = (4EI/L)\theta_A$ at A, and this in turn induces a clockwise moment of $M_B = \frac{1}{2}M_A$ on the member at B. The expression $4EI/L$ is usually called the *stiffness factor*, which is defined as the moment required to be applied at A to cause a rotation of 1 radian at A of a span AB simply supported at A and fixed at B; the number $+\frac{1}{2}$ is the *carry-over factor*, which is the ratio of the moment induced at B to the moment applied at A. Note that the same sign convention is used in the moment-distribution method as in the slope-deflection method.

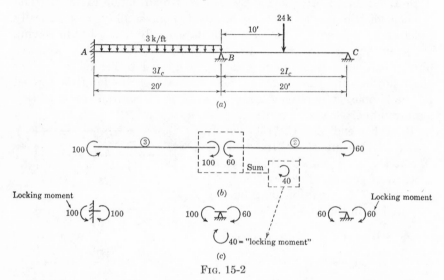

Fig. 15-2

Consider the continuous beam ABC shown in Fig. 15-2a. If the joints A, B, and C are to be restrained against rotation, the moments as shown in Fig. 15-2 must be applied, and these are, in fact, the fixed-end moments on spans AB and BC. The restraining moments required to hold the joints against rotation are (1) 100 kip-ft counterclockwise at A, (2) $100 - 60 = 40$ kip-ft clockwise at B, (3) 60 kip-ft clockwise at C. These restraining moments are sometimes called the "locking" moments to "lock" the joints against rotation. Note that the joint B shown as a free body in Fig. 15-2c is in equilibrium under the action of the fixed-end moments, which are opposite in direction to those acting on the members, and the locking moment. The procedure may be described as follows: First lock all three joints. Then release joint B only. Joint B, now under the action of 40 kip-ft counterclockwise, will rotate a certain amount in the counterclockwise direction, which will in turn induce counterclockwise moments at B to act on BA and BC in amounts proportional to the *stiffness factors* of each, with a sum of 40 kip-ft. The relative stiffness factors of BA and BC are 3 and 2. Thus $3/(3 + 2) = 0.600$ times 40 kip-ft, or 24 kip-ft counterclockwise, will act on BA, and

$$\frac{2}{(2 + 3)} = 0.400 \text{ times } 40 \text{ kip-ft}$$

or 16 kip-ft counterclockwise, will act on BC. The numbers 0.600 and 0.400 are usually called the *distribution factors*. Now lock joint B in its new position, and release joint C, which will rotate a certain amount in the counterclockwise direction. This rotation must be such as to induce a counterclockwise moment of 60 kip-ft to act on CB at C. Joint A is a fixed support; so it need not be released at all. Thus the first cycle of the moment distribution has been completed (see the adjoining moment-distribution table). To summarize, all joints are first locked by locking moments $+100$, -100, $+60$, -60 acting on all members, then joints B and C are released in succession, and the "balancing" moments are 0, $+24$, $+16$, and $+60$ (joint A is a fixed support).

When a balancing moment of $+24$ kip-ft is placed at B of span AB, one-half of this amount, or $+12$ kip-ft, is induced at A on AB. In the

MOMENT-DISTRIBUTION TABLE

Joint...............		A	B		C
Member.............		AB	BA	BC	CB
Distribution factors...		——	0.600	0.400	1.000
Cycle 1	*FEM*	$+100$	-100	$+60$	-60
	Balance	——	$+24$	$+16$	$+60$
Cycle 2	Carry-over	$+12$	——	$+30$	$+8$
	Balance	——	-18	-12	-8
Cycle 3	Carry-over	-9	——	-4	-6
	Balance	——	$+2.4$	$+1.6$	$+6$
Cycle 4	Carry-over	$+1.2$	——	$+3$	$+0.8$
	Balance	——	-1.8	-1.2	-0.8
Cycle 5	Carry-over	-0.9	——	-0.4	-0.6
	Balance	——	$+0.24$	$+0.16$	$+0.6$

and so on, to any desired degree of accuracy

Total end moments (5 cycles)	$+103.3$	-93.16	$+93.16$	0

same manner, one-half of the $+16$ kip-ft, or $+8$ kip-ft, acts at C on CB, and one-half of the $+60$ kip-ft, or $+30$ kip-ft, acts at B on BC. These moments $+12$, 0, $+30$, $+8$ are called the "carry-over" moments. They are *kept out* during the first balancing and now are considered as new locking moments to lock the joints in position after the first rotations. Then joints B and C are released for the second time, and the second rotations at B and C induce balancing moments as shown in the second cycle and carry-over moments as shown in the first line of the third cycle. The same process is repeated for as many cycles as desired to bring the balancing or carry-over moments to very small magnitudes. Thus any degree of accuracy can be obtained, and the work required decreases as the required accuracy decreases. The final, or total, end moments are obtained by adding all numbers in the respective columns.

Thus the moment-distribution method consists in successively locking and releasing the joints; the first locking moments are the fixed-end

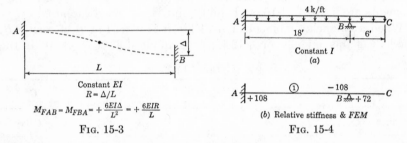

$$M_{FAB} = M_{FBA} = +\frac{6EI\Delta}{L^2} = +\frac{6EIR}{L}$$

Fig. 15-3

(b) Relative stiffness & *FEM*

Fig. 15-4

moments due to the applied loading; after the first balancing, the successive locking moments are the carry-over moments which are induced to act at the other ends of the respective spans by the balancing moments.

The reader is advised to read this article *again* and *again* while working through the rest of this chapter.

15-2. Application of the Moment-distribution Method to the Analysis of Statically Indeterminate Beams. As described in general in the preceding article, the moment-distribution method may be used to analyze statically indeterminate beams due to any applied loading. This method may also be used to analyze statically indeterminate beams due to the yielding of supports. The physical concept involved is that the joints are first locked against rotation and then displaced to conform with the amount of yielding; the locking moments acting on the ends of each member will be the fixed-end moments as derived in Eq. (14-8), which are repeated here and shown in Fig. 15-3. Then the joints are released or balanced; the carry-overs become the next unbalances, the joints are again balanced, and so on. In other words, the fixed-end moments due to the movement of one end relative to the other in a direction perpendicular to the original direction of the member are treated in exactly the same manner as those due to the applied loadings.

Example 15-1.   Analyze the beam shown in Fig. 15-4a by the moment-distribution method.   Draw shear and bending-moment diagrams. Sketch the elastic curve.

SOLUTION.   *Relative stiffness and fixed-end moments* (Fig. 15-4b). Since there is only one span (or member) in this beam, the stiffness of this member can be regarded as unity.

$$M_{FAB} = + \frac{4(18)^2}{12} = +108 \text{ kip-ft}$$
$$M_{FBA} = -108 \text{ kip-ft}$$

When the overhang BC is treated as a cantilever fixed at B, the fixed-end moment at B is $+4(6)^2/2 = +72$ kip-ft because it acts counterclockwise at B on BC.   However, BC has no stiffness and it should *not* be treated as a member.

TABLE 15-1. MOMENT DISTRIBUTION

Joint		A	B	
Member		AB	BA	BC
K		1	1	
Cycle	DF	——	1.000	
1	FEM	+108	−108	+72
	Balance	——	+ 36	
2	Carry-over	+ 18	——	
	Balance	——	——	
Total		+126	− 72	+72
Check:				
Change		+ 18	+ 36	
−½ (change)		− 18	− 9	
Sum		0	+ 27	
θ_{rel} = sum/−K		0	− 27	
		Check		

For explanation of check, see Art. 15-3.

In the moment-distribution table (see Table 15-1) the distribution factor at end A of AB is shown by a dash because at the fixed end A there is no need of releasing or balancing, while the distribution factor at B of BA is equal to 1.000 because member BA is the only member with stiffness entering joint B. In the body of the moment-distribution table, a dash is used for zero. In the first cycle the "unbalance" at joint B is $-108 + 72 = -36$ kip-ft. All of the balancing moment, $+36$ kip-ft, is placed under BA. The moment distribution comes to an automatic stop at the end of the second cycle. The total moments are $+126$, -72, and $+72$ as shown. Whenever a stage of moment distribution comes to an automatic stop, the results are theoretically exact. The checking procedure will be explained in the next article.

The reactions, shear and bending-moment diagrams, and the elastic curve are shown in Example 14-1 and will not be repeated here.

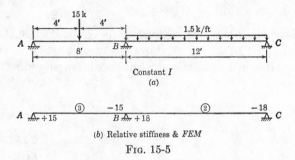

Constant I
(a)

(b) Relative stiffness & FEM

FIG. 15-5

Example 15-2.   Analyze the continuous beam shown in Fig. 15-5a by the moment-distribution method.   Draw shear and bending-moment diagrams.   Sketch the elastic curve.

SOLUTION.   *Relative stiffness and fixed-end moments* (Fig. 15-5b)

AB	$\dfrac{I}{8}(24)$	3
BC	$\dfrac{I}{12}(24)$	2

$$M_{FAB} = +\frac{(15)(8)}{8} = +15 \text{ kip-ft}$$

$$M_{FBA} = -15 \text{ kip-ft}$$

$$M_{FBC} = +\frac{1.5(12)^2}{12} = +18 \text{ kip-ft}$$

$$M_{FCB} = -18 \text{ kip-ft}$$

In the moment-distribution table (see Table 15-2) the distribution factors are first computed. The DF (distribution factor) at joint A (or C) is 1.000 on member AB (or CB) because there is only one member entering the joint. The DF's at joint B at $3/(3 + 2) = 0.600$ on member BA and $2/(3 + 2) = 0.400$ on member BC. In cycle 1, the joints are released and balanced. The "unbalance" at A is $+15.00$; so the balancing moment is -15.00. The "unbalance" at B is

$$-15.00 + 18.00 = +3.00$$

so the balancing moments are $-(0.600)(3.00) = -1.80$ and

$$-(0.400)(3.00) = -1.20 \text{ on } BA \text{ and } BC, \text{ respectively}$$

The "unbalance" at C is -18.00; so the balancing moment is $+18.00$. The carry-overs as shown in the first line of the second cycle are $+\frac{1}{2}$ times the balancing moments placed at the far ends of the respective members in the preceding line. These carry-overs are the new unbalances, which in turn are balanced in the same manner. The process is repeated, keeping all figures to two decimal places. In this example it is observed that the total end moments at all joints are not materially affected by the third (and fourth) cycle of moment distribution. This suggests that no further moment distribution is needed. By adding the moments

TABLE 15-2. MOMENT DISTRIBUTION

Joint...............		A	B		C
Member...........		AB	BA	BC	CB
K...............		3	3	2	2
Cycle	DF	1.000	0.600	0.400	1.000
1	FEM Balance	$+15.00$ -15.00	-15.00 -1.80	$+18.00$ -1.20	-18.00 $+18.00$
2	Carry-over Balance	-0.90 $+0.90$	-7.50 -0.90	$+9.00$ -0.60	-0.60 $+0.60$
3	Carry-over Balance	-0.45 $+0.45$	$+0.45$ -0.45	$+0.30$ -0.30	-0.30 $+0.30$
4	Carry-over Balance	-0.22 $+0.22$	$+0.22$ -0.22	$+0.15$ -0.15	-0.15 $+0.15$
Total...............		0	-25.20	$+25.20$	0
Check: Change........... $-\frac{1}{2}$ (change)....... Sum............... $\theta_{rel} = \text{sum}/-K$.....		-15.00 $+5.10$ -9.90 $+3.30$	-10.20 $+7.50$ -2.70 $+0.90$	$+7.20$ -9.00 -1.80 $+0.90$ Check	$+18.00$ -3.60 $+14.40$ -7.20

For explanation of check, see Art. 15-3.

in the respective columns the total end moments are obtained. These total end moments check with those obtained by the slope-deflection method in Example 14-2.

For reactions, shear and bending-moment diagrams, and the elastic curve, see Example 14-2.

In carrying out the work in the moment-distribution table, it is advisable to first put down all the *signs* in any one line at the same time and then the numerical values. This helps the computer to concentrate on one operation at a time. Note also that at the time of each balancing the sum of the balancing moments placed at the ends of members meeting at one joint must be equal to the total unbalance at that joint. This will ensure that the sum of the total end moments acting on all members meeting at any one joint is zero.

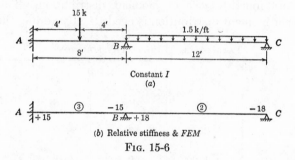

Constant I
(a)

(b) Relative stiffness & *FEM*

Fig. 15-6

Example 15-3.   Analyze the continuous beam shown in Fig. 15-6a by the moment-distribution method.   Draw shear and bending-moment diagrams.   Sketch the elastic curve.

SOLUTION.   *Relative stiffness and fixed-end moments* (Fig. 15-6b)

AB	$\dfrac{I}{8}(24)$	3
BC	$\dfrac{I}{12}(24)$	2

$$M_{FAB} = +\frac{15(8)}{8} = +15 \text{ kip-ft}$$

$$M_{FBA} = -15 \text{ kip-ft}$$

$$M_{FBC} = +\frac{1.5(12)^2}{12} = +18 \text{ kip-ft}$$

$$M_{FCB} = -18 \text{ kip-ft}$$

In the moment-distribution table (see Table 15-3) the distribution factors are first computed.   Since joint A is a fixed support, it can resist any moment assigned to it and therefore need not be released.   In such a case no distribution factor is applicable to A. The distribution factors at joints B and C are determined as previously explained. Joint A requires no balancing; so the balancing moments in all cycles are zero, for which a *dash* is used in the table. The moment distribution is carried out as indicated. By keeping all figures to two decimal places, the table

TABLE 15-3. MOMENT DISTRIBUTION

Joint...............		A	B		C
Member.............		AB	BA	BC	CB
K.................		3	3	2	2
Cycle	DF	——	0.600	0.400	1.000
1	FEM	+15.00	−15.00	+18.00	−18.00
	Balance	——	− 1.80	− 1.20	+18.00
2	Carry-over	− 0.90	——	+ 9.00	− 0.60
	Balance	——	− 5.40	− 3.60	+ 0.60
3	Carry-over	− 2.70	——	+ 0.30	− 1.80
	Balance	——	− 0.18	− 0.12	+ 1.80
4	Carry-over	− 0.09	——	+ 0.90	− 0.06
	Balance	——	− 0.54	− 0.36	+ 0.06
5	Carry-over	− 0.27	——	+ 0.03	− 0.18
	Balance	——	− 0.02	− 0.01	+ 0.18
6	Carry-over	− 0.01	——	+ 0.09	——
	Balance	——	− 0.05	− 0.04	——
7	Carry-over	− 0.02	——	——	− 0.02
	Balance	——	——	——	+ 0.02
8	Carry-over	——	——	+ 0.01	——
	Balance	——	− 0.01	——	——
Total...............		+11.01	−23.00	+23.00	0
Check:					
Change...........		− 3.99	− 8.00	+ 5.00	+18.00
−½ (change)......		+ 4.00	+ 2.00	− 9.00	− 2.50
Sum..............		+ 0.01	− 6.00	− 4.00	+15.50
θ_{rel} = sum/−K....		0	+ 2.00	+ 2.00	− 7.75
		Check	Check		

For explanation of check, see Art. 15-3.

comes to an automatic stop at the end of the eighth cycle. The total end moments obtained by adding the moments in the respective columns check quite closely with those of Example 14-3 in which the slope-deflection solution was used.

For reactions, shear and bending-moment diagrams, and the elastic curve, see Example 14-3.

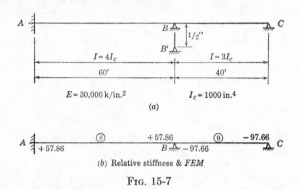

(b) Relative stiffness & *FEM*

FIG. 15-7

Example 15-4. Analyze the continuous beam shown in Fig. 15-7a owing to the effect of a ½-in. settlement at support B by the moment-distribution method. Draw shear and bending-moment diagrams. Sketch the elastic curve.

SOLUTION. *Relative stiffness and fixed-end moments* (Fig. 15-7b)

AB	$\left(\dfrac{4I_c}{60} = \dfrac{I_c}{15}\right)(120)$	8
BC	$\dfrac{3I_c}{40}(120)$	9

$$M_{FAB} = M_{FBA} = +\frac{6EIR}{L}$$

$$= +\frac{6(30,000)(4,000)}{(144)(60)}\left(+\frac{0.500}{720}\right) = +57.86 \text{ kip-ft}$$

$$M_{FBC} = M_{FCB} = +\frac{6EIR}{L}$$

$$= +\frac{6(30,000)(3,000)}{(144)(40)}\left(-\frac{0.500}{480}\right) = -97.66 \text{ kip-ft}$$

The solution of this problem by the moment-distribution method is shown in Table 15-4. For reactions, shear and bending-moment diagrams, and the elastic curve, see Example 14-4.

TABLE 15-4. MOMENT DISTRIBUTION

Joint...............	A	B		C
Member.............	AB	BA	BC	CB
K...................	8	8	9	9

Cycle	DF	——	0.4706	0.5294	1.0000
1	FEM	+57.86	+57.86	−97.66	−97.66
	Balance	——	+18.73	+21.07	+97.66
2	Carry-over	+ 9.36	——	+48.83	+10.54
	Balance	——	−22.98	−25.85	−10.54
3	Carry-over	−11.49	——	− 5.27	−12.92
	Balance	——	+ 2.48	+ 2.79	+12.92
4	Carry-over	+ 1.24	——	+ 6.46	+ 1.40
	Balance	——	− 3.04	− 3.42	− 1.40
5	Carry-over	− 1.52	——	− 0.70	− 1.71
	Balance	——	+ 0.33	+ 0.37	+ 1.71
6	Carry-over	+ 0.16	——	+ 0.86	+ 0.18
	Balance	——	− 0.40	− 0.46	− 0.18
7	Carry-over	− 0.20	——	− 0.09	− 0.23
	Balance	——	+ 0.04	+ 0.05	+ 0.23
8	Carry-over	+ 0.02	——	+ 0.12	+ 0.02
	Balance	——	− 0.06	− 0.06	− 0.02
9	Carry-over	− 0.03	——	− 0.01	− 0.03
	Balance	——	——	+ 0.01	+ 0.03
10	Carry-over	——	——	+ 0.02	——
	Balance	——	− 0.01	− 0.01	——
Total...............		+55.40	+52.95	−52.95	0

Check:					
Change...........		− 2.46	− 4.91	+44.71	+97.66
−½ (change).......		+ 2.46	+ 1.23	−48.83	−22.36
Sum.............		0	− 3.68	− 4.12	+75.30
θ_{rel} = sum/−K.....		0	+ 0.460	+ 0.458	− 8.367
		Check	Check		

For explanation of check, see Art. 15-3.

15-3. Check on Moment Distribution. The moment-distribution table begins with relative stiffness and fixed-end moments (due to the applied loading and/or settlement of supports) and concludes with the required end moments. A check may easily be made to ensure that the correct end moments have been obtained on the basis of the relative stiffness and fixed-end moments at the beginning of the table. Of course, the first obvious check is to see whether or not the moments are balanced at each interior joint and the moment is zero at each exterior simple (or hinged) support; this check is on the conditions of *statics*. There is also a check on the conditions of *geometry*. This may be made by finding the absolute or relative values of the rotation at each joint.

The slope-deflection equations may be written as

$$M_{AB} = M_{TFAB} + \frac{2EI}{L}(-2\theta_A - \theta_B)$$

$$M_{BA} = M_{TFBA} + \frac{2EI}{L}(-2\theta_B - \theta_A)$$

(15-6)

in which M_{TFAB} and M_{TFBA} represent the *total* fixed-end moments due to the applied loading and settlement of supports, or

$$M_{TFAB} = M_{FAB} + \frac{6EIR}{L}$$

$$M_{TFBA} = M_{FBA} + \frac{6EIR}{L}$$

In Eqs. (15-6), the end moments are expressed in terms of the fixed-end moments and the end rotations. Conversely, the end rotations may be expressed in terms of the fixed-end moments and the final end moments. Solving Eqs. (15-6) for θ_A and θ_B,

$$\theta_A = \frac{(M_{AB} - M_{TFAB}) - \frac{1}{2}(M_{BA} - M_{TFBA})}{-3EI/L}$$

$$\theta_B = \frac{(M_{BA} - M_{TFBA}) - \frac{1}{2}(M_{AB} - M_{TFAB})}{-3EI/L}$$

(15-7)

Equations (15-7) give the absolute values of the end rotations. If only the relative values of θ_A and θ_B are desired, the expression $3EI/L$ in the denominator can be replaced by K_{rel}. Thus,

$$(\theta_A)_{\text{rel}} = \frac{(M_{AB} - M_{TFAB}) - \frac{1}{2}(M_{BA} - M_{TFBA})}{-K_{\text{rel}}}$$

$$(\theta_B)_{\text{rel}} = \frac{(M_{BA} - M_{TFBA}) - \frac{1}{2}(M_{AB} - M_{TFAB})}{-K_{\text{rel}}}$$

(15-8)

In Eqs. (15-7) or (15-8), $(M_{AB} - M_{TFAB})$ and $(M_{BA} - M_{TFBA})$ are the changes in moment from the "fixed-end moment" to the "final end moment." Thus the absolute or the relative value of the rotation θ

at any one end of a member is equal to the change in the moment at the near end minus one-half of the change in the moment at the far end and then divided by $-3EI/L$ or $-K_{rel}$. Or

Absolute value of $\theta_{near\ end}$

$$= \frac{(change)_{near\ end} + (-\tfrac{1}{2})(change)_{far\ end}}{-3EI/L} \qquad (15\text{-}9)$$

Relative value of $\theta_{near\ end}$

$$= \frac{(change)_{near\ end} + (-\tfrac{1}{2})(change)_{far\ end}}{-K_{rel}} \qquad (15\text{-}10)$$

Thus either the absolute or relative values of the rotation at the ends of each member may be computed by the use of Eqs. (15-9) or (15-10). A check on the condition of geometry is that the rotations at the ends of all members meeting at one joint must be equal and that the rotation at the fixed support must be zero. It is to be noted that, for the purpose of this check, the use of the relative values of the rotations is recommended. If occasions arise where the absolute values are desired, Eq. (15-9) must be used.

For example, at the end of Table 15-3 for moment distribution in Example 15-3, the check made by applying Eq. (15-10) is shown. The relative values of the rotations at joints A, B, and C are 0, $+2.00$, $+2.00$, and -7.75. The check on the two conditions of geometry are that the rotation at A is zero and that the same value of the rotation at B is obtained from spans BA and BC.

It must be again noted that the above check has nothing to do with the correctness of the values of the relative stiffness and of the fixed-end moments used at the beginning of the moment-distribution solution; i.e., if these are incorrect, the final answers will be correspondingly wrong even though they meet the test of the check.

15-4. Stiffness Factor at the Near End of a Member When the Far End Is Hinged. The stiffness factor has been defined as the moment required to rotate the tangent to the elastic curve at the near end of a member through 1 radian when the far end is *fixed;* for a member with constant cross section, this stiffness factor is $4EI/L$ (Fig. 15-8a). Now, if the far end is hinged instead of being fixed, the moment required to rotate the tangent at the near end through 1 radian will be $3EI/L$ instead of $4EI/L$ (Fig. 15-8b). This may be derived easily by the conjugate-beam method. From Fig. 15-8b,

$$EI\theta_A = R'_A = \frac{M_A L}{3} \qquad \text{or} \qquad M_A = \frac{3EI}{L}\theta_A = \frac{3}{4}\left(\frac{4EI}{L}\right)\theta_A \qquad (15\text{-}11)$$

Thus the stiffness factor at the near end when the far end is hinged is $3EI/L$, or three-fourths of that when the far end is fixed. This stiffness factor, $3EI/L$, will be called the *modified* stiffness factor in subsequent discussions.

In Fig. 15-9 four members AE, BE, CE, and DE meet at a rigid joint E. If A, B, and D are fixed and C is hinged, any unbalanced moment at joint E will cause a certain amount of rotation at joint E, or the unbalance will distribute itself into four parts to act on the ends of EA, EB, EC, and ED in the ratio of K_{AE}, K_{BE}, $\frac{3}{4}(K_{CE})$, and K_{DE}. One-half of the balancing moments placed at E on members EA, EB, and ED will then be carried to A, B, and D; but no carry-over to the hinge

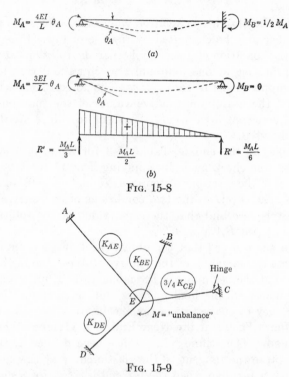

$$M_A = \frac{4EI}{L}\theta_A$$

$$M_B = \frac{1}{2}M_A$$

$$\theta_A$$

(a)

$$M_A = \frac{3EI}{L}\theta_A$$

$$M_B = 0$$

$$\theta_A$$

$$R' = \frac{M_A L}{3}$$

$$\frac{M_A L}{2}$$

$$R' = \frac{M_A L}{6}$$

(b)

FIG. 15-8

A

$$K_{AE}$$

B

$$K_{BE}$$

Hinge

$$\frac{3}{4}K_{CE}$$

C

E

$$M = \text{"unbalance"}$$

$$K_{DE}$$

D

FIG. 15-9

is necessary, because by modifying the stiffness factor of EC to three-fourths of its usual value, provision is made for the fact that the moment at C should always be zero. If, however, the "unbalance" at E is distributed in the ratio of K_{AE}, K_{BE}, K_{CE}, and K_{DE}, one-half of the balancing moment placed at E on EC must be carried over to C and joint C must be balanced in every cycle of moment distribution. The latter procedure has been followed in dealing with the exterior simple support in all examples of Art. 15-2. The alternate procedure involving modification of the stiffness of members with exterior simple or hinged supports will now be shown.

When a continuous beam or rigid frame has one or more exterior simple or hinged supports, the moment-distribution procedure may be performed by the regular method or the modified-stiffness method. In general, the modified-stiffness method maintains accuracy to the same number of significant figures with a shorter moment-distribution table. In applying the modified-stiffness method, the stiffness of the member with an exterior simple or hinged support is changed to three-fourths of its usual value. For such an exterior support, no distribution factor is shown because only one release is needed when there is an initial fixed-end moment. After the first balance, no further carry-overs are brought to the simple support. To prevent a common mistake, it is suggested that a vertical arrow be drawn immediately after the first cycle under the exterior simple support in the moment-distribution table.

The distribution factors are determined from the modified stiffness of members with hinged ends. Upon checking, however, the *unmodified* values of the relative stiffness should be used in $\theta_{rel} = \text{sum}/-K$.

Example 15-5. Using the modified-stiffness procedure, solve Example 15-2 by the moment-distribution method.

SOLUTION

TABLE 15-5. MOMENT DISTRIBUTION

Joint	A	B		C
Member	AB	BA	BC	CB
K	3	3	2	2
Modified K	2.25	2.25	1.5	1.5
Cycle	DF	——	0.600	0.400	——
1 FEM	+15.00	−15.00	+18.00	−18.00
Balance	−15.00	− 1.80	− 1.20	+18.00
2 Carry-over	↑	− 7.50	+ 9.00	↑
Balance	↓	− 0.90	− 0.60	↓
Total	0	−25.20	+25.20	0
Check: Change	−15.00	−10.20	+ 7.20	+18.00
− ½ (change)	+ 5.10	+ 7.50	− 9.00	− 3.60
Sum	− 9.90	− 2.70	− 1.80	+14.40
$\theta_{rel} = \text{sum}/-K$	+ 3.30	+ 0.90	+ 0.90	− 7.20
			Check	

Example 15-6.   Using the modified-stiffness procedure, solve **Example** 15-3 by the moment-distribution method.

SOLUTION

TABLE 15-6. MOMENT DISTRIBUTION

Joint.............		*A*	*B*		*C*
Member...........		*AB*	*BA*	*BC*	*CB*
K................		3	3	2	2
Modified *K*........		3	3	1.5	1.5
Cycle	*DF*	——	0.667	0.333	——
1	FEM	+15.00	−15.00	+18.00	−18.00
	Balance	——	− 2.00	− 1.00	+18.00
2	Carry-over	− 1.00	——	+ 9.00	
	Balance	——	− 6.00	− 3.00	
3	Carry-over	− 3.00	——	——	
	Balance	——	——	——	
Total.............		+11.00	−23.00	+23.00	0
Check:					
Change..........		− 4.00	− 8.00	+ 5.00	+18.00
−½ (change).....		+ 4.00	+ 2.00	− 9.00	− 2.50
Sum.............		0	− 6.00	− 4.00	+15.50
$\theta_{\rm rel}$ = sum/−*K*...		0	+ 2.00	+ 2.00	− 7.75
		Check	Check		

Example 15-7.   Solve the moment distribution in Example 15-4 by the modified-stiffness method.

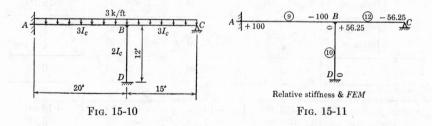

Fig. 15-10   Fig. 15-11

SOLUTION

TABLE 15-7. MOMENT DISTRIBUTION

Joint		A	B		C
Member		AB	BA	BC	CB
K		8	8	9	9
Modified K		8	8	6.75	6.75
Cycle	DF	——	0.5424	0.4576	——
1	FEM	+57.86	+57.86	−97.66	−97.66
	Balance	——	+21.59	+18.21	+97.66
2	Carry-over	+10.80	——	+48.83	↑
	Balance	——	−26.48	−22.35	
3	Carry-over	−13.24	——	——	↓
	Balance	——	——	——	
Total		+55.42	+52.97	−52.97	0
Check:					
Change		− 2.44	− 4.89	+44.69	+97.66
−½ (change)		+ 2.44	+ 1.22	−48.83	−22.34
Sum		0	− 3.67	− 4.14	+75.32
θ_{rel} = sum$/-K$		0	+ 0.459	+ 0.460	− 8.369
			Check	Check	

15-5. Application of the Moment-distribution Method to the Analysis of Statically Indeterminate Rigid Frames. *Case* 1 *Without Joint Movements.* The application of the moment-distribution method to the analysis of statically indeterminate rigid frames wherein no joint movements or "sidesway" is involved is very similar to that of beams as discussed in the previous articles, except that in the case of rigid frames there are frequently more than two members meeting in one joint. In such cases the unbalance at any joint is distributed to the ends of the several members meeting at the joint in the ratio of their relative stiffnesses. There are a number of ways in which the work for the moment-distribution procedure may be arranged, but a tabular form in which all members meeting at the same joint are grouped together is used in this text and is suggested as the most convenient form.

Example 15-8. Analyze the rigid frame shown in Fig. 15-10 by the moment-distribution method. Draw shear and bending-moment diagrams. Sketch the deformed structure.

SOLUTION. $R = 0$ for all members.

Relative stiffness and fixed-end moments (Fig. 15-11)

AB	$\dfrac{3I_c}{20}\,(60)$	9
BC	$\left(\dfrac{3I_c}{15} = \dfrac{I_c}{5}\right)(60)$	12
BD	$\left(\dfrac{2I_c}{12} = \dfrac{I_c}{6}\right)(60)$	10

$$M_{FAB} = +\frac{3(20)^2}{12} = +100 \text{ kip-ft}$$

$$M_{FBA} = -100 \text{ kip-ft}$$

$$M_{FBC} = +\frac{3(15)^2}{12} = +56.25 \text{ kip-ft}$$

$$M_{FCB} = -56.25 \text{ kip-ft}$$

Moment distribution (see Tables 15-8 and 15-9)

TABLE 15-8. MOMENT DISTRIBUTION, MODIFIED-STIFFNESS METHOD

Joint	A	B			C	D
Member	AB	BA	BC	BD	CB	DB
K	9	9	12	10	12	10
Modified K	9	9	9	10	9	10
Cycle *DF*	——	0.3214	0.3214	0.3572	——	——
1 *FEM*	$+100.00$	-100.00	$+56.25$	——	-56.25	——
Balance	——	$+\ 14.06$	$+14.06$	$+15.63$	$+56.25$	——
2 Carry-over	$+\ \ 7.03$	——	$+28.12$	——	↑	$+7.82$
Balance	——	$-\ \ 9.04$	$-\ \ 9.04$	-10.04		——
3 Carry-over	$-\ \ 4.52$	——	——	——	↓	-5.02
Balance	——	——	——	——		——
Total	$+102.51$	$-\ 94.98$	$+89.39$	$+\ 5.59$	0	$+2.80$
Check:						
Change	$+\ \ 2.51$	$+\ \ 5.02$	$+33.14$	$+\ \ 5.59$	$+56.25$	$+2.80$
$-\frac{1}{2}$ (change)	$-\ \ 2.51$	$-\ \ 1.26$	-28.12	$-\ \ 1.40$	-16.57	-2.80
Sum	0	$+\ \ 3.76$	$+\ \ 5.02$	$+\ \ 4.19$	$+39.68$	0
$\theta_{rel} = \text{sum}/-K$	0	$-\ \ 0.418$	$-\ \ 0.418$	$-\ \ 0.419$	$-\ \ 3.307$	0
	Check		Check			Check

TABLE 15-9. MOMENT DISTRIBUTION, REGULAR METHOD

Joint.............	A	B			C	D
Member..........	AB	BA	BC	BD	CB	DB
K...............	9	9	12	10	12	10
Cycle	DF	——	0.2903	0.3871	0.3226	1.000	——
1	FEM	+100.00	−100.00	+56.25	——	−56.25	——
	Balance	——	+ 12.70	+16.94	+14.11	+56.25	——
2	Carry-over	+ 6.35	——	+28.12	——	+ 8.47	+7.06
	Balance	——	− 8.16	−10.89	− 9.07	− 8.47	——
3	Carry-over	− 4.08	——	− 4.24	——	− 5.44	−4.54
	Balance	——	+ 1.23	+ 1.64	+ 1.37	+ 5.44	——
4	Carry-over	+ 0.62	——	+ 2.72	——	+ 0.82	+0.68
	Balance	——	− 0.79	− 1.05	− 0.88	− 0.82	——
5	Carry-over	− 0.40	——	− 0.41	——	− 0.52	−0.44
	Balance	——	+ 0.12	+ 0.16	+ 0.13	+ 0.52	——
6	Carry-over	+ 0.06	——	+ 0.26	——	+ 0.08	+0.06
	Balance	——	− 0.08	− 0.10	− 0.08	− 0.08	——
7	Carry-over	− 0.04	——	− 0.04	——	− 0.05	−0.04
	Balance	——	+ 0.01	+ 0.02	+ 0.01	+ 0.05	——
8	Carry-over	——	——	+ 0.02	——	+ 0.01	——
	Balance	——	——	− 0.01	− 0.01	− 0.01	——
Total.............		+102.51	− 94.97	+89.39	+ 5.58	0	+2.78
Check:							
Change..........		+ 2.51	+ 5.03	+33.14	+ 5.58	+56.25	+2.78
−½ (change)....		− 2.52	− 1.26	−28.12	− 1.39	−16.57	−2.79
Sum............		− 0.01	+ 3.77	+ 5.02	+ 4.19	+39.68	−0.01
θ_{rel} = sum/−K..		0	− 0.419	− 0.418	− 0.419	− 3.307	0
		Check		Check			Check

For reactions, shear and bending-moment diagrams, and the elastic curve, see Example 14-5.

Example 15-9. Analyze the rigid frame shown in Fig. 15-12 by the moment-distribution method. Draw shear and bending-moment diagrams. Sketch the deformed structure.

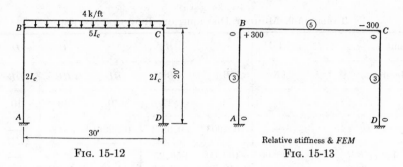

FIG. 15-12   Relative stiffness & *FEM*
                             FIG. 15-13

SOLUTION. $R = 0$ for all members.

Relative stiffness and fixed-end moments (Fig. 15-13)

AB, CD	$\left(\dfrac{2I_c}{20} = \dfrac{I_c}{10}\right) (30)$	3
BC	$\left(\dfrac{5I_c}{30} = \dfrac{I_c}{6}\right) (30)$	5

$$M_{FBC} = +\frac{4(30)^2}{12} = +300 \text{ kip-ft}$$
$$M_{FCB} = -300 \text{ kip-ft}$$

Moment distribution (see Table 15-10)

For reactions, shear and bending-moment diagrams, and the elastic curve, see Example 14-6.

15-6. Application of the Moment-distribution Method to the Analysis of Statically Indeterminate Rigid Frames. *Case 2 With Joint Movements.* The procedure for applying the moment-distribution method to the analysis of statically indeterminate rigid frames in which sidesway or joint movements are involved consists in the following:

1. The joints are first held against sidesway. The fixed-end moments caused by the applied loadings are distributed, and a first set of balanced end moments is obtained.

2. The unloaded frame is then assumed to have a certain amount of sidesway which will cause a set of fixed-end moments. These fixed-end moments are then distributed, and a second set of balanced end moments is obtained.

3. The resulting set of end moments may be obtained by adding the first set and the product of a ratio and the second set, the ratio being determined by use of the shear condition, as will be explained.

TABLE 15-10. MOMENT DISTRIBUTION

Joint		A	B		C		D
Member		AB	BA	BC	CB	CD	DC
K		3	3	5	5	3	3
Cycle	DF	——	0.375	0.625	0.625	0.375	——
1	FEM	——	——	+300.0	−300.0	——	——
	Balance	——	−112.5	−187.5	+187.5	+112.5	——
2	Carry-over	−56.2	——	+ 93.8	− 93.8	——	+56.2
	Balance	——	− 35.2	− 58.6	+ 58.6	+ 35.2	——
3	Carry-over	−17.6	——	+ 29.3	− 29.3	——	+17.6
	Balance	——	− 11.0	− 18.3	+ 18.3	+ 11.0	——
4	Carry-over	− 5.5	——	+ 9.2	− 9.2	——	+ 5.5
	Balance	——	− 3.4	− 5.8	+ 5.8	+ 3.4	——
5	Carry-over	− 1.7	——	+ 2.9	− 2.9	——	+ 1.7
	Balance	——	− 1.1	− 1.8	+ 1.8	+ 1.1	——
6	Carry-over	− 0.6	——	+ 0.9	− 0.9	——	+ 0.6
	Balance	——	− 0.3	− 0.6	+ 0.6	+ 0.3	——
7	Carry-over	− 0.1	——	+ 0.3	− 0.3	——	+ 0.1
	Balance	——	− 0.1	− 0.2	+ 0.2	+ 0.1	——
8	Carry-over	——	——	+ 0.1	− 0.1	——	——
	Balance	——	——	− 0.1	+ 0.1	——	——
Total		−81.7	−163.6	+163.6	−163.6	+163.6	+81.7
Check:							
Change		−81.7	−163.6	−136.4	+136.4	+163.6	+81.7
$-\frac{1}{2}$ (change)		+81.8	+ 40.8	− 68.2	+ 68.2	− 40.8	−81.8
Sum		+ 0.1	−122.8	−204.6	+204.6	+122.8	− 0.1
θ_{rel} = sum/$-K$		0	+ 40.93	+ 40.92	− 40.92	− 40.93	0
		Check	Check		Check		Check

Take, for example, the rigid frame shown in Fig. 15-14a. It is required to analyze this statically indeterminate rigid frame by the moment-distribution method. The given frame shown in Fig. 15-14a is equivalent to the sum of Fig. 15-14b and Fig. 15-14c. In Fig. 15-14b the joints B and C are held against sidesway by the fictitious support at C, the horizontal reaction of which is denoted as H'_C. If the fictitious support at C is removed, the force H'_C would act at joint C. In Fig. 15-14c, Δ'' is the sidesway caused by any arbitrary force H''_C. If H'_C is equal to kH''_C, where k is the unknown ratio, the actual amount of the sidesway, Δ' (Fig. 15-14a), must be equal to $k\Delta''$. Let M'_{AB}, M'_{BA}, M'_{BC}, M'_{CB}, M'_{CD}, and M'_{DC} be the balanced moments obtained by distributing the fixed-end moments due to the applied loading in which joints B and C are permitted to rotate but not to move from their original positions (Fig. 15-14b).

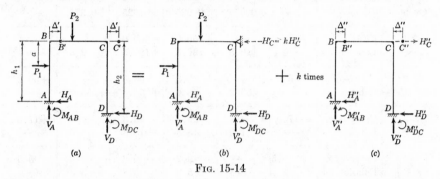

Fig. 15-14

Let M''_{AB}, M''_{BA}, M''_{BC}, M''_{CB}, M''_{CD}, and M''_{DC} be the balanced moments obtained by distributing the fixed-end moments due to any assumed amount Δ'' of the horizontal movement of joint B or C. The shear condition required of the frame shown in Fig. 15-14a is

$$H_A + H_D = P_1$$

Since

$$H_A = \frac{M_{AB} + M_{BA}}{h_1} + \frac{P_1 a}{h_1} \quad \text{and} \quad H_D = \frac{M_{CD} + M_{DC}}{h_2}$$

the shear condition becomes

$$\frac{M_{AB} + M_{BA}}{h_1} + \frac{P_1 a}{h_1} + \frac{M_{CD} + M_{DC}}{h_2} = P_1 \tag{15-12}$$

Also, by superposition,

$$\begin{aligned} M_{AB} &= M'_{AB} + k(M''_{AB}) & M_{BA} &= M'_{BA} + k(M''_{BA}) \\ M_{BC} &= M'_{BC} + k(M''_{BC}) & M_{CB} &= M'_{CB} + k(M''_{CB}) \\ M_{CD} &= M'_{CD} + k(M''_{CD}) & M_{DC} &= M'_{DC} + k(M''_{DC}) \end{aligned} \tag{15-13}$$

By substituting Eqs. (15-13) in Eq. (15-12),

$$\frac{(M'_{AB} + M'_{BA}) + k(M''_{AB} + M''_{BA})}{h_1} + \frac{P_1 a}{h_1}$$
$$+ \frac{(M'_{CD} + M'_{DC}) + k(M''_{CD} + M''_{DC})}{h_2} = P_1 \tag{15-14}$$

The unknown ratio k can then be found by solving Eq. (15-14). Once k

is known, all end moments acting on the frame of Fig. 15-14a may be found from Eqs. (15-13).

Where two or more unknown movements of sidesway are involved, the resulting set of end moments may be expressed as the sum of (1) the balanced end moments by distributing the fixed-end moments due to the applied loading, and (2) the products of an unknown ratio and the balanced end moments found by distributing the fixed-end moments due to a certain amount of the first movement in sidesway, and (3) the products of a *second* unknown ratio and the balanced end moments due to a certain amount of the *second* movement in sidesway, and so on. The unknown ratios are determined from the shear conditions.

The procedure discussed above will be illustrated by the following examples.

Example 15-10. Analyze the rigid frame shown in Fig. 15-15a by the moment-distribution method. Draw shear and bending-moment diagrams. Sketch the deformed structure.

SOLUTION. *Relative stiffness*

AB	$\left(\dfrac{3I_c}{18} = \dfrac{I_c}{6}\right)(12)$	2
BC	$\left(\dfrac{4I_c}{24} = \dfrac{I_c}{6}\right)(12)$	2
CD	$\left(\dfrac{3I_c}{12} = \dfrac{I_c}{4}\right)(12)$	3

Distribution of fixed-end moments due to the applied loading (see Fig. 15-15b)

$$M_{FAB} = + \frac{(10)(12)(6)^2}{(18)^2} = +13.33 \text{ kip-ft}$$

$$M_{FBA} = - \frac{(10)(6)(12)^2}{(18)^2} = -26.67 \text{ kip-ft}$$

$$M_{FBC} = + \frac{3(24)^2}{12} = +144 \text{ kip-ft}$$

$$M_{FCB} = -144 \text{ kip-ft}$$

For distribution of these fixed-end moments, see Table 15-11.

Distribution of fixed-end moments due to sidesway (see Fig. 15-15c)

FIXED-END MOMENTS DUE TO SIDESWAY

	Relative magnitudes			
$M_{FAB} = M_{FBA}$	$+6E(3I_c)\Delta/(18)^2$	$+\frac{1}{3}_{24} \times 1{,}296$	$+4$	$+400$
$M_{FCD} = M_{FDC}$	$+6E(3I_c)\Delta/(12)^2$	$+\frac{1}{144} \times 1{,}296$	$+9$	$+900$

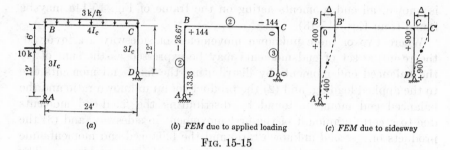

(a)

(b) FEM due to applied loading

(c) FEM due to sidesway

FIG. 15-15

Note that only the relative magnitudes of the fixed-end moments due to an assumed amount of sidesway Δ are required.

For distribution of these fixed-end moments, see Table 15-12.

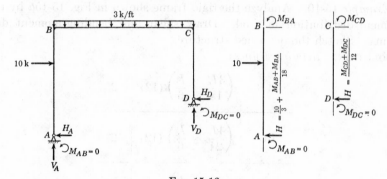

FIG. 15-16

Determination of ratio k. The shear condition (Fig. 15-16) is

$$H_A + H_D = 10$$

or

$$\left(\frac{10}{3} + \frac{M_{AB} + M_{BA}}{18}\right) + \left(\frac{M_{CD} + M_{DC}}{12}\right) = 10$$

Simplifying, $2(M_{AB} + M_{BA}) + 3(M_{CD} + M_{DC}) = +240$

Substituting

$$M_{AB} = 0 + k(0) = 0$$
$$M_{BA} = -99.75 + k(+156.7) = -99.75 + 156.7k$$
$$M_{CD} = +99.70 + k(+227.0) = +99.70 + 227.0k$$
$$M_{DC} = 0 + k(0) = 0$$

in the above equation and solving for the ratio k,

$$2(-99.75 + 156.7k) + 3(+99.70 + 227.0k) = +240$$
$$k = +0.1412$$

TABLE 15-11. DISTRIBUTION OF *FEM* DUE TO THE APPLIED LOADING

Joint................	A	B		C		D
Member.............	AB	BA	BC	CB	CD	DC
K.................	2	2	2	2	3	3
Modified K.........	1.5	1.5	2	2	2.25	2.25

Cycle	DF	——	0.4286	0.5714	0.4706	0.5294	——
1	FEM	+13.33	−26.67	+144.00	−144.00	——	——
	Balance	−13.33	−50.29	− 67.04	+ 67.77	+76.23	——
2	Carry-over	↑	− 6.67	+ 33.88	− 33.52	——	↑
	Balance		−11.66	− 15.55	+ 15.77	+17.75	
3	Carry-over		——	+ 7.88	− 7.78	——	
	Balance		− 3.38	− 4.50	+ 3.66	+ 4.12	
4	Carry-over		——	+ 1.83	− 2.25	——	
	Balance		− 0.78	− 1.05	+ 1.06	+ 1.19	
5	Carry-over		——	+ 0.53	− 0.52	——	
	Balance		− 0.23	− 0.30	+ 0.24	+ 0.28	
6	Carry-over		——	+ 0.12	− 0.18	——	
	Balance		− 0.05	− 0.07	+ 0.08	+ 0.10	
7	Carry-over		——	+ 0.04	− 0.04	——	
	Balance		− 0.02	− 0.02	+ 0.02	+ 0.02	
8	Carry-over		——	+ 0.01	− 0.01	——	
	Balance		——	− 0.01	——	+ 0.01	↓

Total..............	0	−99.75	+ 99.75	− 99.70	+99.70	0

Check:

Change...........	−13.33	−73.08	− 44.25	+ 44.30	+99.70	0
−½ (change)......	+36.54	+ 6.66	− 22.15	+ 22.12	0	−49.85
Sum..............	+23.21	−66.42	− 66.40	+ 66.42	+99.70	−49.85
θ_{rel} = sum/−K....	−11.60	+33.21	+ 33.20	− 33.21	−33.23	+16.62
		Check		Check		

TABLE 15-12. DISTRIBUTION OF *FEM* DUE TO SIDESWAY

Joint.............	A	B		C		D
Member..........	AB	BA	BC	CB	CD	DC
K...............	2	2	2	2	3	3
Modified K.......	1.5	1.5	2	2	2.25	2.25

Cycle	DF		0.4286	0.5714	0.4706	0.5294	
1	FEM	+400.0	+400.0	——	——	+900.0	+900.0
	Balance	−400.0	−171.4	−228.6	−423.5	−476.5	−900.0
2	Carry-over	↑	−200.0	−211.8	−114.3	−450.0	↑
	Balance		+176.5	+235.3	+265.6	+298.7	
3	Carry-over		——	+132.8	+117.6	——	
	Balance		− 56.9	− 75.9	− 55.3	− 62.3	
4	Carry-over		——	− 27.6	− 38.0	——	
	Balance		+ 11.8	+ 15.8	+ 17.9	+ 20.1	
5	Carry-over		——	+ 9.0	+ 7.9	——	
	Balance		− 3.8	− 5.2	− 3.7	− 4.2	
6	Carry-over		——	− 1.8	− 2.6	——	
	Balance		+ 0.8	+ 1.0	+ 1.2	+ 1.4	
7	Carry-over		——	+ 0.6	+ 0.5	——	
	Balance		− 0.3	− 0.3	− 0.2	− 0.3	
8	Carry-over		——	− 0.1	− 0.1	——	
	Balance	↓	——	+ 0.1	——	+ 0.1	↓

Total.............	0	+156.7	−156.7	−227.0	+227.0	0

Check:

	A	B		C		D
Change.........	−400.0	−243.3	−156.7	−227.0	−673.0	−900.0
−½ (change)....	+121.6	+200.0	+113.5	+ 78.4	+450.0	+336.5
Sum............	−278.4	− 43.3	− 43.2	−148.6	−223.0	−563.5
θ_{rel} = sum/−K..	+139.2	+ 21.6	+ 21.6	+ 74.3	+ 74.3	+187.8
			Check		Check	

Combination of the two sets of balanced moments (see Table 15-13)

TABLE 15-13. COMBINATION OF THE TWO SETS OF BALANCED MOMENTS

Joint	A	B		C		D
Member	AB	BA	BC	CB	CD	DC
FEM from Table 15-11	+13.33	− 26.67	+144.00	−144.00	——	——
+0.1412 times FEM from Table 15-12	+56.48	+ 56.48	——	——	+127.08	+127.08
Total FEM	+69.81	+ 29.81	+144.00	−144.00	+127.08	+127.08
Balanced moments from Table 15-11	——	− 99.75	+ 99.75	− 99.70	+ 99.70	——
+0.1412 times balanced moments in Table 15-12	——	+ 22.13	− 22.13	− 32.05	+ 32.05	——
Total balanced moments	——	− 77.62	+ 77.62	−131.75	+131.75	——
Check:						
Change	−69.81	−107.43	− 66.38	+ 12.25	+ 4.67	−127.08
−½ (change)	+53.72	+ 34.90	− 6.12	+ 33.19	+ 63.54	− 2.34
Sum	−16.09	− 72.53	− 72.50	+ 45.44	+ 68.21	−129.42
θ_{rel} = sum/−K	+ 8.04	+ 36.26	+ 36.25	− 22.72	− 22.74	+ 43.14
			Check		Check	

For reactions, shear and bending-moment diagrams, and the elastic curve, see Example 14-10.

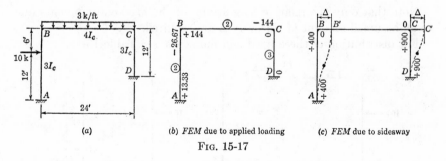

(a)   (b) *FEM* due to applied loading   (c) *FEM* due to sidesway

FIG. 15-17

Example 15-11. Analyze the rigid frame shown in Fig. 15-17a by the moment-distribution method. Draw shear and bending-moment diagrams. Sketch the deformed structure.

SOLUTION. *Relative stiffness*

AB	$\left(\dfrac{3I_c}{18} = \dfrac{I_c}{6}\right)(12)$	2
BC	$\left(\dfrac{4I_c}{24} = \dfrac{I_c}{6}\right)(12)$	2
CD	$\left(\dfrac{3I_c}{12} = \dfrac{I_c}{4}\right)(12)$	3

Distribution of FEM due to the applied loading (see Fig. 15-17*b*)

$$M_{FAB} = +\frac{(10)(12)(6)^2}{(18)^2} = +13.33 \text{ kip-ft}$$

$$M_{FBA} = -\frac{(10)(6)(12)^2}{(18)^2} = -26.67 \text{ kip-ft}$$

$$M_{FBC} = +\frac{3(24)^2}{12} = +144 \text{ kip-ft}$$

$$M_{FCB} = -144 \text{ kip-ft}$$
$$M_{FCD} = M_{FDC} = 0$$

For distribution of these fixed-end moments, see Table 15-14.

Distribution of fixed-end moments due to sidesway (see Fig. 15-17*c*)

FIXED-END MOMENTS DUE TO SIDESWAY

	Relative magnitudes			
$M_{FAB} = M_{FBA}$	$+6E(3I_c)\Delta/(18)^2$	$+\frac{1}{324} \times 1{,}296$	$+4$	$+400$
$M_{FCD} = M_{FDC}$	$+6E(3I_c)\Delta/(12)^2$	$+\frac{1}{144} \times 1{,}296$	$+9$	$+900$

Note that only the relative magnitudes of the fixed-end moments due to an assumed amount of sidesway Δ are required.

For distribution of these fixed-end moments, see Table 15-15.

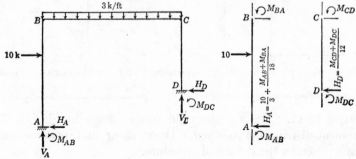

FIG. 15-18

Determination of ratio k. The shear condition (Fig. 15-18) is

$$H_A + H_D = 10$$

or

$$\frac{10}{3} + \frac{M_{AB} + M_{BA}}{18} + \frac{M_{CD} + M_{DC}}{12} = 10$$

Simplifying, $2(M_{AB} + M_{BA}) + 3(M_{CD} + M_{DC}) = +240$

Substituting

$$M_{AB} = -25.12 + k(+342.1)$$
$$M_{BA} = -103.58 + k(+284.2)$$
$$M_{CD} = +109.47 + k(+394.7)$$
$$M_{DC} = +54.73 + k(+647.4)$$

TABLE 15-14. DISTRIBUTION OF *FEM* DUE TO THE APPLIED LOADING

Joint	A	B		C		D
Member	AB	BA	BC	CB	CD	DC
K	2	2	2	2	3	3

Cycle	DF		0.5000	0.5000	0.4000	0.6000	
1	FEM	+13.33	− 26.67	+144.00	−144.00	——	——
	Balance	——	− 58.66	− 58.67	+ 57.60	+ 86.40	——
2	Carry-over	−29.33	——	+ 28.80	− 29.33	——	+43.20
	Balance	——	− 14.40	− 14.40	+ 11.73	+ 17.60	——
3	Carry-over	− 7.20	——	+ 5.86	− 7.20	——	+ 8.80
	Balance	——	− 2.93	− 2.93	+ 2.88	+ 4.32	——
4	Carry-over	− 1.46	——	+ 1.44	− 1.46	——	+ 2.16
	Balance	——	− 0.72	− 0.72	+ 0.58	+ 0.88	——
5	Carry-over	− 0.36	——	+ 0.29	− 0.36	——	+ 0.44
	Balance	——	− 0.15	− 0.14	+ 0.14	+ 0.22	——
6	Carry-over	− 0.08	——	+ 0.07	− 0.07	——	+ 0.11
	Balance	——	− 0.04	− 0.03	+ 0.03	+ 0.04	——
7	Carry-over	− 0.02	——	+ 0.02	− 0.02	——	+ 0.02
	Balance	——	− 0.01	− 0.01	+ 0.01	+ 0.01	——
Total		−25.12	−103.58	+103.58	−109.47	+109.47	+54.73

Check:						
Change	−38.45	− 76.91	− 40.42	+ 34.53	+109.47	+54.73
−½ (change)	+38.46	+ 19.22	− 17.26	+ 20.21	− 27.36	−54.74
Sum	+ 0.01	− 57.69	− 57.68	+ 54.74	+ 82.11	− 0.01
$\theta_{rel} = $ sum$/−K$	0	+ 28.84	+ 28.84	− 27.37	− 27.37	0
	Check	Check		Check		Check

TABLE 15-15. DISTRIBUTION OF *FEM* DUE TO SIDESWAY

Joint	A	B			C		D
Member	AB	BA	BC	CB	CD	DC
K	2	2	2	2	3	3
Cycle DF	—	0.5000	0.5000	0.4000	0.6000	—
1 FEM	+400.0	+400.0	—	—	+900.0	+900.0
Balance	—	−200.0	−200.0	−360.0	−540.0	—
2 Carry-over	−100.0	—	−180.0	−100.0	—	−270.0
Balance	—	+ 90.0	+ 90.0	+ 40.0	+ 60.0	—
3 Carry-over	+ 45.0	—	+ 20.0	+ 45.0	—	+ 30.0
Balance	—	− 10.0	− 10.0	− 18.0	− 27.0	—
4 Carry-over	− 5.0	—	− 9.0	− 5.0	—	− 13.5
Balance	—	+ 4.5	+ 4.5	+ 2.0	+ 3.0	—
5 Carry-over	+ 2.2	—	+ 1.0	+ 2.2	—	+ 1.5
Balance	—	− 0.5	− 0.5	− 0.9	− 1.3	—
6 Carry-over	− 0.2	—	− 0.4	− 0.2	—	− 0.6
Balance	—	+ 0.2	+ 0.2	+ 0.1	+ 0.1	—
7 Carry-over	+ 0.1	—	—	+ 0.1	—	—
Balance	—	—	—	—	− 0.1	—
Total	+342.1	+284.2	−284.2	−394.7	+394.7	+647.4
Check:						
Change	− 57.9	−115.8	−284.2	−394.7	−505.3	−252.6
−½ (change)	+ 57.9	+ 29.0	+197.4	+142.1	+126.3	+252.6
Sum	0	− 86.8	− 86.8	−252.6	−379.0	0
θ_{rel} = sum/−K	0	+ 43.4	+ 43.4	+126.3	+126.3	0
	Check	Check		Check		Check

in the above equation and solving for the ratio k,

$$2(-128.70 + 626.3k) + 3(+164.20 + 1{,}042.1k) = +240$$
$$4{,}378.9k = +4.8$$
$$k = +0.0011$$

Combination of the two sets of balanced moments (see Table 15-16)

TABLE 15-16. COMBINATION OF THE TWO SETS OF BALANCED MOMENTS

Joint	A	B		C		D
Member	AB	BA	BC	CB	CD	DC
FEM from Table 15-14	+13.33	− 26.67	+144.00	+144.00	——	——
+0.0011 times *FEM* from Table 15-15	+ 0.44	+ 0.44	——	——	+ 0.99	+ 0.99
Total *FEM*	+13.77	− 26.23	+144.00	−144.00	+ 0.99	+ 0.99
Balanced moments from Table 15-11	−25.12	−103.58	+103.58	−109.48	+109.47	+54.73
+0.0011 times balanced moments in Table 15-15	+ 0.38	+ 0.31	− 0.31	− 0.43	+ 0.43	+ 0.71
Total balanced moments	−24.74	−103.27	+103.27	−109.90	+109.90	+55.44
Check: Change	−38.51	− 77.04	− 40.73	+ 34.10	+108.91	+54.45
−½ (change)	+38.52	+ 19.26	− 17.05	+ 20.36	− 27.22	−54.46
Sum	+ 0.01	− 57.78	− 57.78	+ 54.46	+ 81.69	− 0.01
θ_{rel} = sum/−K	0	+ 28.89	+ 28.89	− 27.23	− 27.23	0
	Check	Check		Check		Check

For reactions, shear and bending-moment diagrams, and the elastic curve, see Example 14-11.

PROBLEMS

15-1 to 15-22. Analyze the statically indeterminate beams or rigid frames shown in Probs. 14-1 to 14-22 by the moment-distribution method. Wherever applicable, use both the regular and the modified-stiffness methods in moment distribution. Draw shear and bending-moment diagrams. Sketch the deformed structure.

ANSWERS TO PROBLEMS

2-1. $F_4 = 138.42$ lb at $(\theta_4)_x = 53°38'$.

2-2. $F_3 = 24.62$ lb; $F_4 = -104.82$ lb.

2-3. (a) $F_3 = 54.58$ lb, $(\theta_4)_x = 176°44'$; (b) $F_3 = -38.68$ lb, $(\theta_4)_x = 243°16'$.

2-4. $F_4 = 160$ lb at 0.5 ft to left of F_1.

2-5. $V_A = 105$ lb; $V_B = 165$ lb.

2-6. $V_A = 17\frac{2}{3}$ lb; $V_B = 50\frac{1}{3}$ lb.

2-7. $F_4 = 202.2$ lb at $(\theta_4)_x = 2°40'$ passing through $(0, +3.99')$.

2-8. $R_A = 24.68$ kips; $R_B = 21.19$ kips at $(\theta_B)_x = 136°22'$.

2-9. $F_3 = 825$ lb; $F_4 = 606.25$ lb; $F_5 = -168.75$ lb.

2-10. $H_A = 12$ kips to the right; $V_A = 16$ kips upward; $H_C = 12$ kips to the left; $V_C = 11$ kips upward.

2-11. $H_A = 23\frac{1}{3}$ kips to the right; $V_A = 17\frac{2}{3}$ kips upward; $H_C = 17\frac{1}{3}$ kips to the left; $V_C = 24\frac{1}{3}$ kips upward.

3-1. (a) $V = +44$ kips, $M = -234$ kip-ft; (b) $V = +10$ kips, $M = -38$ kip-ft; (c) $V = +4$ kips, $M = -12$ kip-ft.

3-2. (a) $V = -23$ kips, $M = -45.5$ kip-ft; (b) $V = +47.44$ kips, $M = +58.32$ kip-ft; (c) $V = -6.56$ kips, $M = +178.72$ kip-ft.

3-3. $V = +1.68$ kips; $M = +74.88$ kip-ft.

3-4. A as origin, $V = +40 - 3x$, $M = -288 + 40x - 1.5x^2$; B as origin, $V = -14 + 3x$, $M = -54 + 14x - 1.5x^2$.

3-5. A as origin, $V = -6.56$, $M = -6.56x + 309.92$; B as origin, $V = -6.56$, $M = -6.56x + 264$; C as origin, $V = -6.56$, $M = +6.56x + 100$.

3-6. A as origin, $V = +18 - 2.4x + 0.045x^2$, $M = +18x - 1.2x^2 + 0.015x^3$; B as origin, $V = -12 + 0.6x + 0.045x^2$, $M = +12x - 0.3x^2 - 0.015x^3$.

3-7. At support, $V = -P$, $M = -PL$.

3-8. At support, $V = +wL$, $M = -\frac{1}{2}wL^2$.

3-9. At support, $V = +16$ kips, $M = -190$ kip-ft.

3-10. At center, $V = 0$, $M = +\frac{1}{8}wL^2$.

3-11. At center, $V = 0$, $M = +\frac{1}{4}PL$.

3-12. At one-third point, $V = +P$, $M = +\frac{1}{3}PL$.

3-13. At center, $V = \pm\frac{1}{2}P$, $M = +\frac{1}{2}PL$.

3-14. Max $M = +142.2$ kip-ft under 14-kip load.

3-15. At support, $V = +50$ kips, $M = -328$ kip-ft.

3-16. Max $M = +199.01$ kip-ft at 16.187 ft from the right support.

3-17. Max $M = +75.736$ kip-ft at 9.028 ft from the left support.

3-18. See Prob. 3-9.

3-19. See Probs. 3-1 and 3-15.

3-20. See Probs. 3-2 and 3-16.

4-1. $M_A = 0$; $M_B = 38.4$ kip-ft (compression inside); $M_C = 173.4$ kip-ft (compression inside); $M_D = 19.8$ kip-ft (compression inside).

4-2. $M_A = M_C = M_D = 0$; $M_B = 38.4$ kip-ft (compression outside); M at 15-kip load $= 62.55$ kip-ft (compression outside).

4-3. $M_A = M_B = 128$ kip-ft (compression inside); $M_C = 50$ kip-ft (compression inside).

4-4. Max M in $BC = +115.2$ kip-ft at 9.633 ft from B.

4-5. $M_{BA} = 24$ kip-ft (compression inside); $M_{BC} = 373.5$ kip-ft (compression inside); $M_{BD} = 349.5$ kip-ft (compression inside).

4-6. Max M in $CD = 150$ kip-ft (compression outside) at 2.167 ft from C.

4-7. $M_B = 90$ kip-ft (compression outside); $M_C = 54$ kip-ft (compression outside).

4-8. T in rod $= 1,500$ lb.

4-9. Direct stress in $AB = 2.67$ kips tension; direct stress in $BC = 4.80$ kips compression; direct stress in $AD = 6.93$ kips compression; direct stress in $CD = 0.27$ kip tension; shear in $AD = +4.80$ kips.

4-10. $M_C = 41.143$ kip-ft (compression inside); direct stress in $CD = 6.343$ kips compression; shear in $CD = 2.743$ kips.

4-11. $H_A = 0.625$ kip to the right; $V_A = 8.50$ kips upward; $H_D = 8.625$ kips to the left; $V_D = 16.50$ kips upward; $M_B = 29$ kip-ft (compression inside).

5-1. $U_0U_1 = 0$, $L_0L_1 = +16.5$ kips, $U_1U_2 = -24$ kips, $L_1L_2 = +16.5$ kips, $U_2U_3 = -24$ kips, $L_2L_3 = +19.5$ kips, $U_3U_4 = 0$, $L_3L_4 = +19.5$ kips, $U_0L_0 = -6$ kips, $L_0U_1 = -27.5$ kips, $U_1L_1 = 0$, $U_1L_2 = +12.5$ kips, $U_2L_2 = -16$ kips, $L_2U_3 = +7.5$ kips, $U_3L_3 = 0$, $U_3L_4 = -32.5$ kips, $U_4L_4 = -8$ kips.

5-2. $L_0U_1 = -4,507$ lb, $L_0L_1 = +3,750$ lb, $U_1U_2 = -3,606$ lb, $L_1L_2 = +3,750$ lb, $U_2U_3 = -2,704$ lb, $L_2L_3 = +3,000$ lb, $U_1L_1 = 0$, $U_1L_2 = -901$ lb, $U_2L_2 = +500$ lb, $U_2L_3 = -1,250$ lb, $U_3L_3 = +2,000$ lb.

5-3. $L_0U_1 = -40.72$ kips, $L_0L_1 = +28.80$ kips, $U_1U_2 = -31.31$ kips, $L_1L_2 = +28.80$ kips, $U_2U_3 = -19.80$ kips, $L_2L_3 = +29.70$ kips, $U_3U_4 = -20.87$ kips, $L_3L_4 = +13.20$ kips, $U_4L_5 = -18.66$ kips, $L_4L_5 = +13.20$ kips, $U_1L_1 = +18.00$ kips, $U_1L_2 = +1.27$ kips, $U_2L_2 = +23.10$ kips, $U_2L_3 = -16.50$ kips, $U_3L_3 = +6.60$ kips, $L_3U_4 = +9.33$ kips, $U_4L_4 = 0$.

5-4. $AB = +1,118$ lb, $AF = -1,000$ lb, $BC = +1,565$ lb, $FG = -2,000$ lb, $CD = +2,012$ lb, $GK = -4,000$ lb, $DE = +2,460$ lb, $CG = -1,789$ lb, $BF = DH = -894$ lb, $GH = +2,000$ lb, $CF = CH = +1,000$ lb, $EH = +3,000$ lb.

5-5. See Prob. 5-1.

5-6. See Prob. 5-2.

5-7. See Prob. 5-3.

5-8. See Prob. 5-4.

5-9. See Prob. 5-3.

5-10. $U_1U_2 = -4P \tan \theta$, $U_1L_1 = +P$, $U_2U_3 = -4.5P \tan \theta$, $U_2L_2 = -0.5P$, $L_0L_1 = +2.5P \tan \theta$, $U_3L_3 = 0$, $L_1L_2 = +2.5P \tan \theta$, $L_0U_1 = -2.5P \sec \theta$, $L_2L_3 = +4P \tan \theta$, $U_1L_2 = +1.5P \sec \theta$, $U_2L_3 = +0.5P \sec \theta$.

5-11. $U_1U_2 = -4P \tan \theta$, $L_0U_1 = -2P \sec \theta$, $U_2U_3 = -6P \tan \theta$, $U_1L_1 = +2P \sec \theta$, $L_0L_1 = +2P \tan \theta$, $L_1U_2 = -P \sec \theta$, $L_1L_2 = +5P \tan \theta$, $U_2L_2 = +P \sec \theta$, $L_2L_3 = +6P \tan \theta$, $L_2U_3 = 0$.

5-12. See Prob. 5-1.

5-13. See Prob. 5-2.

5-14. See Prob. 5-3.

5-15. See Prob. 5-4.

6-1. Dead panel load $= 2,000$ lb on top chord and 900 lb on bottom chord. Snow panel load $= 1,000$ lb. Wind panel load $= 2,250$ lb. Max caused by $D + S$: $AB = -21,800$; $BC = -17,440$; $CD = -13,080$; $Ab = +19,500$; $bc = +19,500$; $cd = +15,600$; $Bb = +900$; $Dd = +8,700$. Max caused by $\frac{3}{4}(D + S/2 + W_L$ or $W_R)$: $Bc = -4,960$; $Cc = +2,900$; $Cd = -6,270$.

6-2. Dead panel load $= 2,810$ lb. Snow panel load $= 2,430$ lb. Wind panel load $= 3,440$ lb. $(D + E)$ panel load $= 6,060$ lb. Max caused by $D + S$: $AB =$

$-33,060$; $BC = -30,150$; $CD = -27,260$; $DE = -24,350$; $AK = +27,510$; $KL = +23,590$; $LP = +15,720$. Max caused by $\frac{3}{4}(D + S/2 + W_L$ or $W_R)$: $BK = DM = -5,090$; $CK = CM = +4,590$; $CL = -10,190$; $EM = +13,770$; $LM = +9,180$. $(D + E)$ stresses: $AB = -38,240$; $BC = -34,880$; $CD = -31,520$; $DE = -28,160$; $AK = +31,820$; $KL = +27,270$; $LP = +18,180$; $BK = CM = -5,040$; $CK = CM = +4,550$; $CL = -10,090$; $EM = +13,640$; $LM = +9,090$.

7-1. Dead panel load = 2,810 lb. Snow panel load = 2,430 lb. Wind panel load = 3,440 lb on truss and 4,500 lb on column. $(D + E)$ panel load = 6,060 lb. Max caused by $D + S$: $AB = -33,060$; $BC = -30,150$; $CD = -27,260$; $DE = -24,350$; $AK = +27,510$; $KL = +23,590$; $LP = +15,720$. Max caused by $\frac{3}{4}(D + S/2 + W_L$ or $W_R)$; $BK = DM = -5,090$; $CK = +10,640$; $CM = +4,590$; $CL = -13,550$; $EM = +16,800$; $LM = +12,210$; $RK = -35,800$. Min caused by $\frac{3}{4}(D + W_L$ or $W_R)$: $AB = +15,170$; $BC = +16,340$; $KL = -12,850$; $LP = -2,860$; $CK = -24,720$; $CL = +11,090$; $EM = -8,420$; $LM = -9,990$; $RK = +8,240$. For stresses due to $(D + E)$ condition, see answers to Prob. 6-2.

8-1. (a) 45 kips; (b) $+26.67$ kips, -11.67 kips, 300 kip-ft; (c) ±18.75 kips, 337.5 kip-ft.

8-2. (a) 16 kips, $+12.25$ kips, 9 kips, 6.25 kips, 4 kips; (b) 70 kip-ft, 120 kip-ft, 150 kip-ft, 160 kip-ft.

8-3. (a) -13 kips, -16 kips, -19 kips; (b) -57.5 kip-ft, -130 kip-ft, -217.5 kip-ft.

8-4. (a) 30.67 kips, 50.67 kips; (b) $+15$ kips, -18 kips; (c) -11.67 kip-ft to $+201.67$ kip-ft, -73.33 kip-ft to $+193.33$ kip-ft, -185 kip-ft.

8-5. (a) 82.75 kips, 99.625 kips; (b) -30 kips on left and $+57.75$ kips on right of left support, $+34.25$ kips, -18.125 kips, -62.125 kips on left and $+50$ kips on right of right support; (c) -200 kip-ft; -86.25 kip-ft to $+372.5$ kip-ft, -102.5 kip-ft to $+485$ kip-ft, -425 kip-ft.

8-6. (a) $+76.67$ kips, -5.73 to $+51.73$ kips, -16.27 to $+31.6$ kips; (b) $+1,533$ kip-ft, $+2,453$ kip-ft, $+2,760$ kip-ft; (c) $+44$ kips.

8-7. (a) $+46.4$ kips, -5.6 to $+28.8$ kips, ±15.2 kips; (b) $+1,160$ kip-ft, $+1,740$ kip-ft; (c) $+580$ kip-ft, $+1,405$ kip-ft, $+1,650$ kip-ft.

8-8. (a) 189 kips compression, 183 kips tension; (b) 60 kips tension; (c) 29.36 kips compression to 33.93 kips tension, 36.21 kips compression to 31.38 kips tension.

9-1. 81.6 kips, $+56$ kips, $+33.6$ kips, $+10.8$ kips.

9-2. 22.5 kips, 16.25 kips, 162.5 kip-ft.

9-3. 28.75 kips, -150 kip-ft.

9-4. 16 kips, 45 kip-ft.

9-5. 18 kips, 81 kip-ft.

9-6. 13.39 kips, 83.70 kip-ft.

9-7. 38.33 kips, 220.1 kip-ft.

9-8. 58.33 kips, 41.67 kips, 300 kip-ft.

9-9. 43.33 kips; 306.2 kip-ft at 14.13 ft from the left support.

9-10. (a) 43.55 kips, 26.22 kips, 12.08 kips; (b) 358.6 kip-ft; 461.3 kip-ft; (c) 462.6 kip-ft with 15.14 ft of uniform load on the span.

9-11. (a) 93.33 kips, 60.83 kips, 30 kips; (b) 975 kip-ft, 1,250 kip-ft; (c) 1,255.12 kip-ft.

9-12. (a) 1,250 kip-ft, 1,300 kip-ft; (b) 1,300 kip-ft.

9-13. (a) $+68.21$ kips, -3.96 to $+45.71$ kips, -13.46 to $+27.21$ kips; (b) 1,364.17 kip-ft, 2,065.83 kip-ft, 2,269.38 kip-ft; (c) 34.25 kips.

9-14. 682.08 kip-ft, 1,713.75 kip-ft, 2,153.02 kip-ft.

9-15. (a) 105.42 kips compression, 102.25 kips tension; (b) 34.25 kips tension; (c) 22.26 kips compression to 24.36 kips tension, 26.00 kips compression to 23.76 kips tension.

10-1. Panel load = 7.43 kips at top and 19.85 kips at bottom; $L_0L_1 = L_1L_2 = +95.48$; $L_2L_3 = +130.94$; $L_3L_4 = +146.14$; $L_0U_1 = -135.01$; $U_1U_2 = -135.00$; $U_2U_3 = -147.75$; $U_3U_4 = -155.88$; $U_1L_2 = +50.14$; $U_2L_3 = +24.34$; $U_3L_4 = +16.76$; $U_1L_1 = +19.85$; $U_2L_2 = -15.61$; $U_3L_3 = +0.85$; $U_4L_4 = -7.43$.

10-2.

H20-44 loading	40-ft span		80-ft span	
	Truck	Lane	Truck	Lane
Max end shear, kips..................	37.2	38.8	38.6	51.6
Max V at $\frac{1}{4}$ point, kips..............	27.2	26.7	28.6	33.9
Max M at $\frac{1}{4}$ point, kip-ft............	272	231	572	654
Max M at center, kip-ft..............	344	308	744	872
Absolute max M, kip-ft..............	345.96	308	744.98	872

10-3.

H20-S16-44 loading	80-ft span		180-ft span	
	Truck	Lane	Truck	Lane
Max end shear, kips............	63.6	51.6	68.27	83.6
Max V at $\frac{1}{4}$ point, kips........	45.6	33.9	50.27	51.9
Max M at $\frac{1}{4}$ point, kip-ft.......	912	654	2,262	2,551.5
Max M at center, kip-ft........	1,160	872	2,960	3,402
Absolute max M, kip-ft........	1,164.9	872	2,962.2	3,402

10-4.

H20-S16-44	Shear, kips			Moment, kip-ft	
	Panel 0-1	Panel 1-2	Panel 2-3	Point 1	Point 2
Truck..............	52.224	37.824	23.424	1,305.6	1,913.6
Lane................	52.8	33.6	18.4	1,160	1,740

10-5. $L_0L_1 = L_1L_2 = 0$ to $+60.55$; $L_2L_3 = 0$ to $+83.04$; $L_3L_4 = 0$ to $+92.68$; $L_0U_1 = -95.52$ to 0; $U_1U_2 = -85.61$ to 0; $U_2U_3 = -93.70$ to 0; $U_3U_4 = -98.86$ to 0; $U_1L_2 = -13.54$ to $+43.50$; $U_2L_3 = -20.57$ to $+33.33$; $U_3L_4 = -22.09$ to $+33.95$; $U_1L_1 = 0$ to $+38.8$; $U_2L_2 = -22.72$ to $+23.43$; $U_3L_3 = -18.31$ to $+28.35$; $U_4L_4 = 0$.

10-6. (a) 175.68 kips, 11.712 kips per ft; (b) 108.18 kips, 12.821 kips per ft; (c) 1,809.45 kip-ft with wheel 12 at $\frac{1}{4}$ point, 10.723 kips per ft; (d) 2,291.4 kip-ft, 10.184 kips per ft.

10-7. $V_{0-1} = 325.75$ kips, 9.307 kips per ft; $V_{1-2} = 246.55$ kips, 9.588 kips per ft; $V_{2-3} = 176.35$ kips, 9.876 kips per ft; $V_{3-4} = 114.76$ kips, 10.042 kips per ft; $M_1 = 6,515.1$ kip-ft, 9.307 kips per ft; $M_2 = 10,740.3$ kip-ft, 8.950 kips per ft; $M_3 = 13,255.9$ kip-ft, 8.837 kips per ft; $M_4 = 14,317.2$ kip-ft, 8.948 kips per ft.

10-8. $U_1L_1 = 118.0$ kips tension (11.80 kips per ft), $U_2L_2 = 101.3$ kips compression (10.00 kips per ft) or 79.3 kips tension (11.11 kips per ft), $U_2L_3 = 149.0$ kips tension (9.93 kips per ft) or 68.0 kips compression (11.18 kips per ft).

10-9. With or without counters: $U_1U_2 = -76.8$ to -355.2; $U_2U_3 = -86.4$ to -399.6; $L_0L_1 = L_1L_2 = +48.0$ to $+222.0$; $L_2L_3 = +76.8$ to -355.2; $L_0U_1 = -76.8$ to -355.4; $U_1L_1 = +16.0$ to $+103.0$; $U_1L_2 = +35.0$ to $+224.4$. Without counters: $U_2L_2 = +14.8$ to -98.3; $U_2L_3 = -29.2$ to $+115.6$; $U_3L_3 = -8.0$. With counters: $U_2L_3 = 0$ to $+115.6$; $L_2U_3 = 0$ to $+29.2$; $U_2L_2 = -8.0$ to -98.3; $U_3L_3 = -8$ to -30.8.

10-10. (a) Moment $= 120$ kip-ft clockwise acting on ends of cross member; (b) direct stress in upper horizontals $= -10.75$ kips and $+0.75$ kip, direct stress in lower horizontals $= +5.75$ kips and -5.75 kips; (c) direct stress in the cross horizontal member $= -15$ kips and $+5$ kips; (d) direct stress in the cross horizontal member $= -18\frac{1}{3}$, -5, and $+8\frac{1}{3}$ kips.

11-1. $\theta_A = 5.484 \times 10^{-3}$ radian clockwise, $\theta_B = 4.193 \times 10^{-3}$ radian counterclockwise, $\theta_C = 2.580 \times 10^{-3}$ radian clockwise, $\Delta_C = 0.3251$ in. downward.

11-2. $\theta_B = PL^2/2EI$ clockwise; $\Delta_B = PL^3/3EI$ downward.

11-3. $\theta_B = 7wL^3/48EI$ counterclockwise; $\Delta_B = 41wL^4/384EI$ downward.

11-4. $\theta_A = 11wL^3/384EI$ clockwise; $\theta_B = 11wL^3/384EI$ counterclockwise; Δ at center $= 19wL^4/2,048EI$ downward.

11-5. $\theta_A = \theta_B = 7wL^3/256EI_c$, Δ at center $= 93wL^4/12,288EI_c$ downward.

11-6. $\theta_C = 261.6$ kip-sq ft/EI counterclockwise; $\Delta_C = 1,052.8$ kip-cu ft/EI upward.

11-7. See Prob. 11-1. $\Delta_{max} = 0.3695$ in. downward at 8.985 ft from left support

11-8. See Prob. 11-4.

11-9. See Prob. 11-5.

11-10. See Prob. 11-6.

11-11. $\theta_C = 161.6$ kip-sq ft/EI counterclockwise; $\Delta_C = 652.8$ kip-cu ft/EI upward

11-12. See Prob. 11-1.

11-13. See Prob. 11-2.

11-14. See Prob. 11-3.

11-15. See Prob. 11-4.

11-16. See Prob. 11-5.

11-17. See Prob. 11-6.

11-18. See Prob. 11-11.

11-19. $R_A = \frac{3}{8}wL$ upward; $R_B = \frac{5}{8}wL$ upward; $M_B = \frac{1}{8}wL^2$ clockwise.

11-20. $R_A = 8.8487$ kips upward; $R_B = 19.5417$ kips upward; $R_C = 1.9904$ kips downward.

11-21. $R_A = R_C = \frac{5}{16}P$ upward; $R_B = 1\frac{1}{8}P$ upward.

11-22. $M_A = M_B = wL^2/12$; $R_A = R_B = wL/2$.

11-23. $R_A = 11.6558$ kips upward; $R_B = 15.1104$ kips upward; $R_C = 0.3662$ kip downward; $M_A = 37.535$ kip-ft counterclockwise.

11-24. $R_A = 1\frac{9}{56}P$ upward; $R_B = 1\frac{7}{14}$ upward; $R_C = 2\frac{5}{56}P$ upward; $M_C = \frac{3}{28}PL$ clockwise.

11-25. R_4: 0; 0.3672; 0.6875; 0.9141; 1.0000; 0.9141; 0.6875; 0.3672; 0. M_2: 0; $+3.828$; $+8.125$; $+3.361$; 0; -1.641; -1.875; -1.172; 0.

11-26. R_4: 0; $+0.3906$; $+0.7250$; $+0.9469$; $+1.0000$; $+0.8531$; $+0.5750$; $+0.2594$; 0; -0.1312; -0.1500; -0.0938; 0. M_4: 0; -2.500; -4.000; -3.500; 0; -2.875; -3.000; -1.625; 0; $+0.875$; $+1.000$; $+0.625$; 0.

12-1. $\theta_A = \theta_B = 503.66$ kip-sq ft/EI_c clockwise; $\theta_C = 593.89$ kip-sq ft/EI_c counterclockwise; $\theta_D = 1,054.69$ kip-sq ft/EI_c counterclockwise; Δ_H of $A = 30,231$ kip-cu ft/EI_c to the left; Δ_H of B or $C = 21,165$ kip-cu ft/EI_c to the left.

12-2. $\theta_A = 171$ kip-sq ft/EI_c clockwise; $\theta_B = 90$ kip-sq ft/EI_c clockwise; $\theta_C = 99$ kip-sq ft/EI_c counterclockwise; $\theta_D = 195$ kip-sq ft/EI_c counterclockwise; Δ_H of $A = 6{,}504$ kip-cu ft/EI_c to the left; Δ_H of B or $C = 3{,}912$ kip-cu ft/EI_c to the left.

12-3. $\theta_A = 4{,}246.65$ kip-sq ft/EI_c counterclockwise; Δ_H of $A = 36{,}004.5$ kip-cu ft/EI_c to the right; Δ_V of $A = 145{,}933.65$ kip-cu ft/EI_c downward.

12-4. $\theta_A = 26$ kip-sq ft/EI_c counterclockwise; Δ_H of $A = 291$ kip-cu ft/EI_c to the right; Δ_V of $A = 450$ kip-cu ft/EI_c downward.

12-5. $\theta_A = 291$ kip-sq ft/EI_c counterclockwise; Δ_H of $A = 4{,}560$ kip-cu ft/EI_c to the right; Δ_V of $A = 4{,}644$ kip-cu ft/EI_c downward.

12-6. $\theta_A = 450$ kip-sq ft/EI_c clockwise; Δ_H of $A = 4{,}644$ kip-cu ft/EI_c to the left; Δ_V of $A = 14{,}256$ kip-cu ft/EI_c upward.

12-7. See Prob. 12-1.

12-8. See Prob. 12-2.

12-9. See Prob. 12-3.

12-10. See Prob. 12-4.

12-11. See Prob. 12-5.

12-12. See Prob. 12-6.

12-13. $H_A = 4.648$ kips to the right; $V_A = 12.125$ kips upward; $H_D = 2.552$ kips to the right; $V_D = 5.875$ kips upward; $\theta_A = 291.2$ kip-sq ft/EI_c counterclockwise; $\theta_B = 85.3$ kip-sq ft/EI_c clockwise; $\theta_C = 133.7$ kip-sq ft/EI_c counterclockwise; $\theta_D = 148.3$ kip-sq ft/EI_c counterclockwise; Δ_H of $B = \Delta_H$ of $C = 2{,}982$ kip-cu ft/EI_c to the left.

12-14. $M_A = 37.722$ kip-ft clockwise; $H_A = 5.5730$ kips to the right; $V_A = 10.8614$ kips upward; $M_D = 2.230$ kip-ft clockwise; $H_D = 1.6270$ kips to the right; $V_D = 7.1386$ kips upward; $\theta_B = 111.91$ kip-sq ft/EI_c clockwise; $\theta_C = 92.05$ kip-sq ft/EI_c counterclockwise; Δ_H of $B = \Delta_H$ of $C = 347.0$ kip-cu ft/EI_c to the left.

12-15. Horizontal reaction at 6-ft intervals: 0; 0.07034; 0.11439; 0.13076; 0.11808; 0.07495; 0.   Moment at the mid-point: 0; 1.523; 3.598; 6.254; 3.520; 1.426; 0.

13-1. $\Delta_H = 0.19728$ in. to the right; $\Delta_V = 0.91194$ in. downward.

13-2. $\Delta_H = 0.0144$ in. to the left; $\Delta_V = 0.3237$ in. downward.

13-3. Δ_H of $L_0 = 0.144$ in. to the right; Δ_V of $L_0 = 0.697$ in. downward; Δ_H of $L_2 = 0.072$ in. to the right; Δ_V of $L_2 = 0.102$ in. downward.

13-4. $\Delta = 0.051$ in. toward each other.

13-5. $\Delta = 0.051$ in. away from each other.

13-6. See Prob. 13-1.

13-7. See Prob. 13-2.

13-8. See Prob. 13-3.

13-9. See Prob. 13-4.

13-10. See Prob. 13-5.

13-11. $R_3 = 911.94/37.71 = 24.183$ kips.

13-12. $R_5 = 323.70/104.75 = 3.090$ kips.

13-13. Stress in $U_2L_3 = 6.25$ kips compression.

13-14. Stress in $L_2U_3 = 6.25$ kips tension.

13-15. Influence ordinates for R_3: 0; $+0.4861$; $+0.8576$; $+1.0000$; $+0.5859$; 0. Influence ordinates for R_5: 0; -0.0916; -0.1146; 0; $+0.4484$; $+1.0000$.

13-16. Influence ordinates for $U_2L_3 = 0$; $-\frac{5}{24}$; $-\frac{5}{12}$; 0; $-\frac{5}{24}$; $-\frac{5}{12}$.   Influence ordinates for $L_2U_3 = 0$; $+\frac{5}{24}$; $+\frac{5}{12}$; 0; $+\frac{5}{24}$; $+\frac{5}{12}$.

14-1. $R_A = 21.24$ kips upward; $R_B = 48.76$ kips upward; $M_B = 215.2$ kip-ft clockwise.

14-2. $R_B = 50.44$ kips upward; $R_C = 41.56$ kips upward; $M_C = 167.2$ kip-ft clockwise.

14-3. $M_{BA} = -594.34$ kip-ft; $\theta_B = 1{,}481.2$ kip-sq ft/EI counterclockwise.

14-4. $M_{AB} = +853.1$ kip-ft; $M_{BA} = -350.6$ kip-ft; $\theta_B = 668.7$ kip-sq ft/EI counterclockwise.

14-5. $M_{BA} = -629.66$ kip-ft; $M_{CB} = +164.83$ kip-ft; $\theta_B = -1,324.2$ kip-sq ft/EI counterclockwise.

14-6. $M_{AB} = +849.48$ kip-ft; $M_{BA} = -357.83$ kip-ft; $M_{CB} = +28.92$ kip-ft; $\theta_B = -644.7$ kip-sq ft/EI counterclockwise.

14-7. $M_{BA} = +83.705$ kip-ft; $\theta_A = 1.9345 \times 10^{-3}$ radian clockwise; $\theta_B = 0.7441 \times 10^{-3}$ radian counterclockwise; $\theta_C = 2.7530 \times 10^{-3}$ radian counterclockwise.

14-8. $M_{AB} = +105.794$ kip-ft; $M_{BA} = +113.932$ kip-ft; $\theta_B = 0.2604 \times 10^{-3}$ radian counterclockwise; $\theta_C = 2.9948 \times 10^{-3}$ radian counterclockwise.

14-9. $M_{BC} = -125.000$ kip-ft; $M_{CB} = -192.708$ kip-ft; $\theta_A = 2.375 \times 10^{-3}$ radian clockwise; $\theta_B = 1.625 \times 10^{-3}$ radian counterclockwise.

14-10. $M_{AB} = +132.534$ kip-ft; $M_{BA} = +167.411$ kip-ft; $M_{CB} = -213.914$ kip-ft; $\theta_B = 1.1161 \times 10^{-3}$ radian counterclockwise.

14-11. $M_{BC} = +531.50$ kip-ft; $M_{CB} = -293.25$ kip-ft; $\theta_B = 1,760.5$ kip-sq ft/EI_c counterclockwise; $\theta_C = 1,363.3$ kip-sq ft/EI_c clockwise.

14-12. $M_{AB} = +878.08$ kip-ft; $M_{BA} = -300.64$ kip-ft; $M_{CD} = +199.84$ kip-ft; $M_{DC} = -500.08$ kip-ft; $\theta_B = 835.2$ kip-sq ft/EI_c counterclockwise; $\theta_C = 667.2$ kip-sq ft/EI_c clockwise.

14-13. $V_A = 24.632$ kips upward; $H_C = 5.053$ kips to the right; $V_C = 4.263$ kips downward; $M_C = 28.42$ kip-ft counterclockwise; $H_D = 5.053$ kips to the left; $V_D = 39.631$ kips upward; $M_D = 25.26$ kip-ft counterclockwise; $\theta_B = 94.73$ kip-sq ft/EI_c counterclockwise.

14-14. $M_{AB} = M_{BC} = +100$ kip-ft; $M_{BA} = M_{CB} = -100$ kip-ft; $M_{BE} = M_{EB} = 0$; $\theta_B = \theta_C = 0$.

14-15. $H_A = 8.4017$ kips to the right; $V_A = 51$ kips upward; $\theta_A = 268.85$ kip-sq ft/EI_c counterclockwise; $\theta_B = 537.70$ kip-sq ft/EI_c clockwise.

14-16. $H_A = 14.1025$ kips to the right; $V_A = 51$ kips upward; $M_A = 112.82$ kip-ft clockwise; $\theta_B = 451.29$ kip-sq ft/EI_c clockwise.

14-17. $H_D = 5.8938$ kips to the right; $V_D = 42.2686$ kips upward; $H_E = 0$; $V_E = 119.4628$ kips upward; $\theta_D = 188.60$ kip-sq ft/EI_c counterclockwise; $\theta_A = 377.19$ kip-sq ft/EI_c clockwise; $\theta_B = \theta_E = 0$.

14-18. $H_D = 10.3912$ kips to the right; $V_D = 43.3025$ kips upward; $M_D = 83.13$ kip-ft clockwise; $H_E = 0$; $V_E = 117.3950$ kips upward; $M_E = 0$; $\theta_A = 332.53$ kip-sq ft/EI_c; $\theta_B = 0$.

14-19. $H_A = 8.860$ kips to the right; $V_A = 29.523$ kips upward; $H_D = 0.140$ kip to the right; $V_D = 24.477$ kips upward; $\theta_A = 360.4$ kip-sq ft/EI_c counterclockwise; $\theta_B = 118.1$ kip-sq ft/EI_c clockwise; $\theta_C = 227.1$ kip-sq ft/EI_c counterclockwise; $\theta_D = 144.5$ kip-sq ft/EI_c counterclockwise; Δ_H of $B = 3,616$ kip-cu ft/EI_c to the left.

14-20. $H_A = 10.409$ kips to the right; $V_A = 27.428$ kips upward; $M_A = 66.13$ kip-ft clockwise; $H_D = 1.409$ kips to the left; $V_D = 26.572$ kips upward; $\theta_B = 165.3$ kip-sq ft/EI_c clockwise; $\theta_C = 183.8$ kip-sq ft/EI_c counterclockwise; $\theta_D = 47.5$ kip-sq ft/EI_c clockwise; Δ_H of $B = 199$ kip-cu ft/EI_c to the left.

14-21. $H_A = 7.858$ kips to the right; $V_A = 28.836$ kips upward; $H_D = 1.142$ kips to the right; $V_D = 25.164$ kips upward; $M_D = 30.77$ kip-ft clockwise; $\theta_A = 271.1$ kip-sq ft/EI_c counterclockwise; $\theta_B = 153.3$ kip-sq ft/EI_c clockwise; $\theta_C = 232.6$ kip-sq ft/EI_c counterclockwise; Δ_H of $B = 2,333$ kip-cu ft/EI_c to the left.

14-22. $H_A = 11.088$ kips to the right; $V_A = 27.511$ kips upward; $M_A = 74.16$ kip-ft clockwise; $H_D = 2.088$ kips to the left; $V_D = 26.489$ kips upward; $M_D = 15.07$ kip-ft counterclockwise; $\theta_B = 153.8$ kip-sq ft/EI_c clockwise; $\theta_C = 175.8$ kip-sq ft/EI_c counterclockwise; Δ_H of $B = 412$ kip-cu ft/EI_c to the left.

15-1 to 15-22. See answers for Probs. 14-1 to 14-22.

INDEX